Manhattan Project

Bruce Cameron Reed

Manhattan Project

The Story of the Century

 Springer

Bruce Cameron Reed
Emeritus, Alma College
Bedford, NS, Canada

ISBN 978-3-030-45736-5   ISBN 978-3-030-45734-1   (eBook)
https://doi.org/10.1007/978-3-030-45734-1

This Springer imprint is published by the registered company Springer Nature Switzerland AG
The registered company address is: Gewerbestrasse 11, 6330 Cham, Switzerland

For Laurie.
And Fred, Leo, Stella, Cassie, Nyx, and Newton.

Prologue

Nuclear weapons are the most destructive devices ever developed by human beings. A single one can cause a level of destruction to a large city akin to that of an earthquake, tsunami, or hurricane, while also depositing radioactive fallout over a large area. A modest-scale nuclear war could loft enough dust into the atmosphere to cause climatic effects like that of an asteroid strike, posing a truly existential threat to human civilization. Since their development in World War II, thousands of nuclear weapons have been constructed, over 2,000 have been tested, and thousands remain in the arsenals of the world's nuclear powers. At any time, hundreds are deployed on bombers, missiles, and submarines, ready for essentially immediate use.

The development and use of nuclear weapons was one of the pivotal events of human history, and certainly the watershed event of the twentieth century. The bombings of Hiroshima and Nagasaki helped bring World War II to a close. The number of people killed by the bombs was estimated to be about 125,000, with a further 130,000–160,000 injured. A similar number of deaths had occurred a few months earlier in one night in a fire-bombing raid of Tokyo, but that raid involved nearly 300 bombers carrying about 1,700 tons of bombs. In contrast, the Hiroshima and Nagasaki atomic bombs, *Little Boy* and *Fat Man*, each exploded with energies equivalent to over 10,000 tons of conventional explosive.

The ramifications were staggering. A single bomber could now obliterate an entire city with a single bomb. Advances in science and engineering had given humanity the capability of annihilating itself on a global scale if an accidental or deliberate large-scale nuclear war were to break out. Nations that saw themselves as global powers would henceforth depend as much on the depth of their scientific and technological prowess as on traditional military capabilities as a measure of their influence. The Cold War, nuclear proliferation, the possibility of nuclear terrorism, the stockpiles of nuclear weapons currently held by various countries, and public fear of radioactivity and nuclear power are all legacies of 1945.

Little Boy and *Fat Man* were the products of what has come to be known as the Manhattan Project. Carried out by the United States Army between 1942 and 1945, this project was properly known as the Manhattan Engineer District. The scale and cost of Manhattan were just as enormous as its eventual significance. Physics Nobel Laureate Niels Bohr, who contributed much to the understanding of the process of nuclear fission, initially thought the idea of a nuclear weapon so improbable as to state that "It can never be done unless you turn the United States into one huge factory." To some extent, that is exactly what happened. Some 480,000 people were employed in the Project's laboratories and factories at one time or another, about one out of every 250 people in the country at the time. Peak employment reached just over 125,000 in late 1944. By mid-1945 the cost of the Project was approaching $2 billion, an enormous amount for two bombs. The Project's own official record, the *Manhattan District History*, lists nearly 1,000 government agencies, universities, individuals, businesses, and industrial contractors that contributed to the work.

In this book I examine the scientific basis of nuclear weapons, how they function, how the Manhattan Project came to be organized, why the bombs required exotic materials for their construction, why they were so difficult and expensive to make, and the circumstances of the Hiroshima and Nagasaki bombing missions. While thousands of articles and books have been written about Manhattan, my motivation in preparing this volume is to offer readers an authoritative treatment of the Project prepared by a professional physicist who has spent over 25 years studying it.

Any author who takes on the Manhattan Project faces a dilemma: how much detail to cover? The *Manhattan District History* runs to thousands of pages; no single-volume account can hope to be comprehensive. We know what scientific theories and industrial processes proved correct and workable, that the bombs dropped at Hiroshima and Nagasaki functioned properly, that the war ended soon thereafter, and that America entered the Cold War as the world's preeminent power. It is easy to think that this is the way it all had to play out once the United States had committed the resources, money, and political support to developing atomic bombs. But this is not so: *none* of these events were preordained. So many aspects of the Project were so chancy that the entire effort could easily have played *no* role in ending the war. After the discovery of uranium fission in late 1938, it took some of the leading researchers of the time well over a year to appreciate how the subtleties of nuclear reactions might be exploited to make a weapon or a reactor. By the time of the Japanese attack at Pearl Harbor in December, 1941, experimental tests and theoretical analyses were beginning to clarify the situation, but the technical challenges involved in realizing the liberation of nuclear energy on a practical scale looked so overwhelming as to make the idea of a deliverable nuclear weapon seem more a matter of science-fiction than real-world engineering. The administrators, scientists, and engineers who organized and carried out the Project faced immense and unique challenges which required ingenuity and perseverance to surmount. In relating the story of Manhattan, I have aimed for a middle ground which gives a reasonable treatment of the major advances, setbacks, personalities and uncertainties involved, without attempting to dissect every last detail. The mine runs deep.

I have benefited from spoken and electronic conversations, correspondence, suggestions, willingness to read and comment on draft material, and general encouragement from John Abelson, Joseph-James Ahern, John Altholz, Dana Aspinall, Albert Bartlett, Jeremy Bernstein, Alan Carr, David Cassidy, John Coster-Mullen, Steve Croft, Gene Deci, Eric Erpelding, Patricia Ezzell, Charles Ferguson, Henry Frisch, Ed Gerjuoy, Chris Gould, Dick Groves, Robert Hayward, Dave Hafemeister, Art Hobson, Cindy Kelly, William Lanouette, Irving Lerch, Harry Lustig, Mike Magras, Jeffrey Marque, Albert Menard, Tony Murphy, Robert S. Norris, Peter Pesic, Klaus Rohe, Bob Sadlowe, Frank Settle, Ruth Sime, D. Ray Smith, Roger Stuewer, Arthur Tassel, Linda Thomas, Michael Traynor, Mark Walker, Alex Wellerstein, Bill Wilcox, John Yates, and Pete Zimmerman. If I have forgotten anybody, I apologize; you are in this list in spirit. A few of these individuals are, sadly, no longer with us. Alma College interlibrary loan specialists Susan Cross and Angie Kelleher have never failed to dig up any obscure document which I have requested; they are true professionals. Angela Lahee and her colleagues at Springer deserve a big nod of thanks for believing in and seeing through this project.

In addition to the individuals listed above, the fingerprints of a lifetime's worth of family members, teachers, classmates, professors, mentors, colleagues, students, collaborators, and friends are all over these pages; a work like this is never accomplished alone. I thank them all.

Most of all I thank Laurie, who bore with it. Again.

Halifax, NS, Canada   Bruce Cameron Reed
June 2019

Contents

1

The Big Picture: A Survey of the Manhattan Project

1.1   Atoms, Nuclei, and Isotopes

To develop an appreciation of how nuclear weapons are designed and function requires some understanding of the basic properties of atoms, nuclei, isotopes, and the process of nuclear fission. Buckle up for a whirlwind tour of the atomic landscape and how the Manhattan Project came to be.

That atoms can be imagined as being constructed like miniature solar systems is a staple of every high-school chemistry and physics class. The development of this picture began with the discovery of the electron in 1897, and was essentially completed with the discovery of the neutron in early 1932. In the solar system analogy, the Sun is played by nuclei at the centers of atoms, combinations of electrically-positive protons and electrically-neutral neutrons. Surrounding nuclei at various distances are negatively-charged orbiting electrons, the particles which sustain the electric currents that course through our houses, bodies, workplaces, and computers. The numbers of these various particles in a given atom depends on what chemical element the atom identifies with. In any atom, the number of orbiting electrons usually equals the number of protons in the nucleus, making the overall structure electrically neutral. The number of protons in the nucleus is the same for all atoms of the same element, and is the "atomic number" of the element.

A very important refinement to this picture is that different elements can occur in a variety of forms: isotopes. Different isotopes of the same element contain differing numbers of neutrons in their nuclei, but they all have the same electrical and chemical properties because they contain the same number of protons and electrons. The atomic number is always designated by the

© Springer Nature Switzerland AG 2020
B. C. Reed, *Manhattan Project*,
https://doi.org/10.1007/978-3-030-45734-1_1

letter Z. For atoms of life-sustaining oxygen, $Z = 8$, that is, they all contain eight protons in their nuclei and have eight orbiting electrons (unless they have been ionized). But there are three naturally-occurring stable isotopes of oxygen: one contains eight neutrons, another nine, and another ten. In these three types there are consequently totals of 16, 17, or 18 protons plus neutrons in the nuclei: Oxygen-16, Oxygen-17, and Oxygen-18, or just O-16, O-17, and O-18. O-16 is by far the most common type, accounting for over 99.7% of naturally-occurring oxygen, but you can and do breathe the other two types.

Neutrons can be thought of as a sort of nuclear glue that holds nuclei together against the immense repulsive electrical forces that act between the like-charged protons. The total number of neutrons plus protons in the nucleus is known as its atomic weight (a bit of a misnomer), and also as its mass number. This is always designated with the letter A. Adopting N to designate the number of neutrons, it follows that $N = A - Z$. Any isotope can be represented by the shorthand notation $_Z^A X$, where X is the symbol for the element involved, A is the mass number, and Z the atomic number corresponding to the element X. The three oxygen isotopes are then abbreviated as $_8^{16}O$, $_8^{17}O$, and $_8^{18}O$.

All nuclear weapons ultimately derive their power from reactions involving isotopes of the very heavy elements uranium ($Z = 92$) and plutonium ($Z = 94$). Two isotopes of uranium and one of plutonium are involved in the Manhattan Project: U-235 $\left(_{92}^{235}U\right)$, U-238 $\left(_{92}^{238}U\right)$, and Pu-239 $\left(_{94}^{239}Pu\right)$. All U-235 nuclei contain 143 neutrons ($N = A - Z = 235 - 92 = 143$), while all U-238 nuclei contain 146 neutrons. The three-neutron difference plays a huge role. Only U-235 can be used to make a nuclear weapon, but this isotope makes up only about 0.7% of naturally-occurring uranium. The remaining 99.3% is U-238, which is useless for making a first-generation nuclear weapon. Plutonium is an artificially-prepared element, and its mass-239 isotope makes for a very efficient nuclear explosive.

1.2 Fission, Neutrons, and Chain Reactions

The process by which uranium and plutonium release their energy in nuclear weapons is termed nuclear fission. Fission, which is synonymous with "splitting"—the word was in fact borrowed from the process of cell division in biology—was discovered quasi-serendipitously in Berlin in late 1938. This discovery involved the realization that nuclei of uranium atoms could be caused to break apart when struck by incoming neutrons. That the bombarding particles are neutrons is important. Fission cannot be induced by striking a uranium nucleus with one of another element; the repulsive forces between the protons in the nuclei are so great that it is practically impossible to have the nuclei come into contact with each other. But because neutrons are electrically neutral, they experience no repulsion: there is nothing to stop them from striking a target nucleus that lies in their path. It was quickly realized that the disintegrated nucleus loses a small amount of mass, but this mass corresponds to a huge amount of energy thanks to Albert Einstein's famous equation $E = mc^2$. The amount of energy released per fission of a single nucleus proved, atom-for-atom, to be millions of times that released in any known chemical reaction.

It was immediately apparent to researchers that if such reactions could be induced on a large scale, millions of pounds of conventional explosive could be replaced with a few pounds of nuclear explosive. A few weeks after the discovery, it was found that a by-product of each fission was the simultaneous liberation of two or three neutrons from the fissioned nucleus, "secondary" neutrons. These neutrons, if they do not escape the mass of uranium involved, can go on to fission other nuclei and initiate a chain reaction. In theory, this process can continue until all of the uranium is fissioned, releasing an enormous amount of energy.

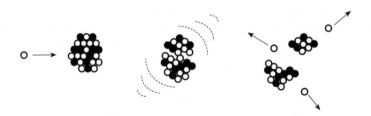

A sketch of the fission process. A nucleus comprising protons (filled circles) and neutrons (open circles) is struck by an incoming neutron (left). The nucleus captures the neutron, becomes agitated (middle), and then fragments into two product nuclei and three neutrons (right). In a real uranium nucleus there are many more protons and neutrons than are drawn here

These discoveries raised a host of questions. Were any other elements fissile? Was there a minimum amount of uranium that would have to be arranged in one place to realize a chain reaction? Did one or both uranium isotopes undergo fission? Could the process be controlled by human intervention to create a power source (a nuclear reactor), or would the result be an uncontrollable explosion? Why hadn't all of the uranium ores in the Earth spontaneously fissioned themselves into oblivion millions of years ago?

1.3 Understanding Fission: Neutrons Fast and Slow

By the time of the outbreak of World War II in September, 1939, physicists had developed theoretical arguments which indicated that only the very heavy elements thorium ($Z = 90$) and uranium were likely to be fissile. The element between thorium and uranium, protactinium, is so rare as to be of no practical use in the nuclear weapons business. Thorium would prove unworkable for either a bomb or a reactor. In the case of uranium, the theory indicated that only the rare U-235 isotope would fission under neutron bombardment, whereas U-238 nuclei would tend to capture incoming neutrons without fissioning. These predictions were confirmed experimentally in early 1940. With the overwhelming preponderance of U-238 in natural uranium, this capture effect promised to literally poison the prospect for a chain reaction using uranium of natural isotopic abundance. To obtain a chain reaction would require isolating a sample of pure U-235 from its sister isotope, or at least processing uranium in some way to isolate a sub-sample with a dramatically increased percentage of U-235. Given that 993 of every 1,000 lb of uranium ore extracted from the Earth will be of the undesirable U-238 isotope, this presents an immense challenge. Isotope enrichment is very difficult, even for a technically advanced country. Since isotopes of any element behave identically so far as their chemical properties are concerned, no chemical reaction can be used to achieve enrichment: only techniques that depend on the mass difference between the two isotopes can be used. In the case of uranium, the three-neutron difference amounts to a mass difference of only about 1% between the two isotopes. Physicists and chemists had developed three workable enrichment techniques, namely centrifugation, mass spectrometry, and a process known as diffusion, but these had been applied successfully only in very limited laboratory settings involving minute samples of much lighter elements where the mass differences between isotopes was relatively large. For uranium, the prospects looked dim to non-existent.

By mid-1940, understanding of the differing responses of the two isotopes of uranium to bombarding neutrons had led to the development of a new idea for obtaining a controlled (not explosive) chain reaction using natural uranium without enrichment. When a nucleus is struck by a neutron, various reactions are possible: the nucleus might fission, it might capture the neutron without fissioning, or it can deflect the neutron just as a billiard ball would an incoming marble. Each result has some probability of occurring, and these probabilities depend on the speed of the incoming neutrons. Neutrons released in fission reactions are extremely energetic, emerging with average speeds of about 45 million miles per hour. For obvious reasons, such neutrons are termed "fast." U-238 nuclei tend to capture fast neutrons emitted in fissions of U-235 nuclei and not subsequently fission. But when a nucleus of U-238 is struck by a neutron traveling at a pokey few thousand miles per hour, it behaves as a much more benign target, with scattering of the neutron being about three times as likely as capture. But—and this is a key point—U-235 nuclei turn out to have an enormous probability for undergoing fission when struck by slow neutrons: over 200 times the slow-neutron capture probability of U-238. This factor is large enough to compensate for the small natural abundance of U-235 to the extent that, in a sample of natural-abundance uranium, a slow neutron is about as likely to fission a nucleus of U-235 as it is to be captured by one of U-238. This speed-sensitive behavior is what makes possible the slow-neutron chain reactions used in power-producing nuclear reactors. In effect, slowing fission-liberated neutrons is equivalent to enriching the abundance percentage of U-235. This point is so important that it bears reiterating: if a neutron emitted in a fission can be slowed, then it has about as good a chance of going on to fission another U-235 nucleus as it does of being uselessly captured by a nucleus of U-238. In actuality, both processes proceed simultaneously within a reactor: U-235 fissions generate energy and liberate neutrons, while U-238 nuclei capture some of the neutrons and become a waste product. In a curious twist, however, this very waste product turned out to be the seed material for producing the bomb-suitable isotope plutonium-239.

The trick to slowing neutrons during the very brief interval between when they are emitted in fissions and when they strike other nuclei is to work not with a single large mass of uranium, but rather to disperse it as small chunks within a surrounding medium which slows neutrons without capturing them. The medium is known as a moderator, and the entire package is a reactor. During the Manhattan Project, the synonymous term "pile" was used in the literal sense of an arrangement of slugs of uranium metal embedded within a heap of moderating material. Ordinary water can serve as a moderator, but, at the time, graphite—like that used in pencils—proved easier to employ. By introducing moveable rods of neutron-capturing material into the pile, the reaction can be controlled by adjusting them as necessary. It is in this way that natural-abundance uranium proved capable of sustaining a controlled nuclear reaction, although not an explosive one. All power-producing nuclear reactors operate via slow-neutron chain-reactions. But you can sleep comfortably. Reactors cannot behave like bombs; the reaction is far too slow, and even if the control rods are rendered inoperative, the reactor will melt rather than blow up. Video footage of the explosion of a reactor at Fukushima, Japan, actually showed a steam explosion involving the reactor's cooling water, not a nuclear explosion.

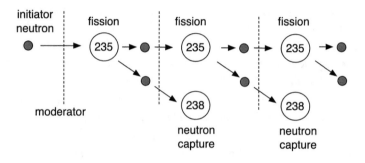

Schematic illustration of a chain reaction utilizing moderated neutrons. Each fission of a U-235 liberates two secondary neutrons, one of which goes on to fission another U-235 nucleus while the other is captured by a nucleus of U-238. From Reed, Atomic Bomb (2015) Fig. 1.2

1.4   Reactors and Plutonium

By mid-1940, it appeared that to make a chain-reaction mediated by fast neutrons—a bomb—would require isolating pure U-235. However, it was soon appreciated that the moderated-neutron concept could be used in an indirect way to make a different fissile material for use in a nuclear weapon. As a reactor operates, neutron capture by U-238 nuclei proceeds alongside fission of U-235 nuclei to about the same degree of probability. On capturing a neutron, a nucleus of U-238 becomes one of U-239. Based on projecting from known patterns of the stability of nuclei, it was predicted that U-239 nuclei might decay within a short time to nuclei of atomic number 94 (later called plutonium), and that such an element might be very similar to U-235 in its fissility properties. If this proved so, then a reactor could be used to "breed" plutonium through neutron capture by U-238 nuclei, while maintaining a self-sustaining reaction via U-235 fissions. The advantage of this would be that the plutonium could subsequently be separated from the mass of parent uranium fuel by chemical processing, which would allow engineers to circumvent the need to develop enrichment techniques. These predictions were soon confirmed on a laboratory scale by creating a tiny sample of plutonium via neutron-bombardment of uranium.

By the time of the Japanese attack at Pearl Harbor in December, 1941, two possible routes to developing a nuclear explosive had been identified: isolate tens of kilograms of U-235, or develop reactors to breed plutonium. U-235 was considered practically certain to make an excellent nuclear explosive, but the tens of kilograms necessary would have to be separated atom by atom from tons of uranium ore. In the case of plutonium, the likely chemical separation techniques were understood by chemical engineers, but nobody had ever constructed an operating reactor or isolated any significant quantity of the new element. Fundamental questions of engineering and physics loomed. Could a large-scale reactor be safely controlled? What if plutonium proved to have some property that rendered it useless as an explosive? With the possibility that German scientists could be thinking along the same lines, the leaders of the Manhattan Project made the only decision that they could in such a circumstance: both methods would be tried.

1.5   The Manhattan Engineer District

The possibility that nuclear fission might have military applications was brought to the attention of President Franklin Roosevelt in the fall of 1939, and support for research was soon organized under the direction of a committee that he ordered assembled. Until mid-1942, this effort was under the authority of various civilian branches of the government, although the work was being conducted in secrecy. By that time, researchers in both Britain and America had reached the conclusion that reactors and weapons could be feasible, but that isolating the relevant materials would require large-scale factories. The only organization capable of carrying out the work with the necessary secrecy was the United States Army, and the project was assigned to the Army's Corps of Engineers in August, 1942. To carry out the work, the Corps established a new administrative entity, the Manhattan Engineer District (MED). In September, overall command of the MED was assigned to Brigadier General Leslie Richard Groves, who had extensive experience with large construction projects.

Left: Brigadier (later Major) General Leslie R. Groves (1896–1970). Right: Robert Oppenheimer (1904–1967) ca. 1944. Sources: http://commons. wikimedia.org/wiki/File:Leslie_Groves.jpg; http:// commons.wikimedia.org/wiki/File:JROppenheimer-LosAlamos.jpg

Manhattan Project work was focused in two major directions: acquiring fissile material (U-235 and Pu-239), and designing and testing possible configurations for actual bombs. Fissile materials were produced at enormous factory complexes located at Oak Ridge, Tennessee (uranium enrichment) and Hanford, Washington (plutonium production reactors). The vast majority of MED funding went into the construction and operation of these facilities. At the same time, a highly-secret bomb-design laboratory was established at Los Alamos, New Mexico. The Los Alamos Laboratory began operating in the spring of 1943, and was directed by theoretical physicist Dr. J(ulius) Robert Oppenheimer of the University of California.

Locations of major Manhattan Project research and production sites. Other sites were located in Montreal and British Columbia, Canada. From Reed (2014a)

In theory, the tasks facing Los Alamos scientists seemed straightforward. Fissile isotopes such as U-235 or Pu-239 possess a so-called critical mass, a minimum mass necessary to achieve a chain reaction. The value of the critical mass depends on factors such as the density of the material, its probability for undergoing fission, and the number of neutrons liberated per fission. Much of the experimental work at Los Alamos was directed toward obtaining accurate measurements of these quantities. With these data, critical masses could be calculated by using mathematical relationships adopted from a well-established branch of physics known as diffusion theory.

1.6 Little Boy, Fat Man, Trinity, Hiroshima, and Nagasaki

By mid-1945, Oak Ridge and Hanford were beginning to produce critical-mass quantities of U-235 and Pu-239. For U-235, the critical mass is about 50 kg. A more efficient explosion can be created if you have more material available than just one critical mass, so imagine that you have 70 kg available. To make a bomb, divide the 70 kg into two pieces, and then arrange to bring them together when you are ready to have the device detonate. This is exactly what was done in the uranium-based Hiroshima "Little Boy" bomb. Ordnance engineers mounted the barrel of a naval artillery cannon inside a bomb casing, and placed one piece of uranium, the "target piece", at the nose end of the barrel. The second piece, the "projectile piece", was placed at the tail end. When radars mounted on the bomb indicated that it had fallen to a pre-programmed detonation height, a conventional powder charge was set off to propel the projectile piece into the target piece. A source of neutrons must be supplied to initiate the chain reaction, but this is the essence of a "gun-type" fission bomb.

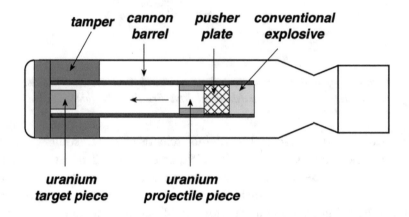

Schematic illustration of a gun-type fission weapon

The Hiroshima bomb contained about 60 kg of U-235, but weighed nearly five tons overall. Much of this was the weight of the cannon, but a significant contributor was that the target end of the cannon was surrounded by a steel tamper weighing several hundred kilograms. The tamper served three essential functions. First, it stopped the projectile piece from flying through the front end of the bomb; unlike in an artillery cannon, the projectile piece has to remain seated around the target piece while the reaction proceeds. Second, the tamper briefly retards the expansion of the assembled bomb core as it detonates, buying a bit more time (microseconds) over which the chain reaction can operate. Third, by making the tamper of a material which reflects escaping neutrons back into the assembled target and projectile pieces, they will have a chance to cause further fissions; this effectively decreases the necessary critical mass. These effects all enhance the efficiency of the weapon by a factor of ten or more over an untamped device, so a tamper is certainly a worthwhile investment despite its being dead weight.

The plutonium bomb was a much more difficult matter. Reactor-produced plutonium proved to fission spontaneously, a completely uncontrollable process. Los Alamos scientists calculated that if they tried to make a gun-type bomb using plutonium, the nuclear explosion would self-initiate before the target and projectile pieces were fully mated, and that the result would be an expensive but very low-efficiency "fizzle" explosion. Two approaches to overcoming this setback looked plausible: either find a way to use less fissile material (fewer spontaneous fissions), and/or assemble the sub-critical pieces more rapidly than could be achieved with the gun mechanism in order to lower the pre-detonation probability. Both approaches were adopted. The critical mass of a fissile material depends on its density; a greater density means a lower critical mass. A mass of material that would be sub-critical at normal density can be made critical by crushing it to a higher density, a feature that lets you get away with using less material than would "normally" be required. This led to the idea of an "implosion" weapon wherein a small subcritical core with a naturally low rate of spontaneous fissions is surrounded with a fast-burning explosive configured to detonate inwards to crush the core to high density in a very short time. For maximum efficiency, the implosion has to be essentially perfectly symmetric: all of the pieces of surrounding explosive need to be triggered within about a microsecond of each other. The feasibility of implosion was considered so uncertain that it was decided to perform a full-scale test of the method. This was the *Trinity* test of July 16, 1945, the world's first nuclear explosion. The test succeeded, and three weeks later the method was put to use in the Nagasaki "Fat Man" bomb. The Little Boy bomb was not tested in advance: by August, 1945, there was only enough U-235 available for one bomb.

Left: The Trinity fireball 25 ms *after detonation. Right: The fireball a few seconds later. Sources: Left:* http://commons.wikimedia.org/wiki/File:Trinity_Test_Fireball_25ms.jpg; *Right: Courtesy of the Los Alamos National Laboratory Archives*

The combat bombs used at Hiroshima and Nagasaki reflected these designs. *Little Boy* was a cylindrically-shaped mechanism that looked like a regular bomb, but *Fat Man* required a bulbous configuration to accommodate its spherical implosion assembly. General Groves' two-billion-dollar gamble paid off in spectacular fashion, and nuclear physics brought the world into a new era.

Left: Little Boy in its loading pit. Right: The Fat Man bomb. Note graffiti on tail. Sources: Left: http://commons.wikimedia.org/wiki/File:Atombombe_Little_Boy_2.jpg; *Right:* http://commons.wikimedia.org/wiki/File:Fat_Man_on_Trailer.jpg

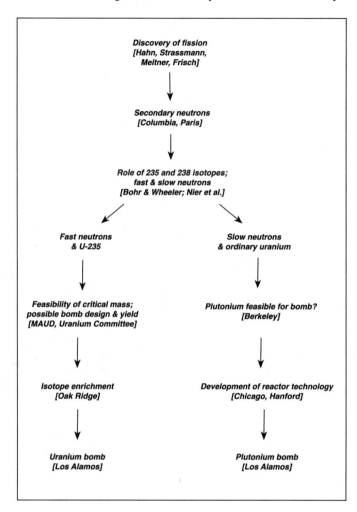

Flowchart summarizing major elements of the Manhattan Project

2

From Atoms to Nuclei: An Inward Journey

The modern scientific picture of atoms as having nuclei of protons and neutrons surrounded by clouds of orbiting electrons emerged between the late 1890s and the early 1930s. Once this understanding was in place, research began shifting toward exploring the internal structure of nuclei, and reactions that could happen with them: nuclear physics. During the 1930s, an immense amount of knowledge on nuclear reactions was accumulated, setting the stage for the discovery of nuclear fission in late 1938. At just the time when events in Europe were heading toward another World War, physicists were beginning to appreciate that fission might hold the key to developing immensely powerful new sources of energy and weapons. What had been a quiet area of academic research was about to assume profound importance in world affairs.

This chapter describes the development of our understanding of atomic structure.

2.1 X-Rays and Radioactivity

In late 1895, German physicist Wilhelm Conrad Röntgen accidentally discovered X-rays. He soon found that his mysterious rays could not only pass through objects such as his hand, but that they also ionized air when they passed through it. This was the first known example of "ionizing radiation." Electrons had not been formally discovered in 1895 (though their existence

© Springer Nature Switzerland AG 2020
B. C. Reed, *Manhattan Project*,
https://doi.org/10.1007/978-3-030-45734-1_2

was strongly suspected); the ionization was caused by X-rays, which are essentially a high-energy form of light, knocking electrons from molecules in the air as they passed through it, leaving the molecules with net electrical charges.

To see X-ray images, Röntgen used a cardboard screen coated with a phosphorescent compound which glowed when struck by the X-rays; when holding his hand in front of the screen, he would be treated to a ghostly image of his bones. That phosphorescence seemed to be involved caught the attention of French mineralogist Antoine Henri Becquerel. Becquerel was an expert on phosphorescence, and wondered if phosphorescent materials such as uranium salts might be induced to emit X-rays if they were exposed to sunlight. While this supposition proved incorrect, investigating it led him, in February 1896, to the accidental discovery of radioactivity. Becquerel observed that samples of uranium ores left on top of wrapped photographic plates would cause them to be exposed, even in the absence of any external illumination. The exposures seemed to be created by the uranium itself; when the plates were unwrapped and developed, images of the samples appeared. Nuclear physics originated with this discovery.

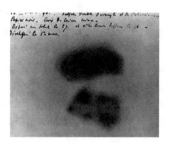

Henri Becquerel (1852–1908) and the first image created by "Becquerel rays" emitted by uranium salts placed on a wrapped photographic plate. In the lower part of the plate a Maltese cross was placed between the plate and the lump of uranium ore. Sources: http://upload.wikimedia.org/wikipedia/commons/a/a3/Henri_Becquerel.jpg; http://upload.wikimedia.org/wikipedia/commons/1/1e/Becquerel_plate.jpg

These exposures are now understood to be due to so-called "alpha" and "beta" particles emitted by nuclei of uranium and other heavy elements as they naturally decay to more stable elements. Some of the decay timescales are fleeting at minutes or seconds only, while others are astronomically long at hundreds of millions or billions of years. In the latter cases it is only because there are so many trillions upon trillions of individual atoms in even a small lump of ore that enough decay within a span of seconds or minutes to leave an image on a film or trigger a Geiger counter. A third type of decay particles, "gamma rays," were subsequently discovered by French chemist Paul Villard in 1900. Gamma-rays are photons just like those by which we see, but of energies about a million times greater than visible-light photons. X-ray photons are intermediate in energy between visible-light photons and gamma-ray photons.

2.2 Marie Curie: Polonium, Radium, and Radioactivity

Becquerel's discovery did not attract much attention at first. Unlike X-rays, which could be created with cathode ray tubes and focused on a target, uranium-generated images required having a sample of ore, tended to be indistinct, and required long exposure times. But the phenomenon of "Becquerel rays" came to the attention of Marie Sklodowski, a native of Poland who had graduated from the Sorbonne (University of Paris) in July 1893 with a degree in physics, ranking first in her class, and then again in mathematics in July, 1894. In 1895 she married Pierre Curie, a physicist at the École Municipale de Physique et de Chimie Industrielles. Seeking a subject for a doctoral thesis, she decided to follow up Becquerel's work, and set up a laboratory in her husband's institution in late 1897.

Marie (1867–1934) and Pierre (1859–1906) Curie; Right: Irène (1897–1956) and Frédéric Joliot-Curie (1900–1958) in 1935. Sources: http://commons.wikimedia.org/wiki/File:Mariecurie.jpg; http://commons.wikimedia.org/wiki/File:PierreCurie.jpg; http://commons.wikimedia.org/wiki/File:Irène_et_Frédéric_Joliot-Curie_1935.jpg

Curie's work began from the observation that X-rays could ionize air. Pierre Curie and his brother had developed a device for detecting minute electrical currents, which Marie used to determine that the amount of electricity generated was directly proportional to the amount of uranium in a sample. On testing other ores, she found that the heavy element thorium also emitted Becquerel rays, although not as many per second as an equal weight of uranium. Further work revealed that samples of pitchblende ore, a blackish material rich in uranium compounds mined in Joachimstal, northwest of Prague, emitted more Becquerel rays than could be accounted for solely by the quantity of uranium that they contained. Inferring that there must be some other active element present, Curie began the laborious and very unpleasant task of chemically isolating it from the tons of ore she had available. Pierre soon abandoned his own research on the properties of crystals in order to join Marie in her work. Between 1898 and Pierre's death in 1906, the Curies would acquire over 23,000 kg of waste Joachimstal ores.

Spectroscopic and chemical analyses of the active substance showed that it was a new, previously unknown element. Christening their find "polonium" in honor of Marie's native country, they published their discovery in the weekly journal of the French Academy of Sciences in July, 1898. In this paper they introduced two new words of scientific jargon: "radioactivity" to designate whatever process deep within atoms was giving rise to Becquerel's ionizing rays, and "radioelement" to designate any element that possessed the property of doing so. The term "radioisotope" is now more commonly used as not all of the individual isotopes of elements that exhibit radioactivity are themselves radioactive.

In December of the same year, the Curies announced that they had found a second radioactive substance, which they dubbed "radium." By the spring of 1902, they had isolated a mere tenth of a gram of radium from ten tons of pitchblende ore, enough for spectroscopic confirmation of its status as a new element. In the summer of 1903 Marie defended her thesis, "Researches on Radioactive Substances," and received her doctorate from the Sorbonne. In the fall of that year the Curies would be awarded half of the 1903 Nobel Prize for Physics; Henri Becquerel received the other half.

2.3 Ernest Rutherford: Alpha, Beta, and Half-Life

In the fall of 1895, New Zealand native Ernest Rutherford arrived at the Cavendish Laboratory of Cambridge University in England on a postgraduate scholarship. His timing was propitious. The Director of the Laboratory was the distinguished experimental physicist Joseph John "J. J." Thomson, who in the fall of 1897 would discover electrons. On stripping electrons away from their parent atoms, Thomson was able to determine their mass and electrical charge by measuring how they were deflected by electrical forces of known strength. Electrons from all parent atoms behaved identically, confirming their existence as the universal negatively-charged particles of matter which orbit nuclei and account for the volumes of atoms. Thomson determined that electrons weighed about one eighteen-hundredth as much as hydrogen atoms, which led him to infer that they must be small components of atoms which resided within some larger overall structure. It is energy released in rearrangements of the outermost electrons of interacting atoms that power our bodies and propel internal-combustion automobiles.

Soon after Rutherford arrived in Cambridge, Röntgen discovered X-rays. As a student, Rutherford had acquired considerable experience with electrical devices, and the Cavendish was well-equipped with Thomson's cathode ray tubes, the core apparatus for generating X-rays. Rutherford began studying their ionizing properties, and when radioactivity was discovered it was natural for him to turn his attention to this new ionizing phenomenon.

Left: Ernest Rutherford (1871–1937) Nobel Prize photo. Right: Seated left to right in this 1921 photo are J. J. Thomson (1856–1940), Rutherford, and Francis Aston (1877–1945), inventor of the mass spectrograph. Sources: https://upload.wikimedia.org/wikipedia/commons/d/df/ Ernest_Rutherford_%28Nobel%29.jpg; *AIP Emilio Segre Visual Archives, Gift of C. J. Peterson*

Rutherford discovered that he could suppress some of the uranium rays by wrapping the samples in thin aluminum foils; adding more layers of foil progressively decreased the activity. He deduced that there appeared to be two types of emissions present, which he termed "alpha" and "beta." Alpha-rays could be stopped easily by a thin layer of foil or a few sheets of paper, but beta-rays were more penetrating; Becquerel later showed that alpha-rays were positively charged, while beta-rays were ordinary negative electrons. By passing both by a magnet, Becquerel further determined that alpha-rays weighed about 7,000 times as much as the electronic beta rays. Magnets exert forces on electrical charges; by observing how the two types are deflected, their masses can be determined.

In the fall of 1898, Rutherford completed his studies at Cambridge and moved to McGill University in Montreal, where he had been appointed as the McDonald Professor of Physics. Over the following three decades he continued his radioactivity research, both at McGill and later back in England, work which would result in a series of groundbreaking discoveries and earn him the 1908 Nobel Prize for Chemistry.

Rutherford's first major discovery at McGill occurred in 1900, when he found that, upon emitting its radiation, thorium simultaneously emitted a product which he named "emanation." Emanation was also radioactive, and, when isolated, its rate of radioactivity was observed to decline in a geometrical progression with time. Specifically, the activity decreased by a factor of one-half for every minute that elapsed. Rutherford had discovered the property of radioactive half-life, a natural random decay process.

Rutherford sought to identify what the thorium emanation actually was, and teamed up with Frederick Soddy, a young Demonstrator in Chemistry. They had expected the emanation to be some form of thorium, but Soddy's analysis revealed that it behaved like an inert gas. This suggested that, remarkably, thorium could spontaneously transmute itself into another element, one of the pivotal discoveries of twentieth-century physics. Soddy initially thought that the emanation was argon (element 18), but it would later come to be recognized as radon (element 86). Various isotopes of thorium, radium, and actinium decay to various isotopes of radon, which themselves subsequently decay until achieving stability when they become atoms of lead.

Frederick Soddy (1877–1956). Source: http://commons.wikimedia.org/wiki/File:Frederick_Soddy_(Nobel_1922).png

Each isotope that decays has its own characteristic half-life. As a standard for comparing radioactivity rates of different substances, Marie and Pierre Curie adopted the rate of decay of a freshly-isolated one-gram sample of radium-226. This isotope, which has a half-life of 1599 years, is a prodigious emitter of alpha-particles. For a one-gram sample, the initial decay rate is about 37 billion nuclear disintegrations per second. This is a huge number, but a gram of radium contains some one billion trillion atoms, and so will maintain for many years essentially the same rate of activity with which it began. In 1910, the International Radium Standard Commission defined one Curie of radioactivity to be exactly 37 billion disintegrations of any type per second. This designation honored Pierre Curie, who had been killed tragically when run over by a heavily-laden horse-drawn wagon while crossing a rainy street in Paris in 1906.

In many situations, a Curie is too large a unit of activity for practical use, so it is more common to encounter millicuries (one-thousandth of a Curie), or even microcuries (one-millionth of a Curie). Household smoke detectors contain about one microcurie (37,000 decays per second) of radioactive material which ionizes a small volume of air around a sensor in order to aid in the detection of smoke particles. The modern unit of radioactivity is the Becquerel; one Becquerel is equal to one decay per second. In this unit, a smoke detector would be rated as having an activity of 37,000 Bq, or 37 kBq. Decay rates are critically important factors in nuclear weapons engineering.

2.4   The Energy of Radioactive Decay

The most fundamental quantity in a physicist's universe is energy. In everyday life, we acquire energy by digesting food, and use it to do work. Machines are no different: energy is supplied by burning fuel or running water and returned as mechanical work, minus what is eroded into useless heat by friction.

Electric bills quantify energy use in kilowatt-hours, and nutrition labels speak of food calories. At the atomic level, however, such units are far too large for quantifying individual reactions. Physicists who study atomic processes adopt a more convenient one, the electron-Volt, always abbreviated eV. There is a formal technical definition of this oddly-named quantity, but an everyday example will make the point: with an ordinary 1.5-V battery, the electrons supplied each emerge with 1.5 eV of kinetic energy. A 9-V battery consists of six 1.5-V batteries connected in series, so their electrons emerge with energies of 9 eV. On an atom-by-atom basis, chemical reactions involve energies of a few eV; when dynamite is detonated, the energy released is 10 eV per reacting molecule.

Nuclear reactions are much more energetic than chemical ones, typically involving energies of *millions* of electron-volts, abbreviated as MeV ("emm-ee-vee"). If a nuclear reaction liberates 1 MeV per atom involved (nucleus, really) while a chemical reaction liberates 10 eV per atom involved, the ratio of the nuclear to chemical energy releases is 100,000. This immense leap is what gives nuclear weapons their compelling power in comparison to ordinary chemical bombs. An ordinary bomb that contains 1,000 lb of chemical explosive could in theory be replaced with a nuclear bomb that utilizes only a fraction of a pound of a nuclear explosive. Even if a nuclear bomb detonates with low efficiency, the destructive energy released can still be immense. Thousands of tons of conventional explosive can be replaced with a few tens of pounds of nuclear explosive. The fission bombs used at Hiroshima and Nagasaki involved reactions which liberated about 170 MeV per reaction, and had efficiencies of about 1% and 20%, respectively. The immense energy latent in nuclei was appreciated well before Einstein wrote $E = mc^2$.

2.5   Isotopes, Mass Spectroscopy, and the Mass Defect

In modern chemistry, an element's location in the periodic table is dictated by the number of protons in the nuclei of its atoms. Each block on the table represents a single element, which might comprise many isotopes.

The concepts of atomic number and isotopy developed over the course of a full century. In 1803, English chemist John Dalton hypothesized that all atoms of a given element are identical to each other and equal in weight. Evidence from chemical reactions indicted that the masses of atoms of various elements seemed to be very nearly equal to whole-number multiples of the mass of hydrogen atoms. This notion was formalized around 1815 by English physician and chemist William Prout, who postulated that all heavier elements are aggregates of hydrogen atoms. He called the hydrogen atom a "protyle," a forerunner of Ernest Rutherford's "proton."

Parts of both Dalton's and Prout's hypotheses would be verified, but, as with any evolving scientific theory, other aspects came to require modification. In particular, something looked awry with Prout's idea in that some elements had atomic weights that were not close to integer multiples of that of hydrogen. An outstanding case was chlorine, whose atoms seemed to weigh 35.5 times as much as hydrogen atoms. This is now understood on the basis that chlorine has two naturally-occurring isotopes: $^{35}_{17}Cl$ and $^{37}_{17}Cl$, which have abundances of about 75 and 25%; the overall weight of 35.5 reflects a percentage-weighted average of 35 and 37.

The concept of isotopy first arose from evidence gathered in studies of natural radioactive decay sequences. Substances that appeared in different decay sequences had similar properties, but could not be separated from each other by chemical means. "Isotope" was coined in 1913 by Frederick Soddy, who had taken a position at the University of Glasgow. By that time, it was appreciated that the masses of nuclei could not be accounted for solely by the masses of their protons. Extra, electrically-neutral mass had to be present, and it was hypothesized to be due to additional protons within nuclei whose positive charges were balanced by the presence of negative electrons somehow contained within the nuclei. Soddy hypothesized that the net positive charge of the nucleus dictated the atom's place in the periodic table, and termed atoms with the same net charge but differing numbers of proton/electron combinations "isotopic elements" which occupied the same place in the table because they would have the same net positive charge. "Iso" comes from the Greek "isos," meaning "equal," and the *p* in *tope* serves as a reminder that it is the number of positive charges which is the same in all isotopes of a given element.

True understanding of the nature of isotopy came with the invention of mass spectroscopy, a technique for making extremely precise measurements of atomic masses. In his 1897 electron work, J. J. Thomson measured the ratio of the electrical charge carried by electrons to their mass by using electric and magnetic fields to deflect them and track their trajectories. In 1907, he modi-fied his apparatus to investigate the properties of positively-charged (ionized) atoms, and so developed the first mass spectrometer. In these devices, electric and magnetic fields were configured to force ionized atoms to travel along separate trajectories which depended on their masses. The ions were directed to strike photographic films, and their landing locations could be analyzed to yield their masses. A vastly scaled-up version of this technique would be used to separate isotopes of uranium during the Manhattan Project.

In 1909, Thomson acquired an assistant, Francis Aston, a gifted instrument-maker. Aston improved Thomson's instrument, and in November, 1912, obtained evidence for the presence of two isotopes of neon, of masses 20 and 22 (taking hydrogen to be of mass 1). The atomic weight of neon was known to be 20.2, and Aston reasoned that this could be explained if the two isotopes were present in a ratio of 9:1.

Aston's research was sidelined by World War I, but he returned to Cambridge in 1919 and began constructing his own series of mass spectrometers. In all, he would discover over 200 naturally-occurring isotopes, including uranium-238. Remarkably, he does not have an element named after him, but he did receive the 1922 Nobel Prize for Chemistry.

Francis Aston (1877–1945).
Source: http://commons.
wikimedia.org/wiki/File:
Francis_William_Aston.jpg

Mass spectrometers are now a staple of every chemistry laboratory, and all still work on the principles pioneered by Thomson and Aston. Fundamentally, these devices utilize an effect known as the Lorentz Force Law, named after Dutch physicist Hendrik Lorentz. The sample to be investigated is first heated in a small oven. The heating will ionize the atoms, some of which escape through a narrow slit. The ionized atoms are then accelerated by an electric field and directed into a region of space where a magnetic field is present, arranged to be perpendicular to the plane of travel of the positively-charged ions. The magnetic field exerts a so-called Lorentz Force on the ions, which causes them to move in circular trajectories whose radii depend on their masses and net charges. For a given net charge, heavier ions will be deflected less strongly than lighter ones, and will enter into larger-radius orbits. The ions streams will be maximally separated after one-half of an orbit, where they are collected on a computerized detector for analysis.

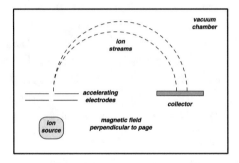

Mass spectroscopy. Positive ions are acceler-ated by an electric field and then directed into a magnetic field which emerges per-pendicularly from the page. Ions of differ-ent mass will follow different circular tra-jectories, with those of greater mass having larger orbital radii. There will be one ion stream for each isotope present

Aston's work revealed that Dalton's conjecture had been partially correct: atoms of the same element behave identically as far as their chemistry is con-cerned, but the presence of isotopes means that not all atoms of the same element have the same weight. Similarly, Aston found that Prout's conjec-ture that the masses of all atoms are integer multiples of that of hydrogen, was also very nearly true if one substitutes "isotopes" for "atoms". But "very nearly" proved to involve some very important physics.

A good example of this "very nearly" is provided by the common form of iron, Fe-56, nuclei of which contain 26 protons and 30 neutrons. In Prout's hypothesis, the mass of an iron-56 atom should be 56 hydrogenic "mass units", if the very tiny contribution of the electrons is neglected. Mass spectroscopy can determine the masses of atoms to remarkable precision, and reveals that the actual weight of iron-56 atoms is 55.934937 mass units. Aston termed the minute shortfall, a mere 0.065063 mass units, the "mass defect". This effect proved to be systemic across the periodic table: all stable atoms proved less massive than would be predicted on the basis of Prout's whole-number hypothesis. The inescapable conclusion was that when protons and neutrons assemble themselves into nuclei, they must give up some of their mass to do so. Mass defects are now quoted in terms of the equivalent energy in MeVs via $E = mc^2$. One mass unit is equivalent to about 930 MeV of energy, so the iron-56 mass defect amounts to some 60 MeV, a substantial amount.

Where does the mass go when nature assembles nuclei? Empirically, stable nuclei have to hold themselves together against the immense mutually repulsive electrical forces of their constituent protons; some sort of nuclear "glue" must be present. To physicists, this glue is known as "binding energy," and is presumed to originate from the missing mass transformed into an attractive form of energy. All stable nuclei show a mass defect. Conversely, some nuclei, such as U-235 and U-238, possess mass surpluses. All mass-surplus nuclei eventually decay to more stable mass-energy configurations.

2.6   Alpha Particles and the Nuclear Atom

In the spring of 1907, Rutherford returned to England to take up a position at Manchester University. When he arrived there he drew up a list of promising research projects, one of which was to pin down the nature of alpha particles. Based on experiments where the number of alphas emitted by a sample of radium had been counted and the electrical charge carried by each had been determined, he had begun to suspect that they were helium atoms that had been stripped of their usual complement of two electrons, that is, that they are bare helium nuclei. However, he needed to trap a sample of alphas for confirming spectroscopic analysis. Working with student Thomas Royds, Rutherford accomplished this in 1909 with one of his typically elegant table-top experiments.

In the Rutherford-Royds experiment, a sample of radon gas, a known alpha-decayer with a half-life of about four days, was trapped inside a very thin-walled glass tube. This tube was placed inside another, thicker-walled one. The space between the two tubes was evacuated, and the radon in the inner tube was allowed to decay for a week. The energetic radon alphas could easily penetrate through the wall of the inner tube; in doing do they would pick up electrons, become neutralized, and then remain trapped in the space between the tubes. The neutralized alphas were then drawn off for analysis, and clearly showed a helium spectrum. In the shorthand notation for describing nuclei, alpha particles are identical to helium-4 nuclei: ^4_2He.

The discovery for which Rutherford is most famous, namely that atoms have nuclei, also had its beginning in 1909. One of the projects on his list was to investigate how alpha particles scattered from atoms when they (the alphas) were directed through a thin metal foil. At the time, the prevailing notion of the structure of atoms, due to J. J. Thomson, was that they comprised clouds of positive electrical material within which were embedded negatively-charged electrons. This picture has often been likened to a plum pudding, with electrons playing the role of embedded raisins. A related line of atomic structure evidence came from chemistry reserach. From the bulk densities of elements and their atomic weights, it could be estimated that individual atoms behaved as if they were about one ten-billionth of a meter in diameter. This length is now known as an Ångstrom, after Swedish spectroscopist Anders Ångstrom. The one-Ångstrom diameter presumably represented the size of the overall cloud of positive material in Thomson's plum-pudding atom.

Rutherford had been experimenting with passing alpha-particles through metal foils since his first days of radioactivity research, and all of his experience indicated that the vast majority of alphas were deflected by only a very few degrees from straight-line paths as they barreled their way through a foil. This behavior was roughly in line with theoretical expectations. Thomson had calculated that the combination of the size of an Ångstrom-scale positively-charged atomic sphere and the kinetic energy of an incoming alpha (itself also presumably about an Ångstrom in size) would be such that an alpha would typically suffer only a small deflection from its initial trajectory. Deflections of a few degrees would be rare, and a deflection of 90° was expected to be so improbable as to never have any practical chance of being observed. In Thomson's atom, collisions between alphas and atoms was not to be imagined as like those between billiard balls, but rather more like diffuse clouds of positive electricity passing through each other. The alphas would presumably strike a number of electrons during the collision, but the effect of the electrons' attractive force on the alphas would be negligible due to the vast difference in their masses; electrons played no part in Rutherford's work.

Rutherford was working with Hans Geiger (of Geiger counter fame), who was looking for a project to occupy undergraduate student Ernest Marsden. Rutherford suggested that they check to see if they could observe any large-angle deflections of alphas on passing them through a thin foil of gold leaf.

Gold was used as the experimental target because it could be pressed into a thin foil only about a thousand atoms thick. Rutherford fully expected a negative result, but to Geiger and Marsden's great surprise, a few alphas, about one in every 8,000, were bounced backward toward the direction from which they came. The number of such occurrences was small, but vastly more than what was expected on the basis of Thomson's plum-pudding model. Rutherford was later quoted as saying that the result was "almost as incredible as if you had fired a 15-in shell at a piece of tissue paper and it came back and hit you."

So unexpected was this result that it took Rutherford over a year to infer that it meant that the positive electrical material within atoms must be confined to much smaller volumes than had been thought to be the case. The alpha-particles—themselves also nuclei—had to be similarly minute; only in this way could the electrical force experienced by an incoming alpha be intense enough to achieve the necessary repulsion to turn it back if it should by chance approach a target nucleus head-on. The vast majority of alpha nuclei sailed through the foil undisturbed, missing gold nuclei by wide margins, but an occasional one would suffer a head-on collision.

The compaction of the positive charge required to explain the scattering experiments was dramatic: down to about a hundred-thousandth of an Ångstrom. But atoms still behaved in bulk as if they were about an Ångstrom in diameter. Both results were experimentally indisputable and had to be accommodated. Rutherford's solution was that the vast majority of the mass in atoms must be contained within minute, positively charged nuclei, with the much less massive electrons orbiting at distances of about an Ångstrom away. Thus was born the high-school "Rutherford atom." Atoms are mostly empty space. A sense of just how empty can be had by thinking of the lone proton that forms the nucleus of an ordinary hydrogen atom as scaled up to being two millimeters in diameter, about the size of an uncooked grain of rice. If this enlarged proton is placed at the center of an American football field, the diameter of the closest electron orbit would be at about the goal lines.

The first public announcement of this new concept of atomic structure came when Rutherford addressed the Manchester Literary and Philosophical Society on March 7, 1911. The formal scientific publication came in July, and directly influenced Niels Bohr's famous atomic model of hydrogen, which was published two years later. Curiously, Rutherford did not use the word "nucleus" in his paper; that term was introduced by Cambridge astronomer John Nicholson in a paper published in November, 1911. Rutherford himself coined "proton" in 1920.

Scattering experiments offered a way to study collisions between nuclei. With this understanding, Rutherford's analysis could be applied to other elements in the sense of using an observed scattering distribution to infer how many protons the target element possessed; this helped to place elements in their proper locations in the periodic table. Elements had theretofore been defined by their atomic weights, but it was the work of researchers such as Rutherford, Soddy, Geiger, and Marsden which showed that an element's chemical identity is dictated by its number of protons.

The atomic weights of elements were still important, however, and still a mystery. Chemical and scattering evidence indicated that the atomic weights of atoms seemed to be proportional to the number of protons their nuclei contained; specifically, atoms of all elements weighed about twice as much or more as could be accounted for on the basis of the numbers of protons alone. This factor of two is what had led to the idea that the additional mass was due to extra protons in the nucleus which for some reason contained electrons within themselves to form electrically-neutral combinations. But by the mid-1920s this concept was becoming untenable: the Heisenberg Uncertainty Principle of quantum mechanics ruled out the possibility of containing electrons within so small a volume as that of an entire nucleus, let alone that of a proton. For many years before its discovery, Rutherford speculated that there likely existed a third, electrically-neutral fundamental constituent of atoms. He would live to see his suspicion proven by one of his own students, James Chadwick. That atoms are built of electrons dancing about nuclei comprised of protons and neutrons is due very much to Rutherford and his collaborators and students.

2.7 Decay Chains

The work initiated by Rutherford and then carried on in many laboratories around the world resulted in the establishment of two golden rules for nuclear reactions, be they natural decay processes or human-induced. Rule number one is that the total number of input nucleons must equal the total number of output nucleons, where the term "nucleon" collectively designates protons and neutrons. While the numbers of protons and neutrons may and usually do change, their sum must be conserved. Rule two is that total electric charge must be conserved. Protons count as one unit of positive charge, and electrons as one unit of negative charge. Beta-decays involve nuclei which create within themselves and then eject an electron to the outside world, and the charges of these electrons must be taken into account in ensuring charge conservation.

But electrons are not considered to be nucleons, and so are not counted when applying the first rule. Electrons are still also known as beta-negative (β^-) particles from Rutherford's early naming of them.

A remarkable feature of these experiments is that in reactions where the input and output nuclei are different, mass is not conserved: the sum of the input masses is not equal to the sum of the output masses. Mass can either be created or lost; what happens depends on what nuclei are involved. The interpretation of this also relates to $E = mc^2$. If mass is lost (sum of the output masses is less than the sum of the input masses), the lost mass will appear as kinetic energy of the output products. Conversely, if mass is gained, then energy must be drawn from somewhere to create mass, and the only source available is the kinetic energy of the bombarding input nucleus. It is common to express the mass gain or loss in units of energy equivalent, almost always in MeV.

Rutherford decoded alpha-decay as a nucleus spontaneously transmuting itself to a more stable mass-energy configuration by ejecting a helium nucleus. In doing so, the original nucleus loses two protons and two neutrons, which means that the remaining nucleus will be two places down in atomic number on the periodic table and have four fewer nucleons in total. Alpha-emission is a common decay mechanism in heavy elements such as uranium. A shorthand notation to designate alpha-decay puts the original nucleus on the left side of a rightward pointing arrow, the resulting lighter nucleus and the alpha particle on the right side of the arrow, and sometimes the half-life for the process below the arrow. For uranium-235:

$$\,^{235}_{92}U \quad \rightarrow \quad \,^{231}_{90}Th + \,^{4}_{2}He.$$

The half-life for this process, about 704 million years, is truly astronomical. In alpha-decays, the total mass of the output products is always less than that of the input particles: nature spontaneously seeks a lower mass-energy configuration. The energy release in alpha-decays is typically 5–10 MeV, the majority of which appears as kinetic energy of the alpha-particle itself.

For beta-decay, experiments showed that the numbers of neutrons in stable nuclei correlates tightly with the number of protons: nature is willing to invest in nuclear glue, but only so much as is necessary to guarantee stability. In cases where nuclei find themselves with more neutrons than necessary to stabilize their number of protons (say, by having captured an incoming neutron), they spontaneously correct the neutron/proton number balance by shedding the extra neutrons. But the shedding must be in accordance with the two golden rules. If a nucleus is too neutron-rich for its number of protons, neutrons spontaneously decay into protons, a process consistent with the first golden rule. But this would represent a net creation of positive electric charge with each decay, leading to a violation of the second rule. To correct the charge balance, Nature creates an ordinary negative electron in each decay to render no net charge created or lost. This is known as β^- decay, and has the net result of moving a nucleus up one place in the periodic table because it gains one proton. An example of this is that if a U-238 nucleus captures a neutron to become U-239, it will undergo beta-decay with a half-life of about 23.5 min to become neptunium-239:

$$^{239}_{92}U \rightarrow {}^{239}_{93}Np + electron.$$

The electron is detected in the outside world when it triggers a Geiger counter. It is usually the product nucleus that is of interest, so it is common to not bother writing down the electron.

Extended sequences of alpha and beta-decays may follow until a decaying nucleus reaches stability. The work of the Curies and Rutherford revealed that three lengthy decay sequences occur spontaneously in Nature. All begin with isotopes of thorium or uranium and terminate with three different isotopes of lead. The original uranium and thorium must have been created in whatever supernova explosion gave rise to the nebula out of which the Sun and solar system formed some five billion years ago; their half-lives are long enough that they can still be found in Earth's crust.

The work accomplished by the first generation of radio-physicists and radio-chemists was staggering. Uranium and thorium ores contain constantly varying amounts of various isotopes which are created by decays of heavier parent isotopes and which themselves decay to lighter products until they arrive at stable configurations. Only by isolating samples of individual elements and subjecting them to mass-spectroscopic and chemical analyses could individual isotopes and decay sequences be worked out.

2.8 Artificial Transmutation

Rutherford's last great discovery came in 1919, the realization that it was possible to set up experiments wherein atoms of a given element could be transmuted into those of another when bombarded by nuclei of yet others. While the idea of transmutation was not new—it is what happens in natural decays—what was new was that transmutations could be induced by human intervention.

The work that led to this discovery was begun around 1915 by Ernest Marsden. As part of a program involving measurements of reaction energies,

Marsden arranged to bombard hydrogen atoms with alpha-particles produced by the decay of radon gas contained in small glass vials. A hydrogen nucleus—a lone proton—would receive a significant kick when struck by an alpha-particle, and be set into motion at high speed. These experiments were done by sealing the alpha source and hydrogen gas inside a small chamber. At one end of the chamber was a small phosphorescent scintillation screen which could be viewed through a microscope. Protons striking the screen would cause a small flash of light, which would be detected by an observer at the eyepiece; this is exactly the routine that had been used in the alpha-scattering experiments. By placing layers of thin metal foils between the screen and the microscope, Marsden could determine the ranges, and hence the energies, of the struck protons. This was routine work that involved using known laws of conservation of energy and momentum to cross-check and interpret measurements.

Breakthroughs favor an attentive mind. Marsden noticed that when the experimental chamber was empty, the radon source itself seemed to give rise to scintillations like those from hydrogen, even though there was no hydrogen in the chamber. The implication seemed to be that protons were arising directly in radioactive decay, a phenomenon never before observed.

Marsden had to leave for New Zealand in early 1915 to take up a professorship at Victoria College in Wellington, although he would soon find himself back in Europe on active duty on the Western Front in the New Zealand Division Signals Company. Rutherford, heavily occupied with research for the British Admiralty, could manage only occasional experiments until World War I came to an end in November, 1918. In 1919 he turned to investigating Marsden's observation.

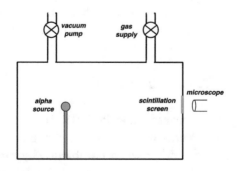

Sketch of the Marsden/Rutherford apparatus for the discovery of artificial transmutation

Rutherford placed a source of alpha particles within a small brass chamber which could be evacuated and then filled with a gas with which he wished to experiment. As reported in his June 1919 discovery paper, he set out to investigate the phenomenon that "a metal source, coated with a deposit of [an alpha-emitter] always gives rise to a number of scintillations on a zinc sulphide screen far beyond the range of the α particles. The swift atoms causing these scintillations… have about the same range and energy as the swift H atoms produced by the passage of α particles through hydrogen. These 'natural' scintillations are believed to be due mainly to swift H atoms from the radioactive source, but it is difficult to decide whether they are expelled from the radioactive source itself or are due to the action of α particles on occluded hydrogen."

In classic scientific style, Rutherford proceeded by investigating various possibilities for the origin of the hydrogen scintillations. No vacuum pump is ever perfect; some residual air would always remain in the chamber no matter how thoroughly it had been evacuated. While hydrogen is normally a very minute component of air, more could be present if the air contained water vapor. Suspecting that the alphas might be striking residual "occluded" hydrogen-bearing water molecules, he began by introducing dried oxygen and carbon dioxide into the chamber, observing, as expected, that the number of scintillations decreased. Surprisingly, however, when he admitted dried air into the chamber, the number of hydrogen-like scintillations increased. This suggested that protons were arising not from the alpha-emitter but rather from some interaction of the alphas with air.

The major constituents of air are nitrogen and oxygen. With oxygen eliminated, Rutherford inferred that nitrogen might be involved. On admitting pure nitrogen into the chamber, the number of scintillations increased yet again. As a final test that hydrogen was not somehow arising from the radioactive source itself, he found that the scintillations persisted on placing thin metal foils close to the source, but with ranges reduced in accordance with what would be expected if the alphas were traveling through the foils before striking nitrogen atoms. The scintillations were clearly arising from within the volume of the chamber, not from the radioactive source itself. As Rutherford wrote, "it is difficult to avoid the conclusion that the long-range atoms arising from collision of α particles with nitrogen are … probably atoms of hydrogen …. If this be the case, we must conclude that the nitrogen atom is disintegrated under the intense forces developed in a close collision with a swift α particle, and that the hydrogen atom which is liberated formed a constituent part of the nitrogen nucleus".

In modern notation, the alpha-nitrogen transmutation is

$$\ _2^4He + \ _7^{14}N \ \rightarrow \ _1^1H + \ _8^{17}O + \gamma.$$

The "γ" indicates that this reaction also releases a high-energy gamma-ray. Gamma-rays have no mass or electrical charge and played no role in interpreting the reaction by via the first golden rule, but they are characteristic products of such reactions, and would play a significant role years later in the discovery of the neutron.

Rutherford and Marsden's discovery opened yet another research horizon. Could alphas induce transmutations in any other elements? What products could be created, and with what efficiencies? What energy gains or losses were involved?

Later in 1919, Rutherford moved from Manchester back to Cambridge University to become Director of the Cavendish Laboratory when that position came vacant upon the retirement of J. J. Thomson. He would remain at Cambridge until his death in October, 1937, nurturing another generation of nuclear experimentalists. Element 104, rutherfordium, is now named in his honor.

2.9   Particle Accelerators and Cyclotrons

Rutherford's discovery of artificial transmutation opened a host of questions, but Nature was not about to give up her secrets without a fight.

A simple interpretation of Rutherford's alpha/nitrogen reaction equation is that if you were to mix helium and nitrogen, say at room-temperature conditions, hydrogen and oxygen would spontaneously result. But this is not so due to an effect known as the Coulomb barrier.

French physicist Charles-Augustin de Coulomb, who performed some of the first quantitative experiments with electrical forces in the late 1700s, determined that electrical charges of the same sign repel each other. Because of this, nitrogen nuclei will inevitably repel Rutherford's incoming alpha particles. It is only because an alpha is born endowed with a significant amount of kinetic energy when its radon parent decays that it can approach a nitrogen nucleus closely enough that stronger but shorter-range inter-nucleon forces can come into play and precipitate transmutation reactions. The motional energy of helium atoms at room-temperature conditions is only a fraction of an electron-Volt, much less than the approximately 4.2 MeV necessary for a helium nucleus to induce a reaction upon striking nitrogen. Decay-generated alphas carry some 5 MeV of kinetic energy, and so can induce the reaction.

In using alphas created in natural decays, it is practical to carry out bombardment experiments with target elements only up to atomic numbers of about 20 (calcium). To try to induce a reaction by striking a uranium nucleus with a 5-MeV alpha would be as hopeless as trying to perturb a bowling ball with a marble. By the mid-1920s, this was becoming a seemingly insurmountable limitation: researchers were literally running out of elements to experiment with. The desire to bombard heavier elements generated a technological challenge: was there any way that the alpha or other particles could be accelerated once they had been emitted by their parent nuclei?

The first practical particle acceleration scheme was published by Norwegian native Rolf Wideröe in a German electrical engineering journal in 1928. The essence of Wideröe's proposal was to place two hollow metal cylinders end-to-end and connect them to a source of voltage which could be rapidly varied between positive and negative. To begin, a stream of protons (say) is directed into the leftmost cylinder, which is initially arranged to be negatively charged. This will attract the protons, which will speed up as they pass through the cylinder.

Left: Rolf Wideröe (1902–1996). Right: Ernest Lawrence (1901–1958). Sources: AIP Emilio Segre Visual Archives; http://commons.wikimedia.org/wiki/File:Ernest_Orlando_Lawrence.jpg

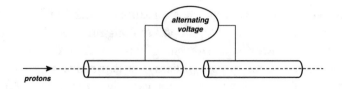

Wideröe's linear acceleration scheme

Just as the bunch of protons emerges from the first cylinder, the voltage polarity is reversed to make the left cylinder positive and the right one negative. The protons then get a push from the first cylinder while being pulled into the second one, which further accelerates them. By placing a number of such units of increasing length end-to-end, substantial accelerations can be achieved. Many of the incoming particles will be lost by crashing into the side of a cylinder or because their speed does not match the cycling of the polarity shifts of the power supplies; only a few might emerge from the last cylinder. But the point was not efficiency; rather, it was to generate some high-speed particles which could surmount the Coulomb barriers of heavy target nuclei. In contrast to Wideröe's table-top model, the longest linear accelerator in the world is now the Stanford Linear Accelerator in California, which can accelerate electrons to 50 billion electron-volts of kinetic energy over a distance of 3.2 km (2 miles).

Wideröe's work came to the attention of Ernest Orlando Lawrence, an experimental physicist at the University of California at Berkeley. Lawrence possessed considerable experience with electrical devices, and by late 1930 he and collaborator David Sloan had built a Wideröe device with which they were able to accelerate mercury ions to kinetic energies of 90,000 eVs. They desired to achieve higher energies, but were daunted by the idea of building an accelerator that would be meters in length. How could the device be made more compact?

Lawrence was well-familiar with how magnetic fields such as used in Francis Aston's spectrometers could be configured to separate ions of different masses by the Lorentz Force Law. Lawrence's inspiration, which he called a cyclotron, also made use of this effect, but in a way that simultaneously incorporated Wideröe's alternating-voltage scheme.

Lawrence connected a voltage supply to two D-shaped metal tanks placed back-to back; such tanks are now known to cyclotron engineers as "Dees." The entire assembly must be placed within a surrounding vacuum tank to avoid deflective effects of collisions of the accelerated particles with air molecules.

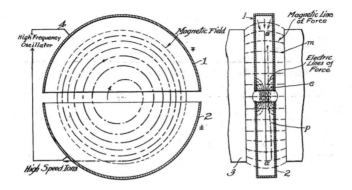

Schematic illustration of Lawrence's cyclotron concept in top and side view, from his patent application. Source: http://commons.wikimedia.org/wiki/File:Cyclotron_patent.png

The source of the ions (usually positive) is placed between the Dees. In the accompanying diagram, the ions are initially directed toward the upper Dee, which is set to carry a negative charge to attract them. If the voltage polarity is not changed and there is nothing to otherwise deflect the ions, they would crash into the outer rim of the Dee. But by placing the Dees between the poles of a powerful electromagnet arranged to make a magnetic field perpendicular to them (into or out of the page), he could invoke the Lorentz effect and force the ions to move in circular paths. The combination of the ions' acceleration toward the charged Dees and the Lorentz force causes them to move in outward-spiraling trajectories. If the magnetic field is strong, the spiral pattern will be tight and the ions will get nowhere near the edge of the Dee in their first orbit. As ions leave the upper Dee, the voltage polarity is switched in order to attract them to the lower Dee. Switching and acceleration continues for a few microseconds until the ions strike a target at the periphery of one of the Dees.

Lawrence and graduate student Nils Edlefsen first reported on the cyclotron concept at a meeting of the American Association for the Advancement for Science in September, 1930, although they had no results available at that time. By May, 1931, Lawrence and student M. Stanley Livingston had a 4.5-in- diameter device in operation, and reported that they were able to accelerate ions of molecular hydrogen to energies of 80,000 eVs using only a 2,000-V power supply. Later the same year, Lawrence achieved MeV energies with an eleven-inch cyclotron. By 1932 he had constructed a 27-in device which achieved an energy of 3.6 MeV, but he had bigger plans yet. As adept at fundraising as he was at electrical engineering, by 1937 Lawrence had constructed a 37-in model capable of accelerating nuclei of "heavy hydrogen" to energies of 8 MeV. (Heavy hydrogen, also known as deuterium, is a form of hydrogen where the nuclei consist of one proton and one neutron bound together: ^2_1H) By 1939, he had brought into operation a 60-in model that required a 220-ton magnet and which could accelerate deuterium nuclei to 16 MeV. In 1942, he brought online his 184-in diameter cyclotron, which is still operating and can accelerate various types of particles to energies exceeding 100 MeV. Along the way, Lawrence established the University of California Radiation Laboratory, the "Rad Lab", which is now the Lawrence Berkeley National Laboratory.

Left: Lawrence's original 4.5-in cyclotron. Middle: Lawrence at the controls of his 184-in cyclotron. Right: M. Stanley Livingston and Lawrence at the 27-in cyclotron. Sources: Lawrence Berkeley National Laboratory, courtesy AIP Emilio Segre Visual Archives

Particle accelerators allowed experimenters to scale the Coulomb barrier and so open up a broad range of energies and targets to further experimentation. Lawrence's ingenuity garnered him the 1939 Nobel Prize for Physics, and a variant of his cyclotron would play a significant role in the Manhattan Project. Today's giant accelerators at the Fermi National Accelerator Laboratory in Illinois and the European Organization for Nuclear Research (CERN) are the descendants of Wideröe's and Lawrence's pioneering efforts.

While Wideröe and Lawrence are now thought of as the fathers of particle acceleration, their efforts were anticipated by an eclectic Hungarian-born inventor, engineer, physicist, and personal friend and sometimes-collaborator of Albert Einstein named Leo Szilard. Szilard had submitted three patent applications for methods of accelerating particles (two in Germany and one in Britain), including both a linear accelerator (1928) and a cyclotron (1929—prior to Lawrence), but he did not pursue them. Later, Szilard would conceive the idea of a chain reaction, and would play a significant role in alerting American government officials to the possibility of nuclear weapons.

Leo Szilard (1898–1964).
Source: http://commons.
wikimedia.org/wiki/File:
Leo_Szilard.jpg

2.10 Enter the Neutron

The discovery of the neutron in early 1932 by Rutherford protégé James Chadwick was a critical turning point in the history of nuclear physics.

Left: Walther Bothe (1891–1957); Right: James Chadwick (1891–1974). Sources: Original drawing by Norman Feather, courtesy AIP Emilio Segre Visual Archives; http://commons.wikimedia.org/wiki/File: Chadwick.jpg

The experiments which led to the discovery of the neutron were first reported in 1930 by German physicist Walther Bothe and his student, Herbert Becker. They were studying the gamma rays produced when light elements are bombarded by energetic alpha-particles, which, as Rutherford had found, typically produces a proton and a gamma-ray. Bothe was an accomplished experimentalist, having worked on alpha-bombardment light-element-disintegration reactions since 1927.

In mid-1930, Bothe and Becker found that boron, lithium, and particularly the light metallic element beryllium gave evidence of gamma-emission under alpha bombardment, but with no protons being emitted. They were certain that some sort of energetic radiation which could penetrate foils of metal was being emitted. But the radiation could not be deflected by a magnetic field, which meant that it must be electrically neutral. Gamma-rays were the only electrically neutral form of foil-penetrating radiation known at the time, so it was natural for them to interpret their results as evidence of gamma-ray emission despite the lack of protons. The energy of their supposed gamma rays was, however, greater than that of those typically released in radioactive decays, which prompted them to propose that they were detecting a new type of gamma-rays, albeit with properties not wildly out of line with known ones.

Bothe & Becker's research was delayed from late 1930 to mid-1931 while they moved to the University of Giessen, where Bothe had been appointed to a professorship. They reported on their work again at a conference on nuclear physics held in Rome in October, 1931, which Marie Curie attended. Crucially, at the same conference, Niels Bohr speculated on the possibility that energy and momentum might not be conserved in nuclear reactions.

Curie reported on the conference to her physicist daughter Irène, who was married to fellow physicist Frédéric Joliot; the two are usually referred to as the Joliot-Curies. They picked up the Bothe and Becker work with the aim of trying to more accurately measure the energies of the presumed gamma-rays. In a paper published in late December, 1931, they reported that beryllium-emitted gammas had energies of 15–20 MeV, substantially more than Bothe and Becker had reported. Just over two weeks later, they reported a crucial new observation: that the presumed gamma-rays, which they termed "beryllium radiation", were capable of knocking protons out of a layer of paraffin wax that had been put in their path.

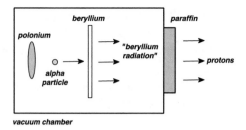

vacuum chamber

The "beryllium radiation" experiment of Bothe, Becker, the Joliot-Curies, and Chadwick

To accommodate the interpretation that alpha-bombardment of beryllium creates gamma-rays with no protons being emitted, the Joliot-Curies hypothesized that the reaction was

$$^4_2He + ^9_4Be \rightarrow ^{13}_6C + \gamma.$$

The energy released in this reaction is about 11 MeV. Polonium decay yields alpha particles with kinetic energies of about 5 MeV, so the emergent γ-ray can at most have an energy of about 16 MeV; this is the origin of the Joliot-Curies' 15–20 MeV estimate. Their presumption was that the gammas were striking protons in the paraffin, setting them into motion. The protons themselves emerged from the paraffin with kinetic energies of about 6 MeV.

At face value, getting a 6 MeV proton from a strike by a particle possessing 16 MeV looks fine, but Chadwick realized that this interpretation was in fact untenable. Suspecting that Bothe & Becker and the Joliot-Curies had stumbled upon the neutron, he immediately set about reproducing and extending their work. His experimental setup involved polonium (the alpha source) deposited on a silver disk one centimeter in diameter placed close to a disk of pure beryllium two centimeters in diameter, with both enclosed in a small vessel which could be evacuated. In comparison to the gargantuan particle accelerators of today, his experiments were literally table-top nuclear physics.

Chadwick key insight was to realize that if a proton was to be accelerated to 6 MeV by being struck by a gamma-ray, conservation of energy and momentum demanded that the gamma-ray would have to possess about 55 MeV of energy, three to four times what it could have from the polonium-beryllium reaction. This surprisingly high energy demand is a consequence of the fact that photons do not possess mass. Einstein's relativity theory had shown that massless particles do carry momentum, but much less than a material particle of the same energy such as a proton or electron. Only an extremely energetic gamma-ray can kick a proton to a kinetic energy of several MeV. The

Joliot-Curies had actually realized that their protons would demand gammas of energy on the order of 50 MeV, but had attributed the discrepancy to the difficulty of accurately measuring energies; it has also been speculated that they may have been influenced by Bohr's thoughts on the possible non-conservation of energy in nuclear processes. But conservation of energy is the most fundamental law in physics, and Chadwick was not willing to cast it aside so quickly.

Before invoking a mechanism involving a hypothetical neutron, Chadwick devised a control experiment to investigate the remote possibility that 55-MeV gammas could be being created in the alpha-beryllium collision. He arranged for the beryllium radiation to strike a sample of nitrogen gas. If the "beryllium radiation" was in fact gamma-rays, he calculated that a nitrogen nucleus should acquire a kinetic energy of about one-half of an MeV. However, the experiment revealed that they were being accelerated to energies of up to nearly 1.5 MeV. This result would in turn require the nitrogen nuclei to have been struck by gamma-rays of about 90 MeV, a result completely inconsistent with the 55 MeV indicated by the proton experiment. Upon having the supposed gamma-rays strike heavier and heavier target nuclei, Chadwick found that "if the recoil atoms are to be explained by collision with a quantum (photon), we must assume a larger and larger energy for the quantum as the mass of the struck atom increases." The absurdity of this situation led him to write that "It is evident that we must either relinquish the application of conservation of energy and momentum in these collisions or adopt another hypothesis about the nature of the radiation." With understated British eloquence, he had utterly demolished the Joliot-Curies' interpretation.

Chadwick then advanced an alternate hypothesis: that if the protons in the paraffin were being struck by electrically neutral material particles of mass similar to that of protons, the kinetic energy of the striking particles need only be on the order of that which the protons acquired in the collision. The collision would then be like one between equal-mass billiard balls, with the incoming one being brought to a stop while the struck one is set into motion with the speed that the incoming one had.

This is the point at which the neutron was born into physicists' menu of particles. In Chadwick's hypothesis, the alpha-beryllium collision leads to the production of carbon and a neutron via the reaction

$$\ {}^{4}_{2}He + {}^{9}_{4}Be \rightarrow {}^{12}_{6}C + {}^{1}_{0}n.$$

${}^{1}_{0}n$ denotes a neutron: it carries no electric charge, but it does count as one nucleon. In this interpretation, a cabrbon-12 atom is produced, as opposed to the Joliot-Curies' carbon-13.

Chadwick initially thought of neutrons as being electrically-neutral combinations of single protons and single electrons, and thus having a mass essentially identical to that of a proton. On this basis he was able to show that the kinetic energy of the ejected neutron would be about 11 MeV, and that a subsequent neutron/proton collision would have no trouble accelerating a proton to a kinetic energy of 6 MeV even after the neutron had found its way out of the beryllium target and through the window of the vacuum vessel on its way to the paraffin. As a check, he calculated that such neutrons striking nitrogen nuclei should set the latter into motion with kinetic energies of about 1.5 MeV, precisely what he had measured.

Chadwick reported his discovery in two papers. The first, titled "Possible Existence of a Neutron," was dated February 17, 1932, and was published in the February 27 edition of *Nature*. An extensive follow-up analysis dated May 10 was published in the June 1 edition of the *Proceedings of the Royal Society of London*; he was awarded the 1935 Nobel Prize in Physics for his discovery. While later experiments showed that the neutron is a fundamental particle in its own right as opposed to a proton/electron composite, that development had no significant effect on his analysis. The proton-neutron-electron picture of atoms is very much a product of the Cavendish Laboratory.

With neutrons, experimenters now had particles that could be used to bombard nuclei without being repelled by them, no matter what the kinetic energy of the neutron or the number of protons in the target nucleus. Neutrons would prove to be the gateway to reactors and bombs, but in 1932 Chadwick anticipated neither development. In the February 29, 1932, edition of the *New York Times*, he was quoted as stating that, "I am afraid neutrons will not be of any use to any one." Thirteen years later he would be in New Mexico to personally witness just what his neutrons could do.

About 18 months after Chadwick's dismissal of the value of neutrons, an idea did arise as to a possible application for them: as links in the progression of a nuclear chain reaction. This notion seems to have occurred inspirationally to Leo Szilard.

Szilard was living in London in the fall of 1933, and happened to read a description of a meeting of the British Association for the Advancement of Science in the September 12 edition of the London *Times*. In an article describing an address to the meeting, Ernest Rutherford was quoted as saying that "We might in these processes obtain very much more energy than the proton supplied, but on the average we could not expect to obtain energy in this way. It was a very poor and inefficient way of producing energy, and anyone who looked for a source of power in the transformation of the atoms was talking moonshine." Szilard reflected on Rutherford's remarks while later strolling the streets of London. From a 1963 interview:

> Pronouncements of experts to the effect that something cannot be done have always irritated me. That day as I was walking down Southampton Row and was stopped for a traffic light, I was pondering whether Lord Rutherford might not prove to be wrong. As the light changed to green and I crossed the street, it suddenly occurred to me that if we could find an element which is split by neutrons and which would emit two neutrons when it absorbed one neutron, such an element, if assembled in sufficiently large mass, could sustain a nuclear chain reaction, liberate energy on an industrial scale, and construct atomic bombs. The thought that this might be possible became an obsession with me. It led me to go into nuclear physics, a field in which I had not worked before, and the thought stayed with me.

It did not take Szilard long to get up to speed in his obsession. Envisioning a chain reaction as a source of power and possibly as an explosive, he filed for patents on the idea in the spring and summer of 1934. His British patent, number 630,726, "Improvements in or relating to the Transmutation of Chemical Elements," was issued on July 4, 1934—the date of Marie Curie's death—and specifically referred to being able to produce an explosion given a sufficient mass of material. To keep the idea secret, Szilard assigned the patent to the British Admiralty. It would be reassigned to him after the war and openly published in 1949.

2.11   Enrico Fermi and Neutron-Induced Radioactivity

Surprisingly, neither the Joliot-Curies nor James Chadwick experimented with using neutrons as bombarding particles. The idea of doing so was instead pursued by a young physicist at the University of Rome, Enrico Fermi. Fermi had established himself as a first-rate theoretician at a young age, publishing his first paper while still a student. As a postdoctoral student in his early twenties he had prepared an important review article on relativity theory, and a few years later he made seminal contributions to statistical mechanics. At 26, he was appointed to a full professorship at the University of Rome, where in 1933 he developed a quantum-mechanically-based theory of beta decay. He was to prove equally gifted as a nuclear experimentalist.

Left: Enrico Fermi (1901–1954). Right: The Fermi family (Laura, Giulio, Nella and Enrico) arrive in America, January, 1939. Sources: University of Chicago, courtesy AIP Emilio Segre Visual Archives; AIP Emilio Segre Visual Archives, Wheeler Collection

Some of Fermi's students and collaborators. Left to right: Oscar D'Agostino (1901–1975), Emilio Segrè (1905–1989), Edoardo Amaldi (1908–1989), and Franco Rasetti (1901–2001). Source: Agenzia Giornalistica Fotovedo, courtesy AIP Emilio Segre Visual Archives

The reticence of Chadwick and the Joliot-Curies to carry out neutron-bombardment experiments was a consequence of the low yields expected. Chadwick estimated that he produced only about 30 neutrons for every million alpha-particles emitted by his polonium. This was because atoms are mostly empty space; most of the alphas do not encounter a target nucleus.

If neutrons interacted with target nuclei with similar improbability, virtually nothing would result. Otto Frisch, one of the co-interpreters of fission, later remarked that "I remember that my reaction and probably that of many others was that Fermi's was a silly experiment because neutrons were so much fewer than alpha particles." But this overlooked the fact that neutrons would not experience a Coulomb barrier. Fermi desired to break into nuclear experimentation, and saw his opening. He began work in the spring of 1934 with a group of students and collaborators that included Edoardo Amaldi, Franco Rasetti, Oscar D'Agostino, and Emilio Segrè, who would later write a very engaging biography of his mentor. In the early 2000's, Italian historians Giovanni Acocella, Francesco Guerra, Matteo Leone, and Nadia Robotti found Fermi's original laboratory notebooks and even some of his neutron sources, and reconstructed a very detailed record of his work.

Fermi's first challenge was to secure a strong neutron source. In this sense he was fortunate in that his laboratory was located in the same building as the Physical Laboratory of the Institute of Public Health, which was charged with controlling radioactive substances in Italy. The Laboratory held many radium sources that had been used for cancer treatments, and Fermi used them as a source of radon gas. Radium decays to radon, which is gaseous at room-temperature conditions. Radon itself then decays with a half-life of about four days, emanating a copious supply of alpha-particles in the process. Fermi harvested the radon into inch-long glass vials which contained powered beryllium. The radon alphas bombarded the beryllium, yielding neutrons as in the Bothe and Becker, Joliot-Curies, and Chadwick experiments. This series of reactions yields neutrons with energies of up to about 10 MeV, enough to escape through the thin walls of the glass vials and bombard a sample of a target element; Fermi estimated that his sources yielded about 100,000 neutrons per second. Samples of the target elements were formed into cylinders which could be placed around the sources to achieve maximum exposure, with the cylinders large enough so that after being irradiated they could be slipped around a handmade Geiger counter.

Fermi's goal was to see if he could induce artificial radioactivity with neutron bombardment. Thinking that the nuclei of a heavy element which were already rich in neutrons would be induced to decay if they should capture additional ones, he first tried platinum (atomic number 78), but fifteen minutes of irradiation gave no discernible signal. He then turned to aluminum, where he hit pay dirt. The reaction involved ejection of a proton from the bombarded aluminum, leaving behind magnesium:

$$\,^1_0 n + \,^{27}_{13}Al \rightarrow \,^1_1 H + \,^{27}_{12}Mg.$$

The magnesium beta-decays back to aluminum with a half-life of about 10 min:

$$\,^{27}_{12}Mg \rightarrow \,^{27}_{13}Al.$$

Guerra and Robotti have pinpointed the date of Fermi's success with aluminum as March 20, 1934. He announced his discovery five days later in the official journal of the Italian National Research Council, and an English-language report dated April 10 appeared in the May 19 edition of *Nature*. By late April, the Rome group had performed experiments on about 30 elements, 22 of which yielded positive results.

Fermi and his co-workers found that, as a rule, light elements exhibited three reaction *channels*: a proton or an alpha could be ejected, or the element might simply capture the neutron to become a heavier isotope of itself and then subsequently decay. In all three cases, the products would undergo beta-decays. With a heavy-element target, the result was typically the latter of these possibilities. Neutron bombardment of gold is characteristic of this pattern:

$$\,^1_0 n + \,^{197}_{79}Au \rightarrow \,^{198}_{79}Au \rightarrow \,^{198}_{80}Hg.$$

The half-life in the last step is just under three days.

By the summer of 1934, Fermi had prepared improved sources, which he estimated were yielding about a million neutrons per second. Based on work with these new sources, he published a stunning result in the June 16, 1934, edition of *Nature*: that his group was producing transuranic elements, that is, ones with atomic numbers greater than that of uranium. Since uranium was the heaviest-known element, this meant that they believed that they were synthesizing new elements, a remarkable claim.

Fermi's assertion was based on the fact that uranium could be activated to produce beta-decay upon neutron bombardment. The results were confusing, however, with evidence for half-lives of 10 s, 40 s, 13 min, and at least two more of up to a day. Whether this was a chain of decays or some mixture of parallel sequences was unknown. Whatever was occurring, however, the initial step was presumably the formation of a heavy isotope of uranium, followed by beta-decay as in the gold reaction:

$$\,^1_0n + \,^{238}_{92}U \rightarrow \,^{239}_{92}U \rightarrow \,^{239}_{93}X,$$

with X denoting a new element with atomic number 93.

The 13-min decay was convenient to work with, and the Rome group managed to chemically isolate its decay product from the bombarded uranium. Analysis showed that the decay product was not any of the elements between lead (atomic number 82) and uranium. Since no natural or artificial transmutation had ever been observed to change the identity of a target element by more than one or two places in the periodic table, it seemed safe to assume that a new element was being created.

Much of the history of nuclear physics is wrapped up in good chemical experimentation techniques. To isolate the product of the 13-min activity, Fermi and his group began with manganese dioxide as a chemical carrier. The rationale for this was that if element 93 was actually being created, it was expected that it would fall in the same column of the periodic table as manganese (Mn; atomic number 25), so the two should have similar chemistry. The Romans' analysis came in for criticism, however, from German chemist Ida Noddack. Noddack was well-regarded; she would be nominated for a Nobel Prize on three occasions. In a paper published in September, 1934, Noddack pointed out that many elements were known to precipitate with manganese dioxide, and that Fermi should have checked for the possibility that elements of lower atomic numbers than that of lead were being produced. In what would prove to be a prescient comment, Noddack remarked that "When heavy nuclei are bombarded by neutrons, it is conceivable that the nucleus breaks up into several large fragments, which would of course be isotopes of known elements but would not be neighbors of the irradiated element." Noddack's breaking up anticipated fission.

1 H																	2 He
3 Li	4 Be											5 B	6 C	7 N	8 O	9 F	10 Ne
11 Na	12 Mg											13 Al	14 Si	15 P	16 S	17 Cl	18 Ar
19 K	20 Ca	21 Sc	22 Ti	23 V	24 Cr	25 Mn	26 Fe	27 Co	28 Ni	29 Cu	30 Zn	31 Ga	32 Ge	33 As	34 Se	35 Br	36 Kr
37 Rb	38 Sr	39 Y	40 Zr	41 Nb	42 Mo	43 Tc	44 Ru	45 Rh	46 Pd	47 Ag	48 Cd	49 In	50 Sn	51 Sb	52 Te	53 I	54 Xe
55 Cs	56 Ba	57 La	72 Hf	73 Ta	74 W	75 Re	76 Os	77 Ir	78 Pt	79 Au	80 Hg	81 Tl	82 Pb	83 Bi	84 Po	85 At	86 Rn
87 Fr	88 Ra	89 Ac															

58 Ce	59 Pr	60 Nd	61 Pm	62 Sm	63 Eu	64 Gd	65 Tb	66 Dy	67 Ho	68 Er	69 Tm	70 Yb	71 Lu
90 Th	91 Pa	92 U	93 Np	94 Pu	95 Am	96 Cm	97 Bk						

A simplified periodic table of the elements with their atomic numbers. Elements 58–71 are "rare earth" elements which have chemical properties similar to each other, as do the "actinide" elements in the bottom row. Otherwise, elements in a given column have similar chemical properties. Elements up to number 118 have now been synthesized and named but play no role in the events described in this book

Ida Noddack (1896– 1978). Source: http:// commons.wikimedia.org/ wiki/File:Ida_Noddack-Tacke.png

While Noddack was ahead of her time in suggesting that heavy nuclei might fission, she evidently did not investigate her own prediction. Ironically, Fermi was probably both inducing fissions and creating transuranic elements. Nuclei of the most common isotope of uranium, U-238, can fission when bombarded by the very energetic neutrons that Fermi was using, but tend to capture slower neutrons and subsequently decay to neptunium and plutonium. That Noddack's idea was not taken seriously has sometimes been construed as an example of scientific sexism, but the reasons were much more prosaic: she offered no supporting calculations of the energetics of such a splitting, and years of experience with nuclear reactions had always yielded products that were near the bombarded elements in atomic number. There was no reason to anticipate such a splitting; Otto Frisch thought Noddack's paper was "carping criticism."

Fermi's next discovery would set the stage for the eventual development of plutonium-based nuclear weapons. In the fall of 1934, his group decided that they needed to more precisely quantify their assessments of activities induced in various elements; previously they had assigned only qualitative "strong-medium-weak" designations. As a standard of activity, they settled on a 2.4-min half-life induced in silver when bombarded by neutrons. However, they soon ran into a problem: the activity induced in silver seemed to depend on where in the laboratory the sample was irradiated, a behavior totally at odds with the scientific touchstone of reproducibility of experiments. In particular, silver irradiated on a wooden table became much more radioactive than when irradiated on a marble-topped one. To try to discern what was happening, Fermi undertook a series of calibration experiments, some of which involved investigating the effects of "filtering" neutrons by interposing layers of lead between the neutron source and the target sample. The layers of lead would reduce the energies of neutrons that passed through them.

Fermi made the key breakthrough on October 22, 1934: "One day, as I came into the laboratory, it occurred to me that I should examine the effect of placing a piece of lead before the incident neutrons. Instead of my usual custom, I took great pains to have the piece of lead precisely machined. I was clearly dissatisfied with something; I tried every excuse to postpone putting the piece of lead in its place. When finally, with some reluctance, I was going to put it in its place, I said to myself: "No, I don't want this piece of lead here; what I want is a piece of paraffin." It was just like that with no advance warning, no conscious prior reasoning. I immediately took some odd piece of paraffin and placed it where the piece of lead was to have been." To his surprise, the presence of the paraffin caused the level of induced radioactivity to increase. Further experiments showed that the effect was characteristic of filtering materials which contained hydrogen; paraffin and water were most effective.

Within a few hours, Fermi had developed a working hypothesis: that by being slowed by collisions with the protons in the paraffin (which contains a lot of hydrogen), the neutrons would have more time in the vicinity of target nuclei to induce a reaction. A head-on collision of a neutron against a proton will essentially bring the neutron to a stop. Since atoms always have random motions due to being at temperatures above absolute zero, the incoming neutrons will never be brought to dead stops, but in practice only a few centimeters of paraffin or water are needed to bring them to an average speed characteristic of the temperature of the slowing medium. This process is now called thermalization, and nuclear physicists define thermal neutrons as having kinetic energy equivalent to a temperature of 25 °C, or 77 °F, not much warmer than the average daily temperature in Rome in October.

The speed of a thermal neutron corresponds to a kinetic energy of about 1/40 of an eV, much less than the 10 MeV of Fermi's radon-beryllium neutrons. Thermal neutrons are also known as "slow" neutrons, while those of MeV-scale kinetic energies are termed "fast". The water or paraffin is now known as a moderator; graphite (crystallized carbon, as is used in pencils) also makes for an effective moderator. Fermi's wooden lab bench, by virtue of its water content, was a more effective moderator than was his marble-topped one. The distinction between fast and slow neutrons proved crucial in the Manhattan Project: when uranium is bombarded by neutrons, what happens depends very critically on the speed of the neutrons. Fast and slow neutrons lie at the heart of why nuclear reactors and bombs function differently, and why a bomb requires enriched uranium to function.

Following Fermi's discovery, his group began subjecting all the elements with which they had previously experimented to slow-neutron bombardment. Extensive results were reported in a paper published in the spring of 1935. For some target elements, the effect was dramatic: activity in vanadium and silver were increased by factors of 40 and 30 over that achieved by unmoderated neutrons. Uranium also showed increased activation, by a factor of about 1.6.

Fermi's hypothesis that slower neutrons have a greater chance of inducing a reaction became quantified in the concept of a reaction cross-section, a measure of the effective cross-sectional area that a target nucleus presents to a bombarding particle that results in a given reaction. Because of quantum-mechanical effects, a target nucleus will generally appear larger to a slower bombarding particle than to a faster one, sometimes by factors of hundreds. Each possible reaction channel for a target nucleus has its own characteristic run of cross-section as a function of bombarding-particle energy. The fundamental unit of cross-section is whimsically termed the "barn," and corresponds to a mere trillionth of a trillionth of a square centimeter, a value characteristic of the geometric cross-sectional area of nuclei. Fermi was exploring what are now called radiative-capture cross-sections, wherein capture of a neutron is followed by a radioactive decay.

Fermi was awarded the 1938 Nobel Prize for Physics for his demonstration of the existence of new radioactive elements produced by neutron irradiation. His wife and children were Jewish, and he and his family used the opportunity of the trip to Stockholm to escape the fascist political situation in Italy by thereafter directly emigrating to America, where he had arranged for a position at Columbia University. They arrived in New York on January 2, 1939.

A few months after Fermi's slow-neutron breakthrough, another discovery was reported: that uranium possesses another, much less abundant isotope than the U-238 that had been assumed to be the sole form of that element. The new isotope was discovered in the summer of 1935 by University of Chicago mass spectroscopist Arthur Dempster. It was not surprising that 235 had escaped earlier detection: Dempster estimated the new isotope to be present to an extent of less than one percent of the abundance of its sister 238 isotope. (Evidence for a third extremely rare isotope of uranium, U-234, had been published in 1912, but this isotope plays no role in the story of the Manhattan Project.) Dempster discovered several isotopes in 1935, but none would change the world like U-235.

3

Fission

Enrico Fermi's claim that he was synthesizing transuranic elements stimulated tremendous interest within the nuclear research community. In addition to Frédéric and Irène Joliot-Curie, the other prominent leaders of that community were Otto Hahn and Lise Meitner at the Kaiser Wilhelm Institute for Chemistry in Berlin. Hahn, a radiochemist, and Meitner, a physicist, had known each other and collaborated on-and-off for 30 years. In 1918 they had discovered the rare element protactinium (atomic number 91), and by the 1930s had accumulated between them years of experience with the chemistry and physics of radioactive elements and their isotopes. Meitner became interested in Fermi's work, and in 1935 convinced Hahn to renew their collaboration in order to sort out exactly how uranium transmuted under slow-neutron bombardment. To bolster their efforts, they brought on board chemist Fritz Strassmann.

© Springer Nature Switzerland AG 2020
B. C. Reed, *Manhattan Project*,
https://doi.org/10.1007/978-3-030-45734-1_3

Left: Lise Meitner (1878–1968) and Otto Hahn (1879–1968) in their laboratory, 1913. Right: Fritz Strassmann (1902–1980). Sources: http://commons.wikimedia.org/wiki/File:Otto_Hahn_und_Lise_Meitner.jpg; AIP Emilio Segre Visual Archives, gift of Irmgard Strassmann

3.1 Stumbling Toward Fission

The Berlin group's chemical procedures followed the same premises that Fermi had used. In the 1930s, heavy elements such as uranium were termed "transition" elements, assigned to the seventh row of the periodic table. Later, they would be recognized as a separate group, the "actinide" elements, whose chemical properties are more similar to each other then they are to the elements above them in the columns of the table in which they reside. Their similar chemistry is now understood as a function of how electron orbits are populated on moving through progressively heavier elements along this row.

It was presumed that any transuranic elements, those with atomic numbers 93, 94, and so on, would have chemical properties similar to the elements above them in the columns of the table in which they were predicted to reside. Those elements are successively rhenium (above 93), osmium (94), iridium (95), platinum (96), and gold (97). The anticipated new elements were given the tentative names eka-rhenium (EkaRe), eka-osmium (EkaOs), and so forth; "eka" is from the Greek for "beyond." In line with this expectation, Hahn, Meitner, and Strassmann separated their induced radioactivities from uranium by precipitating them out of solutions with "carrier" compounds containing the presumed corresponding elements, assuming that any radioactivity that so carried was due to the sought-after transuranic elements. In the case of searching for EkaRe, manganese (atomic number 25) was commonly used as a carrier; the other element in that column, technetium (43) is extremely rare. The chemical separations had to be extremely thorough: the natural alpha radioactivity of uranium of their samples was ten thousand to a hundred thousand times greater than the induced activity being sought; even minute amounts of residual uranium would have masked the induced effects.

By 1937, the situation had become extremely confusing. The Berlin team had identified no less than nine half-lives arising from uranium bombardment, many more than Fermi had detected. These were thought to involve transuranic elements with atomic numbers up to 97. The activities and decay sequences seemed to point to three reaction processes:

$$n + {}_{92}U \rightarrow \left({}_{92}U + n\right) \rightarrow$$
$${}_{93}EkaRe \rightarrow {}_{94}EkaOs \rightarrow {}_{95}EkaIr \rightarrow$$
$${}_{96}EkaPt \rightarrow {}_{97}EkaAu?$$

$$n + {}_{92}U \rightarrow \left({}_{92}U + n\right) \rightarrow {}_{93}EkaRe \rightarrow {}_{94}EkaOs \rightarrow {}_{95}EkaIr?$$

$$n + {}_{92}U \rightarrow \left({}_{92}U + n\right) \rightarrow {}_{93}EkaRe$$

Re, Os, Ir, Pt, and Au are respectively rhenium, osmium, iridium, platinum, and gold. Half- lives for the various steps varied from about 10 s to over 60 h, with, strangely, different half-lives associated with the same elements in the three separate chains. Chemically, the identifications seemed secure, but Meitner struggled to understand the corresponding physics. Neutrons exciting three different energy levels in uranium had never been observed before, and, contrary to all previous experience, the first two sequences appeared to involve "inherited" excited energy levels.

To confound the situation further, in October, 1937, Irène Curie and Paul Savitch in Paris identified an approximately 3.5-h beta-decay half-life resulting from slow-neutron bombardment of uranium, an activity which the Berlin group had not found. Curie and Savitch suggested that the decay might be attributable to thorium, element number 90. If this were true, it would mean that a neutron slowed to the point of possessing less than a single electron-volt of kinetic energy was apparently capable of knocking an alpha-particle out of a uranium nucleus. While such a reaction is energetically possible, the chance of an alpha-ejection could be computed from quantum mechanics, and was found to be extremely unlikely: the half-life for ordinary spontaneous alpha-decay of U-238 is over four billion years. Further, no isotope of thorium was known to have a half-life in the vicinity of 3.5 h.

In Berlin, Meitner asked Strassmann to search for thorium. He did so, but in a way that overlooked the fission product that was actually being created and being mistaken for thorium, of which he found no evidence. Ironically, in 1935 Edoardo Amaldi had tried inducing alpha-emission by neutron bombardment of uranium. But to do so he had to filter out the natural alpha-decay activity of uranium, which he did by wrapping his samples in thin aluminum foils on the rationale that any alphas arising from short half-life decays should be energetic enough to pass through the foils on their way to his Geiger counter, whereas lower-energy naturally-occurring ones would be stopped by the foils. He detected no alphas, but the foils also blocked heavy, slow-moving fission products.

Further work by Curie and Savitch published in September, 1938, indicated that their 3.5-h beta-emitter seemed to have chemical properties similar to that of lanthanum, element 57. Lanthanum is in the same column of the periodic table as actinium, element 89, which is only three places away from uranium, so it seemed sensible to attribute the activity to actinium. But this only made understanding the process more difficult. To get from element 92 to 89 would require the ejection of a lithium nucleus, and if the probability of ejecting an alpha-particle was unlikely to begin with, that of ejecting lithium would be even less. Curie and Savitch's chemistry was indicating thorium or actinium, but all of the known physics of their interpretation seemed improbable. In actuality they may have been detecting the fission product lanthanum-141 (half-life 3.9 h) and/or yttrium-92 (3.54 h); both are in the same column of the periodic table as actinium.

A few months before this complication arose, Lise Meitner had been forced to flee Berlin. Born into a Jewish family in Austria, her Austrian citizenship had protected her against German anti-Semitic laws, but that protection ended with the German annexation of Austria in March, 1938. On July 13, she fled to Holland with only 10 Marks in her purse and the clothes on her back. She eventually made her way to Sweden, where she was given a position at the Nobel Institute for Experimental Physics, but she was not too warmly received nor offered much support. She continued to collaborate with Hahn and Strassmann by letter, but her career was ruined and her pension was stolen by the Nazi government.

After the war, Hahn largely disavowed Meitner's contributions, to the point of suggesting that her considerations of physics impeded the chemical discovery of fission. Fritz Strassmann had a different opinion: "What does it matter that Lise Meitner did not take *direct* part in the discovery? ... [She] has been the intellectual leader of our team and therefore she was one of us, even if she was not actually present for the 'discovery of fission.'" Hahn would be awarded the 1944 Nobel Prize for Chemistry for the discovery; Meitner and Strassmann did not share in the recognition, although element 109, Meitnerium, is now named in her honor. Meitner and Hahn's lives ran in eerie parallel: she was born four months before and died three months after him.

3.2 Breakthrough

In Berlin, Hahn and Strassmann continued to look for Curie and Savitch's 3.5-h activity, with Hahn chronicling their progress and frustrations in letters to Meitner. By October 25, he had become convinced of the existence of the 3.5-h activity, but suspected that a radium isotope (element 88) might be involved. By early November, he was convinced that two or three radium isotopes were being created, a situation which implied ejection of two alpha-particles from uranium, an even more improbable process than Curie and Savitch's actinium proposal. On November 8, Hahn and Strassmann sent a paper to the journal *Naturwissenschaften* reporting half-lives for three radium and three actinium isotopes. But this this interpretation was soon eclipsed by other developments. They refined their chemical techniques, and by mid-December had come to the startling conclusion that what they had thought were isotopes of radium were actually isotopes of barium, element 56. Radium and barium are in the same column of the periodic table, so it is not surprising that there would have been some confusion as to which was being detected. The idea that uranium could be splitting off a nucleus of more than half of its own weight was astounding.

Hahn wrote to Meitner during the evening of Monday, December 19, to seek her opinion of whether slow-neutron bombardment of uranium might yield a product of much lower atomic weight, writing (in translation) "… But we are coming steadily closer to the frightful conclusion: our Ra-isotopes do not behave like Ra but like Ba. … I have agreed with Strassmann that for now we shall tell only *you*. Perhaps you can come up with some sort of fantastic explanation. We know ourselves that it *can't* actually burst apart into Ba." In another letter on the 21st, he remarked that "How beautiful and exciting it would be just now if we could have worked together as before." So far at least, Meitner and physics were still considered part of the team.

Hahn's first letter reached Meitner in Stockholm on Wednesday, December 21. She replied at once: "Your radium results are very startling … At the moment the assumption of such a thoroughgoing breakup seems very difficult to me, but in nuclear physics we have experienced so many surprises, that one cannot unconditionally say: it is impossible." Hahn would receive her reply on December 23, but, anxious not to be scooped, he and Strassmann did not wait for it before submitting a paper to *Naturwissenschaften* on December 22. The paper was authored by Hahn, who hedged by referring to "alkaline earth" elements (of which barium is one), although he did specifically mention the barium finding. The paper would be published on January 6, 1939.

3.3 A Walk in the Snow

On December 23, Meitner traveled from Stockholm to spend Christmas with some friends in the town of Kungälv, near Göteborg on the west coast of Sweden. Her nephew, Otto Frisch, a nuclear physicist and another refugee from Austria, was then working at Niels Bohr's Institute for Theoretical Physics in Copenhagen. He traveled to Sweden to spend Christmas with his aunt, arriving also around the 23rd.

Otto Frisch (1904–1979). Source: Photo-graph by Lotte Meitner-Graf, London, courtesy AIP Emilio Segre Visual Archives

The next morning—Christmas Eve day—Meitner and Frisch went for a walk in the snow, he on skis and she keeping up on foot. He tried to interest her in an experiment he was planning, but she drew him into a discussion of Hahn's letter. Hahn's *Naturwissenschaften* manuscript had also been forwarded to her from her home in Stockholm. As Frisch later related in his memoires, they sat down on a tree trunk and began to calculate on scraps of paper. Working from a theoretical model of nuclei that had been developed some years previously by Russian-American physicist George Gamow and Niels Bohr which envisioned nuclei as acting like drops of liquid, they knew that uranium nuclei with their many protons are near the limit of intrinsic stability beyond which no additional number of neutrons can inhibit them from spontaneously breaking up. Uranium nuclei are somewhat like wobbly drops, liable to fragment in response to a modest provocation such as the impact of a neutron. If a uranium nucleus were to break into two halves, the resulting fragments would experience a mutually repulsive Coulomb force and fly away from each other at high speeds.

Meitner had nuclear mass data committed to memory, and quickly calculated that if a uranium nucleus split into two equal halves, the two fragments would total to a mass less than that of the original nucleus by about one-fifth of the mass of a proton, equivalent to an energy of about 200 MeV. This enormous energy would appear in the outside world in the form of the kinetic energy of the fission fragments. Frisch provided a supporting estimate by calculating that two fragment nuclei of intermediate masses would repel each other with an energy about that which Meitner had calculated. Thus was the process of fission conceived in a snowy Swedish forest.

Hahn's discovery reaction did not involve an equal splitting, but rather the production of barium (atomic number 56) and krypton (36). Also accompanying fission is typically the instantaneous release of two or three neutrons:

$$\frac{1}{0}n + \frac{235}{92}U \rightarrow \frac{141}{56}Ba + \frac{92}{36}Kr + 3\left(\frac{1}{0}n\right).$$

Yet unknown to Hahn, Strassmann, Meitner, or Frisch was that the reaction involved U-235 as opposed to the more common U-238 isotope. Curiously, the masses of fission fragments are rarely equal; nature prefers fragments with about 90 and 140 nucleons, now understood as a consequence of the nuances of inter-nucleon forces.

Fission was unlike any reaction known in 1938. Since alpha and beta decays or neutron bombardments had always led to changes in the atomic number of the bombarded element by at most one or two, researchers concentrated on looking for "nearby" products, never (except for Ida Noddack) anticipating such a radical departure from past experience. The energy released, about 170 MeV, is enormous even by the violent standards of nuclear reactions. The liberated neutrons would be the initiators of Leo Szilard's chain reaction.

A less obvious consequence of this reaction is that since fission products are typically very neutron-rich for the numbers of protons they possess, they will undergo a series of successive beta-decays until they achieve stability. It is the *products* of fission that are responsible for radioactive fallout persisting long after the destructive effects of the energy release have made themselves felt: fission bombs are not simply scaled-up ordinary chemical bombs. Spent reactor fuel rods also contain fission products, but the products remain stuck in them unless the rods are reprocessed.

How did Fermi and his collaborators fail to discover high-energy fission fragments in 1934? The culprit was the nature of their radon-beryllium neutron sources. In addition to being an alpha-emitter, radon is a fairly prolific gamma-ray emitter, and these gamma-rays caused unwanted background signals in Geiger counters when the latter were placed near the neutron sources. To circumvent this, the Rome group adopted the procedure of irradiating target samples away from their counters and then literally running them down a hallway to a detector in a room removed from the neutron source. Since the goal of the experiments was to detect delayed effects (the induced half-lives were often on the order of minutes), this procedure would not affect their results in that sense. But any high-energy fission fragments that might have been detected would have been brought to rest by the time the sample arrived at the detector. What the Romans had attributed to decays of transuranic elements were beta-decays from the fission fragments, although transuranics were no doubt also being created. Fermi never expected fission to happen and so never considered that his experimental arrangement might be biasing him against detecting it: retrospect is always perfect.

The energy released in fission is fantastic. Complete fission of a kilogram of uranium releases energy equivalent to exploding 17,000 kg of conventional explosive. A thousand kilograms is a metric ton (formally, tonne), so the 17,000 kg is spoken of in weapons circles as 17 kt. The vast majority of this energy is carried off in the form of kinetic energy by the barium and krypton fission products, but the neutrons carry off on average about 2 MeV each, a number which proved to be important. Some 30 different elements are produced by uranium fission; it is no wonder that Hahn, Meitner, and Strassmann had observed a confusing plethora of decay chains. Their detection of the barium-krypton fission channel must have been a result of their use of barium chemistry.

Lise Meitner must have experienced a storm of emotions. On one hand was the realization that a rich new area of physics was opening up; on the other was the revelation that phenomena which for several years she had attributed to transuranic elements were likely to have been the products of neutron-rich fission fragments decaying toward stability. Only the 23-min decay of their proposed family of three reactions (the last one) would prove to be correct; the other two would prove to be confounding mixtures of decaying fission products which appeared to be decay chains. But those years of "transuranic" research had paved the way to the discovery of fission.

In another of the confluences of events that seem to characterize nuclear history, it was also on December 24 that Enrico Fermi and his family set out for America from Southampton, England. Fermi would know nothing of these developments until he met Niels Bohr some three weeks later in New York. Since his Nobel Prize was awarded for the presumed discovery of elements 93 and 94, the discovery of fission would prompt him to add a footnote to the published version of his Nobel lecture.

In the meantime, Otto Hahn had also begun thinking that what he had previously assumed to be transuranic elements might be lighter-element fission products, and on the 27th he phoned the editor of *Naturwissenschaften* to append a comment to this effect to his and Strassmann's paper. The next day he wrote Meitner again, pleading with her to consider whether the energetics of the proposed splitting made sense. Back in Stockholm for New Year's Eve, she wrote Hahn that "We have read and considered your paper very carefully; *perhaps* it is energetically possible for such a heavy nucleus to break up." On New Year's Day, Frisch returned to Copenhagen, promising to keep in telephone contact with his aunt as they drafted a paper based on the work they had begun during their walk a few days earlier.

In his memoirs, Frisch relates that in all the excitement, he and Meitner overlooked the possibility of a chain reaction. A Danish colleague, Christian Møller, suggested to him that the fission fragments might contain enough energy to each eject a neutron or two, which might go on to cause other fissions. That the fragments would be neutron-rich in comparison to stable nuclei of the same atomic number made this possibility very real. Frisch's immediate response was that if this were the case, no deposits of uranium ores should exist as they would have blown themselves up long ago. But he then realized that this argument was naïve: ores contained other elements which might capture neutrons, and many might simply escape before causing other fissions. Leo Szilard's vision of a chain-reaction had taken its first steps toward reality.

Meitner wrote Hahn again on January 3 to congratulate him and Strassmann, and to express her frustration at having to watch developments from afar: "I am now almost *certain* that the two of you really do have a splitting to Ba and I find that to be a truly beautiful result, for which I most heartily congratulate you and Strassmann ... And believe me, even though I stand here with very empty hands, I am nevertheless happy for these wondrous findings."

In early 1939, the focus of fission research shifted to Copenhagen. On January 3rd, Frisch caught up with Niels Bohr to apprise him of the situation. Bohr was preparing to depart to spend a semester at the Institute for Advanced Study in Princeton, New Jersey; their conversation was brief. According to Frisch, Bohr's reaction was to hit himself on the head and exclaim "Oh what idiots we have all been. Oh but this is wonderful! This is just as it must be! Have you and Lise Meitner written a paper about it?" Bohr promised not to disclose the discovery until their paper had been prepared. The next day, Frisch also informed Hahn of his and Meitner's work.

Bohr and Frisch conversed again on January 6 to review the calculation of the near-instability of uranium. Frisch also discussed the situation with theoretical physicist George Placzek, who recalled that Irène Curie had told him in the fall of 1938 that she had found light elements from uranium bombardment, but had not trusted herself enough to publish it. The next morning, Frisch met Bohr just before his departure, handing him a draft of the paper he was coauthoring with Meitner.

Curiously, neither Hahn and Strassmann nor Frisch thought of setting up an experiment to detect the expected high-energy fission fragments. It was Placzek who encouraged Frisch to do so, which he did on Friday, January 13. The fragments were immediately detected, and Frisch became credited with being the first person to set up an experiment to deliberately demonstrate and detect fission. He is also credited with coining the term "nuclear fission," after having asked an American biologist working in Copenhagen, William Arnold, what term was used for the process of cell division: "binary fission."

Extending Hahn and Strassmann's work, Frisch also tested thorium, which proved to act like uranium in that it would fission under bombardment by *fast* (unmoderated) neutrons, but to act unlike uranium in that it did *not* do so when bombarded with *slow* neutrons. Frisch wrote two papers. The first was co-authored with Meitner, and described their Christmastime insights, while the second described his own experiments. Both were sent to *Nature* on January 16; the joint paper was published on February 11, and the experimental one on February 18. The uranium/thorium asymmetry would prove to be a crucial observation a few weeks later when Niels Bohr worked to understand the underlying physics of the process.

3.4 Fission Arrives in America

Bohr sailed to America, accompanied by his son Erik and collaborator Léon Rosenfeld of the University of Liège in Belgium. Bohr had a blackboard installed in his stateroom, and during the voyage he and Rosenfeld, battling seasickness, began to develop a theoretical understanding of fission. They arrived in New York on the afternoon of Monday, January 16, where they were met by Enrico and Laura Fermi. Bohr remained in New York while Rosenfeld left directly for Princeton. But Bohr had not told Rosenfeld of his promise to Frisch to keep the news quiet until his and Meitner's paper had been submitted for publication, and Rosenfeld spilled the beans that evening at a meeting of the Princeton Physics Journal Club. When Bohr heard that the word was out, he drafted his own note to *Nature* on January 20 to assert Meitner and Frisch's priority; it would be published on February 25.

Bohr speculated that in cases of ordinary non-fission reactions, the energy of the bombarding particle must become distributed in the target nucleus as vibrations in a manner resembling an agitated drop of liquid. If a large part of the energy should come to be concentrated on some particle such as a proton or alpha-particle at the surface of the nucleus, then that particle will be ejected, just as a molecule can evaporate from a drop. In contrast, he reasoned that in a fission reaction, the distribution of energy would have to result in a mode of vibration of the nucleus that involved a considerable deformation of the surface. Bohr deduced, purely qualitatively as yet, that in heavy nuclei the energy sufficient to distort the nucleus to the point where it would form two lobes which would repel each other and lead to fission must be of about the same value as the energy necessary to cause the escape of a single particle from a lighter nucleus, a few MeV. The concept of a requisite deformation energy would soon find quantitative expression as the so-called fission barrier.

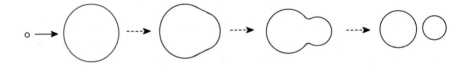

Schematic representation of steps in the progression of droplet fission. An initially spherical nucleus (left) is perturbed by a bombarding neutron, and begins to distort. Lobes form which force each other apart due to electrical repulsion, leading to fission. Sketch is not to scale. The volume of the nucleus is conserved in the process: in reality, the sum of the volumes of the final two spheres is equal to that of the initial nucleus plus that of the incoming neutron

The first demonstration of fission in America occurred at Columbia University. On Wednesday, January 25, Bohr, while on his way to attend a conference in Washington, stopped at Columbia to find Fermi. Fermi was out, but Bohr encountered one of his graduate students, Herbert Anderson. As Anderson told the story, Bohr approached him, grabbed him by the shoulder, and whispered "Young man, let me explain to you about something new and exciting in physics." Anderson, who was preparing a thesis on neutron scattering, instantly understood the significance of what Bohr related.

Enrico Fermi and his reactor group at the University of Chicago. This photograph was taken December 2, 1946, on the fourth anniversary of the operation of the CP-1 reactor. Back row (l-r): Norman Hilberry, Samuel Allison, Thomas Brill, Robert Nobles, Warren Nyer, Marvin Wilkening. Middle row (l-r): Harold Agnew, William Sturm, Harold Lichtenberger, Leona Woods, Leo Szilard. Front row (l-r): Enrico Fermi, Walter Zinn, Albert Wattenberg, Herbert Anderson. Source: http://commons. wikimedia.org/wiki/File:ChicagoPileTeam.png

Bohr went on his way, and Anderson went to find Fermi, who had already heard the news through a contact at Princeton. Fermi had to leave for Washington as well, and so was not present that evening when Anderson set up an experiment to detect fission fragments with an ionization chamber he had prepared for his thesis work. Ionization pulses caused by the fragments were readily apparent, and the experiment was witnessed by Professor John Dunning. Anderson states that Dunning telegraphed the news to Fermi in Washington, but it is not clear if he actually did so. In Paris the next day, Frédéric Joliot read Hahn and Strassmann's paper, and also detected evidence of fission.

Left: Edward Teller (1908–2003) in 1958, as Director of the Lawrence Livermore National Laboratory. Right: George Gamow (1904–1968). Sources: http://commons.wikimedia.org/wiki/File:Edward_Teller_(1958)-LLNL-restored.jpg; *AIP Emilio Segre Visual Archives; AIP Emilio Segre Visual Archives, Physics Today Collection*

While word of the discovery had been spreading at Princeton and Columbia since Bohr and Rosenfeld's arrival, the public coming-out of the news came on January 26. Fermi and Bohr were in Washington to attend the Fifth Washington Conference on Theoretical Physics. These conferences were co-hosted by George Washington University (GWU) and the Carnegie Institution of Washington (a private research foundation), and were mainly organized by George Gamow and Edward Teller, both then at GWU. The topic of the 1939 meeting was to be low-temperature physics, but that agenda quickly found itself derailed. The conference opened at two p.m. on the afternoon of Thursday, the 26th. Gamow opened the proceedings by introducing Bohr, who related Hahn and Strassmann's discovery and Meitner and Frisch's interpretation. The news electrified the fifty-odd participants, some of whom left to perform their own experiments. Today, a plaque outside Room 209 of GWU's Hall of Government commemorates Bohr's announcement.

The next deliberate demonstration of fission in America seems to have occurred on Saturday morning, January 28, at Johns Hopkins University in Baltimore. Apparently tipped off by a colleague attending the conference, R. D. Fowler and R. W. Dodson tested both uranium and thorium, verifying Frisch's observation that slowing neutrons with paraffin increased the fission rate in uranium but had no effect on that for thorium. That evening at the Carnegie Institution, Richard Roberts and colleagues Lawrence Hafstad and Robert Meyer demonstrated fission with Bohr, Fermi, Rosenfeld, Teller, and others in attendance. The *New York Times* reported on the discovery in its edition of Sunday, January 29, noting that scientists at the Washington meeting thought that it might be twenty or twenty-five years before the phenomenon could be put to use. In Berkeley, Luis Alvarez, a member of Ernest Lawrence's Radiation Laboratory staff, read of the discovery in the *San Francisco Chronicle* and passed the word to his graduate student, Philip Abelson, who verified the finding on January 31. Abelson detected iodine as a decay product of tellurium, which was itself a direct fission product; over the following few weeks he identified a number of other products. Alvarez also verified that slow neutrons were more effective in causing fission than fast ones. The Johns Hopkins, Carnegie, and Berkeley reports all appeared in the February 15, 1939, edition of the *Physical Review*. The Columbia group's first paper did not appear until the March 1 edition, but contained the first quantitative measurements of cross-sections for both fast and slow neutrons.

As Leo Szilard and no doubt many others realized, there would have to be on average at least one neutron liberated per fission if there was to be any hope of sustaining a chain reaction. As Herbert Anderson later wrote, "Nothing known then guaranteed the emission of neutrons. Neutron emission had to be observed experimentally and measured quantitatively." A number of research teams began looking for secondary neutrons, and proof of their existence was not long in coming. On March 16, two independent groups at Columbia submitted reports to *The Physical Review* reporting their discovery, with both estimating about two neutrons emitted per each captured. Leo Szilard, a member of one of the groups, recalled later his reaction upon detecting the neutrons: "That night, there was very little doubt in my mind that the world was headed for grief." Confirming evidence for secondary neutrons soon came in from Europe: in Paris on April 7, Hans von Halban, Frédéric Joliot, and Lew Kowarski submitted a paper to *Nature* (published April 22) in which they reported an average of 3.5 ± 0.7 neutrons liberated per fission. The modern value is about 2.4.

The other Columbia group, comprising Anderson, Fermi, and H. B. Hanstein (whose given names now seem lost to history), also reported on what was probably the first nuclear "pile" experiment conducted in America. Their setup involved placing a neutron source inside a spherical glass bulb 13 cm in diameter, which was placed within a water-filled cylindrical tank 90 cm high by the same diameter. Foils of rhodium placed in the water at various positions were used to detect the neutron flux via beta-activity induced in the rhodium. The induced activity was measured when the bulb contained only the neutron source, and then again with the neutron source plus uranium oxide. With the oxide present, a 6% increase in activity was detected. In a follow-up paper dated July 3 (published August 1), Anderson, Fermi, and Szilard reported on the second Columbia pile, which involved using 52 cylindrical metal cans 5 cm in diameter by 60 cm high which were filled with a total of about 200 kg uranium oxide. The cans were placed within the same tank as the earlier experiment, which this time was filled with 540 L of a solution of manganese sulfate; induced activity in the manganese served as the neutron detector. A neutron source was placed in the center of the tank, and the activity induced in the solution was measured both when the cans were empty and full of oxide. The activity was about 10% greater when the cans were full, which prompted them to write that "From this result we may conclude that a nuclear chain reaction could be maintained in a system in which neutrons are slowed down without much absorption until they reach thermal energies ...." From the fall of 1939 onwards, Fermi and his collaborators wrote 47 papers describing experiments which would culminate in the world's first self-sustaining nuclear reaction in late 1942.

3.5   A Walk at Princeton

The observation that the likelihood of uranium to fission depended on the speed of bombarding neutrons and that uranium and thorium differed in their responses to slow-neutron bombardment catalyzed several crucial revelations on the part of Niels Bohr in early 1939.

Sometime in January, George Placzek arrived at Princeton. Over breakfast with Bohr and Rosenfeld one morning at the University's Faculty Club, the conversation turned to fission. Bohr expressed relief that physics was now rid of purported transuranic elements, but Placzek protested that the situation was more confused than ever. Both uranium and thorium were known to have strong radiative-capture non-fission probabilities for slow neutrons, which suggested that transuranic-inducing transmutations were actually somehow proceeding concurrently with fission. Also, why did uranium fission under slow-neutron bombardment, but not thorium? While walking back to his office, Bohr had his revelation.

Again working with haste (and, apparently, with Placzek's uncredited help), Bohr wrote up and sent off a paper to the *Physical Review* in which he argued that it was likely the rare isotope uranium-235 that must be responsible for *slow-neutron* fission in that element, while simultaneously explaining both how transuranic elements could be being produced and why thorium did not exhibit slow-neutron fission. Dated February 7, 1939, it was published in the February 15 edition alongside the reports of fission being detected in various American laboratories.

Bohr's argument comprised two interlinked components. The first involved the liquid-drop model of nuclei. In this new paper, Bohr linked this argument to some earlier experiments of Meitner, Hahn, and Strassmann wherein they examined the radiative-capture response of uranium to neutrons of varying speeds. This work had revealed a rich forest of very strong capture probabilities for neutrons of energies of from a few to thousands of electron-Volts. Meitner, Hahn, and Strassmann had concluded that these high probabilities, known technically as resonances, were likely attributable to the abundant isotope, U-238; it is this process, which is the third of their three reaction sequences, that leads to a transuranic element. However, the resonance captures were apparently not associated with any significant fission probability. This led Bohr to infer that nuclei of the 238 isotope must in fact be very stable, and also to conclude that if 238 does not fission under intermediate-energy neutron bombardment, it would certainly not be expected to do so under slow-neutron bombardment. This left only U-235 as the suspect for slow-neutron fission.

As for thorium, Bohr appealed to an empirical phenomenon known as nuclear parity. Nuclear physicists classify the "parity" of nuclei according to the evenness or oddness of the number of protons and neutrons that they contain, always expressed in the order protons/neutrons, or Z/N. In this scheme, uranium-235 is an even/odd nucleus (Z = 92, N = 143), whereas uranium-238 is classified as even/even (Z = 92, N = 146). The only stable form of thorium is $^{232}_{90}$Th, so it too is of even/even parity. Nature has a definite preference for even/even arrangements: of 266 known stable isotopes, 159 are even/even, 103 are even/odd or odd/even, and only four are odd/odd. This distribution is now understood as a manifestation of the nature of the forces between pairs of nucleons: nuclei wherein every nucleon has a partner with which to pair-bond will enter into a more stable mass-energy state than one in which unpaired nucleons are present.

Bohr argued that if it is the even/odd 235 isotope that is responsible for slow-neutron fission in uranium, we might *not* expect to see slow-neutron fission in thorium as it lacks an isotope of such parity, exactly as Otto Frisch and others had observed. From an energy perspective, Bohr's argument was that when an even/odd nucleus such as U-235 captures a neutron and becomes even/even, it will find itself in a more excited energy state and hence more prone to fission than would an even/even nucleus.

The second-to-last paragraph of Bohr's paper presented an important hypothesis concerning fast-neutron fission, a speculation which was likely largely overlooked at the time with all the attention being devoted to slow neutrons. Quantum-mechanical theory indicated that as the energy of bombarding neutrons increases (that is, as they become faster), the fission cross-section should generally decrease. Since U-238 did not fission under intermediate-energy neutron bombardment, it would certainly not be expected to do so when struck by *fast* ones because of the lower cross-section to be expected at higher energies. Conversely, Bohr argued, U-235 might have a chance of sustaining fast–neutron fission in view of its apparently very large cross-section for slow neutrons, if it had sufficient remaining cross-section for fast neutrons despite the expected general decrease in cross-section with increasing neutron energy.

While Bohr did not remark on what might happen if U-235 could be separated from U-238 and bombarded with fast neutrons, the possible implications of this question had not gone unappreciated. Philip Morrison, a student of Robert Oppenheimer, recalled that "when fission was discovered, within perhaps a week there was on the blackboard in Robert Oppenheimer's office a drawing … of a bomb."

An apparent glaring inconsistency in Bohr's argument is that the quantum-mechanical theory would in fact suggest a greater chance of fissioning U-238 with lower-energy neutrons, contrary to what he concluded. This would be resolved in a more detailed analysis published a few months later.

3.6 Bohr and Wheeler

Bohr's insight that it was likely the rare 235 isotope that was responsible for slow-neutron fission of uranium was but the first step in an extensive chain of experimental and theoretical investigations into the fission process that unfolded over the following year. Verification of the U-235 hypothesis would come in early 1940, but in the meantime Bohr remained busy analyzing the physics of the roles played by different isotopes under fast and slow neutron bombardment.

Upon his arrival in America, Bohr began collaborating with John Wheeler, a young Assistant Professor at Princeton University. The two had known each other since 1934, when Wheeler began a postdoctoral year at Bohr's Institute for Theoretical Physics in Copenhagen. In the September 1, 1939, edition of *The Physical Review*—the very day that World War II began—they published an extensive analysis of the energetics of fission.

John Wheeler (1911–2008). Source: AIP
Emilio Segrè Visual Archives

Their calculations were complicated, but revealed that fission essentially boiled down to a competition between two factors: the energetics involved in neutron capture by nuclei, and the energy necessary to distort a given nucleus to the point of inducing fission.

The first factor is heavily influenced by the odd/even parity of the bombarded nucleus. When a nucleus captures an incoming neutron, its internal structure rearranges itself in such a way that a slight amount of mass is lost. The lost mass is transformed into energy, which leaves the nucleus in an agitated state. More specifically, when a nucleus transforms from being even/odd to being even/even, about 6.5 MeV of agitation energy appears; this is the case with U-235. On the other hand, if the struck nucleus is even/even and transforms to even/odd, about 5 MeV appears; this is the case with U-238 and Th-232. The difference of 1.5 MeV is everything.

As to the competing factor, Bohr and Wheeler found that *any* otherwise stable nucleus can be induced to fission under neutron bombardment, but that any specific isotope possesses a characteristic "fission barrier" or "activation energy", a minimum energy which has to be supplied to deform the nucleus sufficiently to induce the fission process to begin. The necessary activation energy can be supplied by a combination of any kinetic energy carried in by the bombarding neutron, and the 5 or 6.5 MeVs acquired through the resulting parity transformation. For elements lighter than uranium, the barrier energies can be enormous, up to over 50 MeV for those near the middle of the periodic table. Since neutrons emitted in fissions have kinetic energies of only about 2 MeV, there can be no possibility of inducing a chain reaction with such elements even with the contribution of the parity-transformation energy. For heavier elements, the barriers are still too large to get a chain reaction, but on reaching uranium at the end of the periodic table, they have fallen to on the order of a few MeVs.

Bohr and Wheeler imagined fission not as an instantaneous process, but one wherein the incoming neutron and target nucleus first combine to form an intermediate "compound" nucleus. U-235, upon capturing a neutron, becomes the compound nucleus U-236; U-238 becomes U-239. For the transformation of U-235 to U-236, the parity-energy release is about 6.6 MeV, while for that of U-238 to U-239 it is about 4.8 MeV. The corresponding fission-activation energies are respectively about 5.0 and 6.2 MeV. If the bombarding neutrons are slow and bring essentially no kinetic energy into the reactions, then a nucleus of U-236 formed from U-235 will find itself with an agitation energy that exceeds the fission activation threshold by about 1.6 MeV, while one of U-239 will fall about 1.4 MeV short of the fission threshold. *Any bombarding neutron, no matter how little kinetic energy it has, can induce fission of U-235.* To fission U-238 by neutron bombardment requires supplying neutrons of kinetic energy about 1.4 MeV or greater.

Bohr and Wheeler found, however, the issue of the unsuitability of U-238 as a weapons material is more subtle than this description lets on. The average kinetic energy of secondary neutrons liberated in the fission of uranium nuclei is about 2 MeV, and about half of them have energies greater than the approximately 1.4 MeV activation energy of the U-238 → U-239 compound nucleus. At first glance, it would appear that U-238 should make a viable weapons material. But Nature throws a complicating factor in the way.

When a neutron strikes and is scattered away by a target nucleus, the collision may happen in one of two ways: elastically or inelastically. In elastic scattering, the kinetic energy of the neutron/nucleus system is conserved, with the neutron typically being slowed only minutely. In contrast, if the collision is inelastic, the neutron loses considerable kinetic energy, with the lost energy again leaving the struck nucleus in an agitated state. Scattering experiments revealed that fast neutrons striking U-238 nuclei are about eight times as likely to be inelastically scattered as they are to induce fissions. So much energy is robbed from bombarding neutrons in inelastic collisions that they are promptly slowed to below the fission threshold, and so become useless for maintaining a chain reaction. The coup de grace is then that U-238 nuclei have a large radiative-capture cross-section for neutrons of energy less than about 1 MeV. It is this inelastic scattering effect that resolved the apparent contradiction in Bohr's earlier paper. In short, the non-utility of U-238 as a weapons material is due to a parasitic combination of inelastic scattering and a fission threshold below which it has an appreciable non-fission capture cross-section for slowed neutrons. The presence of even small amounts of U-238 in a *fast-neutron* environment will suppress any chain reaction. The net result of Bohr and Wheeler's analysis was that only U-235 could sustain a fast-neutron chain reaction, and it is for this reason that this isotope must be isolated from its more populous sister isotope if one desires to build a fast-fission uranium bomb.

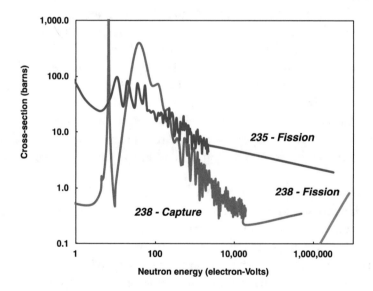

*Approximate fission cross-section for U-235 and neutron-capture and fission cross-sections U-238. The curves have been heavily averaged across the resonance-capture spikes. Only for energies exceeding about 1 MeV does U-238 begin to show an appreciable fission cross-section. Adapted from Reed, "The Manhattan Project," Physica Scripta **89** (2014) 108003, Fig. 6*

For yet heavier elements, the numbers become more favorable for fission, but this gain is offset by the fact that, with few exceptions (plutonium-239 being one), these synthetic elements are so intrinsically unstable that they quickly decay by spontaneous fission or alpha-decay; they are not realistic options if a nuclear weapon is to have any sensible shelf-life. There is a very narrow "window of fissility" at the heavy end of the periodic table.

Despite its non-fissility, U-238 did play a crucial role in the Manhattan Project. Upon capturing a neutron, a U-238 nucleus becomes a U-239 compound nucleus, which sheds its excess energy in a series of two beta-decays of half-lives of about 23 min and 2.3 days before settling down as plutonium-239:

$$^{239}_{92}\text{U} \rightarrow {}^{239}_{93}\text{Np} \rightarrow {}^{239}_{94}\text{Pu}.$$

Like U-235, Pu-239 is even/odd, and is fissile under slow-neutron bombardment: neutron capture by Pu-239 releases a parity-transformation energy of 6.5 MeV, but the fission activation energy is only about 6.0 MeV.

Bohr and Wheeler did not comment on the possibility of Pu-239, but this line of reasoning did occur to Louis Turner of Princeton University in early 1940. Turner speculated that if Pu-239 should prove to be a reasonably stable decay product of neutron bombardment of U-238, it could open an alternate route to obtaining bomb-quality, fast-neutron-fissile material. He wrote up his speculation in a brief paper dated May 29, 1940, which he submitted to the *Physical Review*. In accordance with wartime censorship guidelines, he voluntarily withheld publication until after the war; it appeared in April, 1946.

If natural uranium can maintain a slow-neutron chain reaction due to the large fission cross-section of U-235 for such neutrons, why not build a slow-neutron bomb that incorporates a neutron moderator? In theory, this could be done. However, it would be utterly impractical: essentially, you would have to transport a reactor and drop it on an adversary. Moreover, the neutrons would be so slow that the reaction would grow at a rate not much faster than an ordinary chemical reaction: the result would be that the device would heat itself up, melt, and disperse, which would allow neutrons to escape and cause the reaction to shut down. A slow-neutron bomb would create an expensive fizzle, not a bang. To maintain a fast-neutron reaction, the isotopic abundance ratio must be changed to increase the fraction of U-235 in order that fission can compete against capture by U-238. The break-even point is a U-235 abundance of about 66%, much greater than the natural abundance of 0.7%.

Even at 66%, fission would be only just as probable as capture. Bomb-grade uranium is defined as 90% or more U-235.

Bohr and Wheeler's work provided a solid foundation for understanding the fission process and its possibilities. But in 1939 a huge gulf lay between theoretical understanding and any possible practical energy-producing applications. That gulf could only be filled with further experimental data on cross-sections and secondary neutron numbers, and consideration of large-scale techniques for isotope separation. But while nobody could yet speak definitively regarding the prospects for a chain reaction or a bomb, that did not mean that the possibilities could not be considered.

3.7   Criticality

As soon as secondary neutrons were discovered, physicists began to consider the conditions necessary for achieving a chain reaction, at least in theory.

Even if there has been a fission to begin a possible chain reaction in a material like U-235, the secondary neutrons that are liberated are not guaranteed to cause subsequent fissions. Some will inevitably reach the surface of the sample and escape, particularly if it is small. As the size of the sample increases, the probability that a neutron will escape before causing a secondary fission naturally diminishes. The escape probability never goes strictly to zero unless the sample becomes infinitely large, but it will eventually become low enough that neutrons become more likely to cause fissions than they are to escape. The key concept is that of a *critical size* and thus a corresponding *critical mass*: the minimum mass of U-235 (or Pu-239) that has to be assembled in one place in order to have a self-sustaining reaction which in principle continues until all of the material has fissioned, or, more likely, heats itself up and disperses.

Technically, criticality is said to occur if the number of neutrons within the sample is increasing with time. Whether or not this condition is fulfilled depends on the density of the fissile material, its cross-sections for fission and scattering, and the number of secondary neutrons emitted per fission. To analyze the flights of neutrons in a reactor or bomb core requires the use of time-dependent diffusion theory, which was a long-established branch of thermodynamics by 1939.

First to publish was French physicist Francis Perrin, in the May 1, 1939, edition of the weekly journal of the French Academy of Sciences. Perrin applied diffusion theory to an assemblage of natural-abundance uranium in its oxide form, assuming fast neutrons. With rough estimates for some of the relevant parameters, he arrived at an enormous critical mass, about 40,000 kg, or 44 U. S. tons. He also analyzed how this figure could be reduced by surrounding the material with a tamper to reflect escaped neutrons back into the fissile material. Perrin's 44-ton result has no real relevance for a bomb, which requires essentially pure U-235, but he did establish the relevant diffusion physics and introduced the notion of tampering.

Soon behind Perrin was German physicist Siegfried Flügge of the Kaiser Wilhelm Institute for Chemistry in Berlin, who published a much lengthier analysis in the June 9 edition of *Naturwissenschaften*. Also considering uranium oxide, Flügge deduced the astounding figure that if all of the uranium in one cubic meter of such material were to fission, the energy released could raise a cubic kilometer of water to a height of 27 km, three times the height of Mount Everest. Flügge had assumed that both uranium isotopes fissioned; if fission of only U-235 is considered, the correct height is much less, about 370 m, but is still impressive. Flügge did not estimate a critical mass, but did give a figure for the critical radius of greater than 50 cm, again based on estimated parameters.

The first truly coherent analysis of criticality was published by Rudolf Peierls of the University of Birmingham (England) in October, 1939. Peierls was an outstanding theoretical physicist who had been born in Germany and emigrated to England in 1933. Like Otto Frisch, Peierls was Jewish; both men would come to be concerned about fission research being done in Germany. Peierls' presence in Birmingham was due to another important personality in Manhattan history, Australian native Marcus Oliphant, another protégé of Ernest Rutherford. In 1937, Oliphant was appointed head of the physics department at Birmingham. One of his first faculty recruits was Peierls, to whom he offered a permanent Professorship. Peierls leapt at the opportunity, especially as it carried a salary over twice what he had been receiving in a temporary position at Cambridge. Peierls moved to Birmingham in 1937, and became a naturalized British citizen in February, 1940. With war looming in Europe in mid-1939, Oliphant made another valuable acquisition: Otto Frisch, who was then still in Copenhagen. Circumventing bureaucratic formalities, Oliphant invited Frisch over for a summer vacation and then found him work as an auxiliary lecturer.

Left: Genia (1908–1986) and Rudolf (1907–1995) Peierls in New York 1943. Right: Marcus Oliphant (1901–2000). Sources: Photograph by Francis Simon, courtesy AIP Emilio Segre Visual Archives, Francis Simon Collection; AIP Emilio Segre Visual Archives, Physics Today Collection

Oliphant's strategic disregard for channels manifested itself various valuable ways. While working on radar research for the British Admiralty, he wanted to tap into Peierls' extensive knowledge of electromagnetic theory. But as an enemy alien, Peierls could have no contact with classified work. Oliphant simply posed his questions as purely academic exercises, with both aware of the fiction. Oliphant would later play a seminal role in prodding American physicists to accelerate their country's fission-bomb efforts.

In his memoirs, Peierls described how he read Perrin's paper and realized that he could refine the calculation. Given the potential military applications, he had some doubts about openly publishing his analysis, and consulted Frisch on the advisability of doing so. Confident that Bohr had shown that an atomic bomb was not a realistic proposition, Frisch saw no reason for Peierls not to publish. A few months later, they would find themselves in a very different position. Curiously, Peierls developed explicit formulas for estimating the critical mass, but did not bother to substitute any numbers into his expressions: cross-sections were still experimentally uncertain. Had he been in possession of even approximately correct values for the relevant data, his paper may well never have appeared in the open literature.

3.8   Bohr Verified

Niels Bohr's suggestion that it was the rare 235 isotope of uranium that was responsible for slow-neutron fission begged for experimental verification. The only sure way to test the idea would be to isolate pure, separated samples of U-235 and U-238, and subject them both to neutron bombardment. The only practical method of isotope separation known at the time was mass spectroscopy, and the task of preparing the samples came into the hands of a superb practitioner of that art, Alfred Nier of the University of Minnesota.

Alfred Nier (1911–1994) Source: University of Minnesota, courtesy AIP Emilio Segre Visual Archives

Nier had come to the attention of uranium physicists with a paper he had published in the January 15, 1939, edition of *Physical Review* in which he reported a measurement of the abundance of a third, extremely rare isotope of uranium, U-234. This isotope is present to the extent of only about one atom per every 18,000 of U-238 in natural uranium, but Nier's mass spectrometer, which he constructed himself, was sensitive enough to make the detection. Nier met Enrico Fermi at an American Physical Society meeting held in Washington in April, 1939, at which time Fermi encouraged him to try to separate samples of uranium isotopes in order to test Bohr's theory. Busy with teaching and other projects, Nier did not take up the challenge until prodded again by Fermi in October. In order to achieve sufficient separation, he had to build a new spectrometer, which he completed in February, 1940. His first successful runs were carried out on February 28 and 29, and he glued the samples to a letter which he posted by airmail special delivery to Columbia University, where they would be subjected to slow-neutron bombardment in the Physics department's cyclotron.

Nier's samples were truly miniscule. He did two separation runs, of durations 10 and 11 h, which he predicted to yield 0.17 and 0.29 micrograms of U-238—if all of the ions stuck to the collector. The amounts of U-235 would have been 1/140 as much, or about 1.2 and 2.1 nanograms. To collect a full kilogram at a rate of 2.1 nanograms per 11 h would require some 600 million *years* of continuous operation, a testament to Niels Bohr's opinion of the impracticality of a U-235 bomb.

At Columbia, the U-235 samples clearly showed evidence for slow-neutron fission, while the U-238 samples showed none at all. Despite the minute sample sizes, the Columbia team was also able to use them to estimate the slow-neutron fission cross-section for U-235. The results were reported in a paper published in the March 15, 1940, edition of the *Physical Review*, which listed Nier, Eugene Booth, John Dunning, and Aristide von Grosse as authors. Their paper closed with the observation that "These experiments emphasize the importance of uranium isotope separation on a larger scale for the investigation of chain reaction possibilities in uranium." The concept of isotopes, barely 25 years old, was about to assume enormous importance.

Unfortunately, Nier's samples were too small to test for fast-neutron fission. In a follow-up paper published a month later, results from further tests with larger samples were reported. Enough material was available to allow testing U-238 for fission by both slow and fast neutrons; it was verified to fission only under fast-neutron bombardment. The slow-neutron fissility of U-235 was again verified, but the new samples of U-235 were still too small to test for fast-neutron fissility. Nier later wrote that had his budget been a few hundred dollars richer, he could have afforded better vacuum pumps, which would have allowed him to obtain a sample of U-235 large enough for the fast-neutron test.

The Nier/Columbia work received some remarkable public exposure. In the May 5, 1940, edition of the *New York Times*, science reporter William Laurence—who would later witness the *Trinity* test and the Nagasaki bombing—was wildly optimistic in stating that the prospect of nuclear power was perhaps just a few months to a year distant, but otherwise gave a fairly clear description of the Columbia work, the Bohr/Wheeler nuclear parity argument, the role of slow neutrons in sustaining a chain reaction, and the fact that one pound of U-235 would be equivalent to some 15 kt of conventional explosive. According to Laurence, "reliable sources" indicated that the Nazi government in Germany had ordered its greatest scientists to concentrate their energies on the uranium issue. Laurence's work reached a wider audience in September with a similar article in the popular weekly magazine

The Saturday Evening Post. Some of his prose was overdone, such as describing U-235 as "… a veritable Prometheus bringing to man a new form of Olympic fire …", but did give a reasonable description of the work to date, and elaborated on possibly using diffusion techniques to isolate U-235 and how nuclear power could be used to drive ships and submarines. When he came into command of the Manhattan Project, General Groves attempted to have all copies of the *Post* with Laurence's article withdrawn from libraries across the country; one source reports that a copy of the article was found in a German laboratory at the end of the war. In a like vein, John O'Neill reported on uranium in the June 1940 edition of *Harper's Magazine*, explaining the effects of fast and slow neutrons, the neutron-capture effects of U-238, how a chain-reaction could work, and the difficulty of isotope separation. While some of O'Neill's speculations were over-the-top (nuclear-powered automobiles would put gasoline stations out of business), he did raise the possibility of explosives: "But … if we use too pure a sample of Uranium 235 the process may take place at such a rapid rate that all the energy … may be given off … before control processes can become operative. If this condition were brought about … we should then have not an atomic power source but an atomic energy explosive." Curiously, Groves does not seem to have attempted to have had O'Neill's article impounded.

The Nier/Columbia work verified Bohr's hypothesis, although it did leave open the question of the fast-neutron fissility of U-235. But even as Nier and his collaborators were undertaking their work, Otto Frisch and Rudolf Peierls were considering that very question.

3.9   An Atomic Bomb Might Be Possible

It is rare for a scientific manuscript to have a direct impact on world affairs, but one that did is the "Frisch-Peierls Memorandum" of March, 1940. This document was responsible for initiating British investigations which resulted in the conclusion that nuclear weapons were not only feasible, but could be built in time affect the outcome of the war. The British efforts would have a significant impact on the opinions of American scientists in the summer of 1941, and would strongly influence a report delivered to Franklin Roosevelt later that year.

At Birmingham, Otto Frisch was also barred from war research, and had plenty of time to pursue his own interests. Aware of Bohr's prediction regarding the role of U-235 in slow-neutron fission, he began to contemplate how the theory might be tested. Months before Alfred Nier and his collaborators performed their experiments, Frisch concluded that one approach would be to prepare a sample of uranium in which the proportion of U-235 had been artificially enriched. If Bohr was correct, then the enriched sample should show an increased rate of fission under slow-neutron bombardment when compared to an unenriched one. Frisch began to research isotope enrichment methods, and zeroed in on the thermal-diffusion method. This was also known as the Clusius-Dickel method, after the two German scientists, Klaus Clusius and Gerhard Dickel, who had only recently developed and successfully applied it to enriching neon and chlorine isotopes. Frisch had the Birmingham glassblower prepare a diffusion tube; the experiment did not succeed, but his attention soon became drawn in a much more compelling direction.

Frisch received an invitation from the Royal Society for Chemistry to write a review article on radioactivity and subatomic phenomena. In his memoirs, Frisch relates that the winter of 1940 was unusually cold and snowy in Birmingham, and that he prepared the report while wrapped in a winter coat, sitting before a fire in a room which did not get warmer than 42 °F during the day and fell to below freezing at night. Ironically, Frisch opened his review with the statement that "The year 1940 has produced no spectacular progress in nuclear physics. The "boom" in papers about nuclear fission … has almost faded out." Much of the report concerned decay products of various bombardment reactions, with only a brief mention given to the Columbia verification of Bohr's speculation that U-235 was responsible for slow-neutron fission. The possibility of a chain reaction is raised in one lone sentence, only to be dismissed.

Frisch later wrote that he truly believed that an atomic bomb was impossible when he prepared the report, but this assertion might have been some after-the fact revisionism. Whatever the sequence of events, writing it apparently caused his thoughts to turn back to his enrichment idea. He began to wonder if, in the event that he could produce enough pure or highly enriched U-235, might it be possible to make a truly explosive chain reaction based on fast neutrons as opposed to slow ones? Making a rough estimate of the fission cross-section of U-235 and using Peierls' critical-size formula, he estimated, to his surprise, a critical mass of about a pound. On extrapolating from the expected efficiency of a single Clusius-Dickel tube, he and Peierls then estimated that a cascade of 100,000 such tubes might be sufficient to produce enough U-235 for a bomb in a matter of weeks. As Frisch wrote, "At that point we stared at each other and realized that an atomic bomb might after all be possible."

In his own memoirs, Peierls states that Frisch approached him in February or March of 1940 with the question: "Suppose someone gave you a quantity of pure 235 isotope of uranium—what would happen?" In Peierls' telling, they then worked out the critical mass together, arriving at the one-pound figure. They then went on to estimate, with what Peierls described as a "back of the proverbial envelope" calculation, how much energy the reaction might liberate before the uranium dispersed itself. The result was equivalent to thousands of tons of ordinary explosive. In Peierls' memory, they said to themselves that "Even if this plant costs as much as a battleship, it would be worth having."

Alarmed that German scientists might be thinking along the same lines, Frisch and Peierls felt it their duty to inform the British government of the possibility of atomic weapons, but in a way that would keep the idea secret in case German researchers hadn't yet thought of it. (They had: see Chap. 8). They decided to prepare a memorandum, which Peierls typed up himself. They kept only one carbon copy.

Frisch and Peierls actually prepared two memoranda. The first, titled "Memorandum on the Properties of a Radioactive Super-bomb," was a brief qualitative description intended for government officials. The second, titled "On the construction of a "super bomb" based on a nuclear chain reaction in uranium," ran to seven pages and was more technical. Both documents still make for fascinating reading. The non-technical one lays out in a few pages all of the key factors concerning how such a bomb might operate, as well as the associated strategic implications. After describing why there exists a critical mass and how such a device could be triggered by rapidly bringing together two otherwise perfectly safe sub-critical pieces of uranium, they described some of the military implications: "As a weapon, such a bomb would be practically irresistible. There is no material or structure that could be expected to resist the force of the explosion." On the ethics of nuclear warfare: "the bomb could probably not be used without killing large numbers of civilians, and this may make it unsuitable as a weapon for use by this country." As to civil defense and deterrence strategy, "no shelters are available that would be effective and could be used on a large scale. The most effective reply would be a counter-threat with a similar bomb. Therefore it seems to us important to start production as soon and as rapidly as possible, even if it is not intended to use the bomb as a means of attack." Unknowingly, they had drafted a credible script for the later Cold War.

The estimate of about a pound for the critical mass appears in the technical memorandum. This would prove to be a serious underestimate caused by an overestimate by a factor of about eight for the fast-neutron fission cross-section for U-235. The critical mass is approximately proportional to the inverse-square of the cross-section, so an error of a factor of eight translates to an error of a factor of 64; the true critical mass is more on the order of 100 lb. This can be reduced by use of a tamper, but they did not explore this refinement.

While Frisch and Peierls' critical mass estimate was erroneous, their underlying physics was entirely sound. The technical memorandum contained only one formula, an expression for the energy liberated by a bomb containing a core of a given fissile mass. What is remarkable about this expression is that it can be shown to be exactly equivalent to one adopted at Los Alamos in 1943; Peierls must have worked out the diffusion theory on the back of his proverbial envelope. Presenting their result in this way exemplifies what generations of physics professors have told their students: work out your problem analytically first, and then substitute numerical values at the end of the derivation. That way, it is easy to recompute the result if a numerical value changes as a result of new experimental data. Frisch and Peierls were surely aware that the numbers they adopted were approximations at best which would be refined through further work. The Frisch-Peierls energy formula also showed very directly why the energy that would be liberated in a slow-neutron bomb would not be worth the trouble of making such a device. The energy attainable before the core blows itself apart turns out to be proportional to the square of the neutron speed. Slow neutrons move at speeds about one ten-thousandth of that of fast neutrons, which means a release of a hundred-millionth of the energy of a fast-neutron bomb—not much more than the equivalent of a few pounds of conventional explosive.

Towards the end of the technical memorandum, Frisch and Peierls emphasized a crucial qualitative difference between fission bombs and ordinary explosives: that in addition to the destructive effect of the explosion itself, the blast would distribute highly radioactive fission products over a wide area, plus material from the bomb casing rendered radioactive by neutron capture. They estimated that a bomb would generate radioactivity equivalent to hundreds of *tons* of radium, enough to render the devastated area too dangerous to enter for several days following the explosion. The tons figure is accurate, but much of the radioactivity gets carried off into the atmosphere by wind currents and dispersed over a much larger area than they had anticipated.

At the time they prepared their memoranda, Peierls had only recently been naturalized, and Frisch was still an enemy alien; they were unsure how to get their ideas to appropriate officials. They took their documents to Oliphant, who forwarded them to Sir Henry Tizard, Chairman of the government's Committee on the Scientific Survey of Air Warfare. British historian Ronald

Clark found the non-technical memorandum among Tizard's papers years later, and deduced that the documents reached him on March 19, 1940, just four days after the publication date of the Nier-Columbia verification of U-235 being responsible for slow-neutron fission.

3.10 MAUD

Sir Henry Tizard had already raised the issue of fission as a weapon with George P. Thomson, the son of J. J. Thomson of electron-discovery fame and a professor of physics at Imperial College, London. When Hans von Halban and his collaborators published their measurement of approximately three neutrons emitted per fissioning uranium nucleus, Thomson begun to consider the possibility of achieving a chain reaction if a sufficient mass of uranium could be brought together. Tizard was initially skeptical that a practical uranium bomb could be made, but had to take the possibility seriously.

James Chadwick initially also very much doubted the idea of a uranium bomb, but began to reconsider with the publication of the Bohr-Wheeler analysis in September, 1939. In October, he was contacted by Professor Edward Appleton, Secretary of the Department of Industrial and Applied Research, who asked if Chadwick thought the possibility of a uranium bomb merited concern. Chadwick reported back in early December that he would initiate experimental work, and began readying his cyclotron at the University of Liverpool to make cross-section measurements. Privately, he was expressed concerns to colleagues that British laboratories seemed disorganized, and that leadership in physics would shift to the United States—a fear which would prove well-founded. Thomson was also pursuing uranium work. By February, 1940, he had almost come to the conclusion that atomic energy was not worth pursuing as a war effort; he had tried to achieve a chain reaction, but had been unsuccessful.

It was against this background that that the Frisch-Peierls memorandum reached Tizard, who asked Thomson to convene a committee to investigate the matter. Thomson served as chair; the members included, among others, Chadwick and Oliphant. As enemy aliens, Frisch and Peierls were barred from serving on the committee, and for a while at least had no idea what happened to their memoranda. Frustrated, they sent Thomson a ten-page memo on the "uranium problem" in late July; he arranged a compromise whereby they could serve as consultants. When the work of the committee was split into a Policy Committee and a Technical Committee in March, 1941, they were allowed to serve on the latter.

Thomson's group named itself the MAUD Committee. This unusual designation had a curious provenance. In April, 1940, Germany occupied Denmark. As this was happening, Niels Bohr sent a telegram to Otto Frisch through Lise Meitner, the six concluding words of which were "Tell Cockcroft and Maud Ray Kent." Cockcroft was John Cockcroft of Cambridge University, but the meaning of "Maud Ray Kent" was a mystery. One theory was that by changing the "y" to an "i", "Maud Ray Kent" became an anagram for "radium taken." Another interpretation was that no expense be spared to separate uranium: "make ur day nt". Somebody suggested MAUD as a cover name, and the appellation stuck. Officially, it had periods between the letters (M.A.U.D.; sometimes interpreted as "Military Application of Uranium Disintegration"), but is commonly written in the simplified form. The mystery was not resolved until after Bohr escaped from Denmark to Sweden in late 1943, and then made his way to England: Maud Ray lived in Kent, and had at one time served as a governess for his children.

The MAUD committee held its first meeting on April 10, 1940, in the committee room of the Royal Society in London. Within weeks, the Battle of Britain would be in full fury. Thomson began to take the idea of a bomb seriously, and on the 16th wrote to Chadwick to say that the concept "is not so impossible when you come to look into it." By the summer of 1940, research under MAUD auspices was underway at the universities of Liverpool (cross-section measurements), Birmingham (uranium chemistry), Cambridge and Oxford (separation methods), and at Imperial Chemical Industries. Peierls spent the summer studying isotope separation methods, and reported in September that the most promising approach looked to be gaseous diffusion through a mesh of fine holes; experiments along this line were being conducted by another refugee scientist, Franz Simon, at Oxford. By December, Simon's group was far enough along to estimate parameters for an actual production plant. For an output of 1 kg of U-235 per day, some 70,000 square meters (17 acres) of diffusion membrane would be required; the plant would cover some 40 acres and consume 60 MW of power. Estimates of the cost of construction and the necessary number of operators would prove far too low, but the important thing was that, in Britain at least, thoughts on atomic bombs were moving toward practical engineering considerations.

As enrichment techniques were being considered, James Chadwick's cross-section measurements were tending toward confirming Frisch and Peierls' theoretical analysis. Chadwick's initial skepticism turned to gnawing worry. From a 1969 interview: "I remember the spring of 1941 … I realized then that a nuclear bomb was not only possible—it was inevitable. … I had many sleepless nights. … And I had then to start taking sleeping pills. It was the only remedy. I've never stopped since then. It's been 28 years, and I don't think I've missed a single night in all those 28 years." Peierls too became convinced that a bomb was possible, writing that "there is no doubt that the whole scheme is feasible ... and that the critical size for a U sphere is manageable." He reported his conclusion to a meeting of the MAUD committee on April 9, and in early summer the committee began to prepare its final report to Tizard.

American physicists had not yet begun thinking seriously about engineering practicalities, but were far ahead of their British counterparts in another important aspect of the problem: considering plutonium as a possible bomb fuel.

3.11 Plutonium

At about the time that Otto Frisch and Rudolf Peierls were re-evaluating the possibility of uranium bombs, Louis Turner at Princeton University was conceiving his idea of using U-238 to breed potentially-fissile plutonium-239. If neutron capture by U-238 did generate such a product and it proved stable, it could be separated from the bombarded uranium by ordinary chemical means, and so provide a path for extracting nuclear energy from U-238. Leo Szilard would remark in a 1946 address that "With this remark of Turner, a whole landscape of the future of atomic energy arose before our eyes in the Spring of 1940 and from then on the struggle with ideas ceased and the struggle with the inertia of Man began." Szilard's frontier was being opened in California.

One of the first confirmations of fission had occurred at Ernest Lawrence's Radiation Laboratory at Berkeley, and work on elucidating the nature of that process continued there. In the March 1, 1939, edition of the *Physical Review*, Edwin McMillan reported on an experiment where a thin foil of uranium was placed against a stack of aluminum foils and then exposed to neutrons from a cyclotron. Fission products ejected from the uranium were collected in the aluminum foils, from which they could be chemically extracted and their decay schemes analyzed.

Edwin McMillan (1907–1991), Emilio Segrè, and Glenn Seaborg. Source: AIP Emilio Segre Visual Archives, Segrè Collection

McMillan observed that following the neutron bombardment, the uranium itself (not the fission products) appeared to be exhibiting two beta-decays, with half lives of approximately 25 min and 2 days. He attributed the 25-min decay to an isotope of uranium formed by neutron capture, a suggestion that had initially been made by Meitner, Hahn, and Strassmann in 1937—the third of their three decay sequences.

In June, Emilio Segrè confirmed that the 25-min decayer (by then refined to 23 min) was indeed U-239, and also determined that since the 2-day decayer could be chemically separated from uranium, it must be a different element. Segrè suspected that the product of the 23-min decay was a long-lived isotope of element 93, and that the 2-day source was likely a fission product. If the 23-min decay product was truly an isotope of element 93, it would mean that a transuranic element had finally been synthesized. A snag in Segrè's interpretation, however, was that the 2-day source behaved anomalously in that it remained stuck in the bombarded uranium as opposed to being ejected as fission products normally were; this needed to be clarified.

The next installment in the story is a brief paper prepared a year later by McMillan and Philip Abelson. Their paper was dated May 27, 1940, just two days before Louis Turner's speculation on the possible fissility of element 94. However, unlike Turner's paper, McMillan and Abelson's report was published promptly, in the June 15 edition of the *Physical Review*. They reported that the 2-day source (refined to 2.3 days) did not in fact behave like a fission product, and that there was a clear relationship between the decay of the 23-min substance and the growth of the 2.3-day decayer. The conclusion seemed inescapable that the latter was an accumulating decay product of the former, with the reaction and decay scheme appearing to be exactly as Turner had speculated:

$$\sideset{_0^1}{}{n} + \sideset{_{92}^{238}}{}{U} \rightarrow \sideset{_{92}^{239}}{}{U} \rightarrow \sideset{_{93}^{239}}{}{X} \rightarrow \sideset{_{94}^{239}}{}{Y}.$$

It is surprising that McMillan and Abelson published their result; it is hard to imagine that they would have been unaware of the possibility of 239 as a weapons material. James Chadwick was so upset with their publication that he placed an official protest through the British Embassy.

McMillan and Abelson's work came to the attention of Glenn Seaborg, who in the summer of 1939 had been appointed as an Instructor of Chemistry at Berkeley after completing his Ph.D. there. Seaborg resolved to search for the product of the 2.3-day decay of element 93, which he suspected might be an isotope of element 94. McMillan had detected indications of a long-lived alpha-activity building up in a sample of purified 93; perhaps they were a decay signature of element 94. Seaborg teamed up with fellow faculty member Joseph Kennedy and graduate student Arthur Wahl, who would study element 93 for his doctoral thesis. With access to Lawrence's 60-in. cyclotron, the group was able to create samples of element 93 by bombarding targets containing uranium compounds. The ultimate goal was to detect element 94 by its own alpha-decay. This is now known to have a half-life of 24,100 years and recreates, ironically, U-235:

$$\sideset{_0^1}{}{n} + \sideset{_{92}^{238}}{}{U} \rightarrow \sideset{_{92}^{239}}{}{U} \rightarrow \sideset{_{93}^{239}}{}{Np} \rightarrow \sideset{_{94}^{239}}{}{Pu} \rightarrow \sideset{_{92}^{235}}{}{U}.$$

Two bombardment methods were used. In one, which was first used on August 30, 1940, uranium was bombarded directly with thermalized neutrons, presumably initiating the above sequence. In the second, used for the first time around December 14, 1940, uranium was directly exposed to accelerated deuterons. Various reaction channels are possible in this case, but a representative one is to produce two neutrons and neptunium-238, which decays through plutonium-238 to uranium-234:

$$\sideset{_1^2}{}{H} + \sideset{_{92}^{238}}{}{U} \rightarrow 2\left(\sideset{_0^1}{}{n}\right) + \sideset{_{93}^{238}}{}{Np} \rightarrow \sideset{_{94}^{238}}{}{Pu} \rightarrow \sideset{_{92}^{234}}{}{U}.$$

The half-lives in the last two steps are about 2 days and 88 years. While this method generates plutonium as well, it gives rise to the short-lived Pu-238 isotope, not Turner's Pu-239 isotope generated by direct neutron bombardment of uranium. But this deuteron-bombardment reaction was important, as it was Pu-238 that Seaborg and his group first isolated; this process is considered to be the discovery reaction for plutonium.

By October, Kennedy had developed a counter capable of detecting alpha particles in the presence of background beta decays, and by late November Wahl had perfected a technique for isolating very pure samples of element 93. Evidence for the 2.1-day decay of Np-238 was detected just before Christmas, 1940. By early January, 1941, Wahl had proven that the alpha-emitting material was definitely not element 93, and by the end of that month the group felt sufficiently confident to prepare a paper announcing the discovery of element 94 based on the fact that the 88-year alpha decayer could be chemically separated from both uranium and element 93. Dated January 28, 1941, the paper was withheld from publication until April 1946, but it established priority for the discovery.

Seaborg's more serious goal was to produce sufficient Pu-239 to test its slow-neutron fissility. On February 23, 1941, bombardment of a 1.2-kg sample of uranium-nitrate-hydride was commenced, which proceeded intermittently until March 3. On March 6, the sample of element 93 extracted from this bombardment gave a beta-decay count estimated at 76 millicuries, equivalent to a mass of about 0.3 micrograms. The sample was allowed to sit for three weeks, by which time, with its half-life of only 2.3 days, essentially all of the 93 would have decayed to element 94.

The first test of element 94's slow-neutron fissility was carried out on March 28. The new element did indeed seem to be slow-neutron fissile, with a cross-section estimated to be about one-fifth that of U-235. The sample geometry was poor, however (it was too thick), and since it was covered in a drop of glue the true cross-section was likely greater. By May 12, Wahl had succeeded in further purifying and thinning the minute sample of element 94, and a second slow-neutron experiment was begun on the 17th. This time the result was a cross-section 1.7 times greater than that for U-235, in fair accord with the present value of about 1.3.

The slow-neutron fissility was reported in a paper dated May 29, 1941, which also remained under wraps until 1946. On May 19, Seaborg related the result to Ernest Lawrence, who passed it on Nobel Laureate Arthur Compton at the University of Chicago. Compton had just finished preparing a report to the National Academy of Sciences on possible military applications of atomic fission, which would have to be amended: if element 94 bred from U-238 was as fissile as it seemed, Seaborg and his team had just increased the amount of potential bomb material by over 100 times.

Plutonium is one of the most bizarre elements known. As described by former Los Alamos National Laboratory Director Siegfried Hecker, it seems almost at odds with itself. With little provocation, its density can change by as much as 25%; it can be brittle or malleable; expands when solidifying from a liquid; tarnishes within minutes; reacts vigorously with oxygen, hydrogen, and water; its corrosion products can spontaneously combust in air; and its own alpha-decay causes self-irradiation damage that can fundamentally change its crystalline properties. In the spring of 1944, metallurgists at Los Alamos would discover that it exhibits five different crystalline structures between room temperature and its melting point; six such forms are now known. These various forms all have different densities and mechanical properties, which can affect alloying properties and corresponding critical masses. To top it off, plutonium is, as Glenn Seaborg put it, "fiendishly toxic, even in small amounts." This bizarre new element, which Nature had not seen fit to place upon the planet in any but trace amounts, would assume an outsize importance in world affairs.

4

Organizing: Coordinating Government and Army Support 1939–1943

4.1 Fall 1939: Szilard, Einstein, the President, and the Uranium Committee

The first formal contact between nuclear scientists and American government representatives occurred on March 17, 1939, when, at a meeting set up by Columbia University Dean of Science George Pegram, Enrico Fermi met with naval officers in Washington to explain the possibilities of using chain reactions as power sources or in bombs. One of the officers present was Admiral Stanford Hooper, technical assistant to the Chief of Naval Operations; also present was Ross Gunn, a civilian physicist working for the Naval Research Laboratory who would later become involved with the liquid thermal diffusion project for uranium enrichment. Despite Fermi's downplaying of the possibilities and the group's inherent skepticism, they allocated $1,500 to Columbia to support his nuclear-pile research.

By 1939, Leo Szilard was living in New York, where, although independently wealthy, he maintained a part-time appointment at Columbia. Szilard was much more alarmed than Fermi at the possibility of fission being turned into a weapon, and felt that responsible government officials needed to be alerted to the issue. He discussed the matter with fellow émigré Eugene Wigner, a brilliant theoretical physicist and chemical engineer who had been on the faculty of Princeton University since 1930. In 1936, Wigner had predicted that scientists would figure out how to release nuclear energy; he would later make significant contributions to reactor engineering.

© Springer Nature Switzerland AG 2020
B. C. Reed, *Manhattan Project*,
https://doi.org/10.1007/978-3-030-45734-1_4

Eugene Wigner (1902–1995), at the time of his receiving the Nobel Prize (1963). Source: http://commons.wikimedia.org/wiki/File:Wigner.jpg. *Right: In this 1946 photo, Albert Einstein and Leo Szilard re-enact the preparation of a letter to President Roosevelt. Source: Courtesy Atomic Heritage Foundation,* http://www.atomicheritage.org/mediawiki/index.php/File:Einstein_Szilard.jpg

Both Szilard and Wigner had grown up in Hungary, and had witnessed the rise of European totalitarianism. On the rationale of denying Germany access to uranium ores, they decided to warn the government of Belgium of the issue: some of the world's richest uranium deposits were in the Congo, then a colony of Belgium. But how could two refugee Hungarian scientists living in America deliver such a warning? On recalling that their friend Albert Einstein was close to Belgium's queen mother, they decided to enlist his help. On July 16, 1939, six years to the day before the *Trinity* test, Szilard and Wigner drove to Einstein's summer home on Long Island. Szilard explained the possibility of an explosive chain reaction, which apparently came as a revelation to Einstein.

Wigner suggested that a letter written by refugees on a security issue to a foreign government might not be appropriate, so it was decided that Einstein—the only one of them famous enough to be recognized—would prepare a letter to the Belgian ambassador, along with a covering letter to the State Department. Einstein drafted a letter in German, which Wigner translated, had typed up, and sent to Szilard. A few days later, however, Szilard came into contact with Alexander Sachs, an economist with the Lehman Brothers financial firm. Sachs had also trained as a biologist, and was a close friend of and advisor to President Franklin Roosevelt. Sachs suggested that a better approach would be a letter directly to the President, and he offered to deliver one personally.

Sachs is little-known outside academic Manhattan Project circles, but one of the most valuable sources of information on the early history of the Project is a "Documentary Historical Report," which he prepared in August, 1945. This remarkable 27-page document covers the period from the Szilard/Einstein letter to mid-1940, when the project was placed under the oversight of the National Defense Research Committee. Sachs wrote in a peculiarly convoluted manner, but was an exceptionally perceptive observer of the world situation.

Szilard, this time accompanied by Edward Teller, visited Einstein again on July 30 to revise their original work. Einstein dictated another letter, which addressed not only the issue of Congolese uranium ores but also the possibility of a significantly destructive new type of bomb:

> Albert Einstein
> Old Grove Rd.
> Nassau Point
> Peconic, Long Island
> August 2nd 1939

F. D. Roosevelt
President of the United States
White House
Washington, D.C.

Sir:

Some recent work by E. Fermi and L. Szilard, which has been communicated to me in manuscript, leads me to expect that the element uranium may be turned into a new and important source of energy in the immediate future. Certain aspects of the situation which has arisen seem to call for watchfulness and, if necessary, quick action on the part of the Administration. I believe therefore that it is my duty to bring to your attention the following facts and recommendations:

In the course of the last four months it has been made probable—through the work of Joliot in France as well as Fermi and Szilard in America—that it may become possible to set up a nuclear chain reaction in a large mass of uranium, by which vast amounts of power and large quantities of new radium-like elements would be generated. Now it appears almost certain that this could be achieved in the immediate future.

This new phenomenon would also lead to the construction of bombs, and it is conceivable—though much less certain—that extremely powerful bombs of a new type may thus be constructed. A single bomb of this type, carried by boat and exploded in a port, might very well destroy the whole port together with some of the surrounding territory. However, such bombs might very well prove to be too heavy for transportation by air.

The United States has only very poor ores of uranium in moderate quantities. There is some good ore in Canada and the former Czechoslovakia, while the most important source of uranium is Belgian Congo.

In view of the situation you may think it desirable to have more permanent contact maintained between the Administration and the group of physicists working on chain reactions in America. One possible way of achieving this might be for you to entrust with this task a person who has your confidence and who could perhaps serve in an inofficial capacity. His task might comprise the following:

(a) to approach Government Departments, keep them informed of the further development, and put forward recommendations for Government action, giving particular attention to the problem of securing a supply of uranium ore for the United States;

(b) to speed up the experimental work, which is at present being carried on within the limits of the budgets of University laboratories, by providing funds, if such funds be required, through his contacts with private persons who are willing to make contributions for this cause, and perhaps also by obtaining the co-operation of industrial laboratories which have the necessary equipment.

I understand that Germany has actually stopped the sale of uranium from the Czechoslovakian mines which she has taken over. That she should have taken such early action might perhaps be understood on the ground that the son of the German Under-Secretary of State, von Weizsäcker, is attached to the Kaiser-Wilhelm-Institut in Berlin where some of the American work on uranium is now being repeated.

Yours very truly,
Albert Einstein

Sachs obtained a meeting with Roosevelt on October 11. In a summarizing cover letter, he explained how the discovery that uranium could be split by neutrons could lead to the creation of a new source of energy, the possibility of creating "tons" of radium for use in medical treatments, and the "eventual probability of bombs of hitherto unenvisaged potency and scope." He urged that with the danger of a German invasion of Belgium, it was imperative that arrangements be made with the mining firm of Union Minière du Haut-Katanga, which had its head office in Brussels, to make available supplies of uranium to the United States. He also urged acceleration of experimental work in America, arguing that since such work could no longer be carried out within the limited budgets of university physics departments, "public-spirited executives in our leading chemical and electrical companies could be persuaded to make available certain amounts of uranium oxide and quantities of graphite, and to bear the considerable expense of the newer phases of the experimentation." Sachs also suggested that Roosevelt designate an individual or committee to serve as a liaison between scientists and the government.

Unlike Winston Churchill, Roosevelt was not known to be particularly curious about science, but, like Churchill, he did appreciate its importance in military power and in improving society generally. After hearing Sachs out, Roosevelt allegedly remarked, "Alex, what you are after is to see that the Nazis don't blow us up." Roosevelt ordered his Secretary, General Edwin M. Watson, to act as the White House's liaison on the issue, and to work with the Director of the National Bureau of Standards, Lyman J. Briggs, to put together an advisory committee.

President Roosevelt signs the dec-laration of war against Japan, December 8, 1941 Source: http://commons.wikimedia. org/wiki/File:Franklin_Roo sevelt_signing_declaration_of_ war_against_Japan.jpg

There are actually two versions of Sachs' meeting with Roosevelt. In Sachs' 1945 history of the project, he refers only to the October 11 meeting. But an article in the March 14, 1950, edition of *Look* magazine written by Nat Finney indicates that the President was not convinced at that meeting that he should embark on what could be a very costly endeavor. In response, Sachs asked if he could meet with the President again the next morning over break-fast, a request which Roosevelt granted. Finney claims that Sachs spent that night in his hotel room and wandering around a park, trying to think of an argument that would convince Roosevelt. He presented his argument as a story, allegedly that of how American steam-engine inventor Robert Fulton tried to convince Napoleon Bonaparte to build a fleet of steamships with which to invade England. Napoleon supposedly scoffed at such a radical idea, and thereby lost his opportunity to change the course of history. Roosevelt is said to have remained silent for a couple minutes, and then scribbled a note to an aide. The aide disappeared for a few moments and then returned with a bottle of Napoleon brandy, from which the President and Sachs drank a toast while Roosevelt indicated that he would take action.

In any event, Sachs met with Briggs later on the 12th to assemble an Advisory Committee on Uranium, which came to be known as the Uranium Committee. The initial members were Briggs himself as Chair, plus Colonel Keith Adamson of the Army and Commander Gilbert C. Hoover of the Navy; Adamson and Hoover were ordnance experts whom Sachs had briefed just prior to meeting with the President. The name, membership, organizational structure, and responsibilities assigned to this committee would change many times over the course of the war.

Some of the Manhattan Project's administrators, at the Bohemian Grove meeting of September, 1942. Left to Right: Major Thomas Crenshaw, Robert Oppenheimer, Harold Urey, Ernest Lawrence, James Conant, Lyman Briggs, Eger Murphree, Arthur Compton, Robert Thornton (Univ. of California), Col. Kenneth Nichols. Source: Lawrence Berkeley National Laboratory, courtesy AIP Emilio Segrè Visual Archives

April, 1940. Left to right: Ernest Lawrence, Arthur Compton, Vannevar Bush, James Conant, Karl Compton, Alfred Loomis. Source: http://commons.wikimedia.org/ wiki/File:LawrenceComptonBushConan tComptonLoomis.jpg

The committee held its first meeting at the Bureau of Standards on October 21. Einstein did not attend, but Enrico Fermi, Leo Szilard, Edward Teller, and Eugene Wigner were invited; also present were physicists Fred Mohler of the Bureau of Standards and Richard Roberts of the Carnegie Institution.

Despite the skepticism of the military officers present as to the possibility of revolutionary new weapons or sources of power, Briggs argued that the world situation and American interests must be taken into account in what he called "the equation of probabilities." The War and Navy Departments contributed $6,000 for the purchase of four tons of graphite, paraffin, cadmium, and other supplies so that Fermi could carry out neutron-capture experiments at Columbia. The committee also appointed a Science Advisory Sub-Committee, comprising Harold Urey (Columbia University) , Gregory Breit (University of Wisconsin), George Pegram, Merle Tuve (Carnegie Institution), Jesse Beams (University of Virginia), and Ross Gunn. Urey was recognized as a world leader in techniques of isotope separation; in May 1940 he would be granted a contract to investigate application of thermal diffusion, chemical separation, and centrifugation to enriching uranium. Breit was an outstanding theoretical physicist, and Beams was conducting research on high-speed centrifuges.

Leo Szilard and Enrico Fermi spent considerable time over the summer of 1939 considering how a chain-reacting mass of uranium and graphite might be configured. Szilard, again ahead of his time, followed up with a memorandum to Briggs on October 26, urging the purchase of 100 metric tons of graphite and 20 metric tons of uranium oxide in order to get experiments underway as soon as possible. This was not done at the time, and, as the Manhattan Project progressed, Szilard was to experience no end of frustration with what he saw as bureaucratic inertia and foot-dragging.

Briggs' committee reported to Roosevelt on November 1 with a brief two-page letter. After opening with a technical summary of the process of fission, they related that a chain reaction could prove to be a power source for submarines, and noted that if a nuclear reaction should be explosive, "it would provide a possible source of bombs with a destructiveness vastly greater than anything now known." The letter recommended that four tons of graphite be procured for experiments, which, if successful, would lead to a requirement for 50 tons of uranium oxide; no mention was made of the $6,000 allocated to Columbia. They also recommended that the main committee be enlarged by the addition of Karl Compton, President of the Massachusetts Institute of Technology (and brother of physics Nobel Laureate Arthur Compton), Sachs, Einstein, and Pegram. Also added to the group at some point before the summer of 1940 was Admiral Harold. G. Bowen, Director of the Naval Research Laboratory.

Watson acknowledged Briggs's report on November 17, indicating that the President would keep it on file for reference. Not until February 8, 1940, did Watson follow up, asking Sachs and Briggs if there was anything new to report. Briggs replied on February 20 to indicate that the $6,000 authorized the preceding October had been transferred to Columbia, and that he was waiting to be informed of results of the work.

Through the fall and winter of 1939/40, scientists had not been idle, however. Sachs' Historical Report lists several areas of experimental and theoretical research that were underway: slow neutron reactions; fast neutron reactions; uranium isotope studies; isotope separation by diffusion, centrifugation, and other means; and production of uranium metal. Groups were active at Columbia, Princeton, the Carnegie Institution, Harvard, Yale, MIT, the University of Virginia, and George Washington University; he did not mention the work on creating plutonium that was underway at Berkeley under the direction of Glenn Seaborg.

In response to Watson's request for an update, Sachs responded on February 15 that he felt that the tone of the November 1 report had been too academic and that possible practical applications should have been emphasized; he promised another letter from Einstein. Einstein's letter, dated March 7, indicated that work on fission was being accelerated in Germany, and that Szilard had prepared a manuscript on how to set up a chain reaction. Sachs transmitted the letter to Roosevelt on March 15, about the time that the Frisch-Peierls memorandum began its journey up the chain of command in England.

Watson replied to Sachs on March 27 that the Briggs Committee was awaiting a report on work being carried out at Columbia. Sachs had occasion to meet with Roosevelt in early April, and reiterated the importance of having Belgian ores shipped to the United States as well as the urgency of having government or foundation funds to promote long-term research planning. Roosevelt and Watson both sent letters to Sachs on April 5, asking that another meeting be organized. Sachs encouraged Einstein to attend; he demurred, but did write Briggs on April 25 to express his conviction that the scale and speed of uranium work should be increased, and seconded a proposal by Sachs that a "Board of Trustees" be formed to solicit funds to support the work.

The pace of activity began to pick up in the spring of 1940. The Uranium Committee held its second meeting at the Bureau of Standards on Saturday, April 27, by which time it had been verified that U-235 was responsible for slow-neutron fission. Briggs reported to Watson on May 9 that the committee was not prepared to recommend a large-scale experiment to attempt a chain reaction until the results of experiments being conducted at Columbia were in, which was expected to be within a week or two. In the meantime, Fermi and Szilard were beginning to conceive of a reactor wherein a three-dimensional lattice of blocks of uranium would be distributed within a moderator.

On May 10, the day that Germany invaded Belgium and Winston Churchill became Prime Minister of Great Britain, Sachs drafted a memorandum to himself which recorded that the next stages of the work would be to carry out a survey of nuclear constants (cross-sections and the like) to narrow down limits of experimental error, and then to undertake a "large-scale" experiment to demonstrate whether or not a chain reaction could be set up and maintained. The cost of these steps was estimated at up to $50,000 and $500,000, respectively. Dissatisfied that work was being impeded by organizational difficulties, he wrote to FDR the next day to again raise the idea of a non-profit corporation to raise funds to support research.

Sachs learned from Pegram that Fermi and Szilard had found the neutron capture cross-section of carbon to be encouragingly small, and on May 13 wrote to Briggs with this news and a plea that the project needed to be accelerated while being kept secret. (A small capture cross-section would mean less possibility of losing the neutrons necessary to maintain a chain reaction.) Two days later, Sachs wrote to Watson to apprise him of the situation, and to suggest that the President establish a "Scientific Council of National Defense" which would be invested with authority to develop defense-related technical projects. Sachs followed up with another letter to Watson on May 23, wherein he reiterated many of the same points. Word must have got back to Briggs; on June 5 he authorized Sachs to approach Union Minière to gather information on ore stocks, costs, and anticipated mine extraction rates.

In addition to drawing the attention of government officials to the prospects for nuclear energy, émigré European physicists were also instrumental in alerting the American scientific community to the need to censor publication of developments that could become of military importance. At a meeting of the Division of Physical Sciences of the National Research Council in April, 1940, Gregory Breit suggested the formation of a committee to control publication in all American scientific journals, a concept completely at odds with the historic practice of open scientific publication and debate. Various subcommittees were set up to deal with publications in a number of fields; the first one, chaired by Breit, was devoted to considering uranium fission. Well before formal military involvement in nuclear fission, scientists had begun to police their own publication practices.

Left: Gregory Breit (1899–1981) at the 1939 meeting of the American Physical Society. Right: Vannevar Bush (1890–1974). Sources: Photo by Esther Mintz, courtesy AIP Emilio Segre Visual Archives, Esther Mintz Collection; Harris and Ewing, News Service, Massachusetts Institute of Technology, courtesy AIP Emilio Segre Visual Archives

Alexander Sachs drifts out of the story at this point, but one last inclusion in his Historical Report deserves mention. This is a five-page memo to himself prepared on April 20, 1940, under the convoluted title "Import of War Developments for Application to National Defense of Uranium Atomic Disintegration." This document opens with the observation that superior technology had enabled Nazi forces to overrun a number of European countries, and that other countries which had not brought their defenses up to the

German level could expect the same fate. He then remarked that uranium research may prove as important to national defense as the most advanced chemical and electrical research then being undertaken. Anticipating that a chain reaction would be successfully demonstrated and that war between America and Japan was likely, Sachs argued that nuclear-propelled American naval vessels, particularly aircraft carriers armed with aircraft carrying nuclear bombs, could easily extend their range to Japan without the need for refueling. This remarkable analysis was written over 19 months before Pearl Harbor, and over two years before Enrico Fermi first achieved a chain reaction.

4.2   The National Defense Research Committee; Reorganization I

In 1940, the Uranium Committee underwent a significant change of venue within governmental administration, as well as a change in membership. On June 27, President Roosevelt established the National Defense Research Committee (NDRC), which was charged with supporting and coordinating research conducted by civilian scientists which might have military applications. The NDRC was the brainchild of Vannevar Bush, whom Roosevelt appointed to be its Director. A veteran of many years of government science administration, Bush had earned a joint Ph.D. from the Massachusetts Institute of Technology and Harvard University in 1917. During World War I, he had worked with the National Research Council on the application of science to warfare, including development of submarines. After the war, Bush joined the department of Electrical Engineering at MIT, where he served as a faculty member. In 1932, he moved up to be Dean of Engineering, at which post he remained until 1938. While at MIT he developed, among other things, an early computer known as the differential analyzer. In 1939, he became President of the Carnegie Institution of Washington, as well as Chairman of the National Advisory Committee for Aeronautics (NACA), a forerunner of NASA. These positions enabled him to direct research toward military applications, and gave him a conduit for providing scientific advice to government officials.

During World War I, Bush had observed the lack of cooperation between civilian scientists and the military, and was determined that such inefficiency not repeat itself in the war which was engulfing Europe and would likely eventually involve America. In 1939, he began thinking of a federal-level agency to coordinate research, an idea he discussed with fellow NACA member James B. Conant, a distinguished chemist and President of Harvard University. Bush also ran the concept past his MIT colleague Karl Compton, as well as Frank Jewett, President of the National Academy of Sciences. Bush secured a meeting with President Roosevelt for June 12, 1940, and soon had his agency, which entered into official existence fifteen days later. Conant, Compton, and Jewett were made members of the new Committee, along with Richard Tolman, Dean of the graduate school at the California Institute of Technology. Compton was assigned responsibility for radar, Conant for chemistry and explosives, Jewett for armor and ordnance, and Tolman for patents and inventions. Funded by and reporting directly to the President, the NDRC was remarkably free of bureaucratic interference. In addition to its involvement in the Manhattan Project, the NDRC and its successor agency, the Office of Scientific Research and Development (OSRD), were involved with the development of key wartime technologies such as radar, sonar, proximity fuses, synthetic rubber, and the Norden bomb sight.

On June 15, Briggs received a letter from President Roosevelt informing him that the Uranium Committee was being absorbed into the NDRC. On July 1, Briggs summarized the work of his Committee in a letter to Bush. Fermi's measurements of neutron capture in carbon looked promising as far as eventually obtaining a chain reaction was concerned, and the Science Advisory Subcommittee felt that there was justification to pursue work in methods of separating U-235 and further measurements towards determining the feasibility of a chain reaction in natural uranium. For research into separation methods, $100,000 had been made available by the Army and Navy to investigate centrifuges and thermal diffusion; this work was being administered by the NRL. Briggs recommended that the NDRC provide $140,000 to advance measurements of properties of uranium. An NDRC meeting held the next day included a resolution that the Committee on Uranium be constituted as a special committee of the NDRC, with membership of Briggs (Chair), Beams, Breit, Gunn, Pegram, Sachs, Tuve, and Urey. Einstein, Bowen, Adamson, and Hoover had been dropped from the October, 1939, incarnation of the group, but the minutes indicate that Bowen would continue to follow the activities of the committee. Bowen was apparently present at the meeting, however, as the minutes record that he related that the Navy was coordinating isotope separation projects to the tune of $102,300. In addition, an Executive Committee of the Committee on Uranium was formed, comprising Briggs, Gunn, Pegram, Tuve, and Urey. It was also voted to approve in principle the $140,000 measurements program proposed by Briggs.

With the NDRC in the picture, the pace of work on the uranium project began to pick up. Between the fall of 1940 and the time of the Japanese attack at Pearl Harbor, the NDRC/OSRD had let contracts totaling about $300,000 for fission and isotope-separation research to various universities, industrial concerns such as the Standard Oil Development Company (SODC), the National Bureau of Standards, and private organizations such as Carnegie Institution. At the time the NDRC was established, the British MAUD committee was just beginning its work in response to the Frisch-Peierls memorandum; Edwin McMillan and Philip Abelson had just isolated a minute sample of element 93; and Louis Turner was speculating that neutron bombardment of U-238 might lead to a fissile form of element 94.

In the spring of 1941, however, Vannevar Bush began to receive complaints that the pace of the uranium committee's work was too slow. On March 17, Karl Compton wrote to Bush, referring to a presentation just two weeks earlier by Briggs on what Compton called the "#92 project". While it looked as if the project was moving ahead, there appeared to be a number of disquieting aspects: the English were "apparently farther ahead than we are," there was reason to believe that the Germans were very active in this area, and "very few of our own nuclear physicists are being put to work on the project and even those who are working on it are decidedly restive". Compton related that the committee practically never met, that its conduct was extremely slow, that the work was being conducted in such secrecy that it was preventing people from knowing what was going on in areas closely related to their own, and that Briggs was "slow, conservative, methodical and accustomed to operate at peace-time government bureau tempo". Eugene Wigner, who was working on the theory of chain reactions, described dealing with the Briggs Committee as like "swimming in syrup."

Compton proposed to let in on the project a group of the ablest theoretical physicists, and raised the question of whether the NDRC should take a more vigorous role as opposed to acting as a passive administrator. He further related that he and Ernest Lawrence had spoken that morning, and suggested that Bush appoint Lawrence as his deputy to explore and report on the situation, or, alternatively, assign Briggs a deputy to work full time on the project. Bush responded on the 21st to indicate that that he had met with Lawrence and that he had called Briggs with the suggestion that Lawrence serve as a temporary consultant; the latter two were to meet that day.

Bush also felt that he needed some independent advice on the uranium issue. On April 19, he asked Frank Jewett to appoint a committee under NAS auspices to review possible military aspects of fission.

This would be the first of three such committees; their reports would have far-reaching consequences.

4.3   May 1941: The First NAS Report

Jewett's committee was chaired by Arthur Compton, Dean of Science at the University of Chicago. The other members were William D. Coolidge, who earlier in his career had made significant improvements to X-ray tubes and who had just retired as director of research at General Electric Research Laboratories; Ernest Lawrence; MIT theoretical physicist John Slater; Harvard physicist and future Nobel Laureate (1977) John Van Vleck; and retired Bell Telephone Laboratories Chief Engineer Bancroft Gherardi. Unfortunately, Gherardi was terminally ill and was unable to participate; he passed away in August, 1941.

Compton's group met with Briggs, Breit, Gunn, Pegram, Tuve, and Urey in Washington on April 30, held a second meeting in Cambridge, Massachusetts, on May 5, and submitted their report to Jewett on May 17. Their seven-page document addressed the question of whether uranium research merited greater funds, facilities, and pressure in the light of then-current knowledge and the probability of applications in connection with national defense. The primary recommendation was that a strongly intensified effort should be spent on the problem during the following six months. While the committee felt that it would seem unlikely that nuclear fission could become of military importance within less than two years, they did comment that a chain reaction could become a determining factor in warfare if it could be produced and controlled.

The report listed three possible military applications of uranium fission: production of violently radioactive materials to be used as missiles "destructive to life in virtue of their ionizing radiation"; as a power source for submarines and other ships; and violently explosive bombs. Discussion of the latter concentrated mistakenly on *slow*-neutron fission of U-235, but it was predicted that the time required to separate an adequate amount of uranium would be from three to five years. It was pointed out, however, that element 94 could potentially be produced in abundance in a chain reaction. One day after the report was submitted, Emilio Segrè and Glenn Seaborg succeeded in isolating the sample of plutonium-239 with which they measured that isotope's slow-neutron fission cross-section.

While acknowledging that separation of a sufficiently large quantity of U-235 could become "a most important aspect of the problem," the bulk of the report was devoted to considering what resources would be needed for achieving a chain reaction. The most urgent requirements for the following six months were support for an intermediate-scale uranium/graphite experiment, a pilot plant for producing heavy water (for possible use as a moderator), investigating beryllium as a moderating agent, and maintaining work on isotope separation. The total cost was estimated at $350,000. If graphite proved to be a useable moderator, the cost of a full-scale experiment to produce a chain reaction was estimated to be upwards of $1 million. If beryllium and the heavy-water projects looked favorable, further support should be extended to both, with additional total costs also pushing toward $1 million. The report praised Briggs' committee, but suggested that a subcommittee be formed to plan and carry through the research programs, to confer on developments as they occurred, to see that information was made available to those who needed it, and to report as appropriate to the main committee.

Concerns with the report began to surface almost immediately. On May 28, Jewett solicited input from Robert Millikan, expressing concern that fundamental practical aspects of securing a chain reaction may have been minimized by physicists who were enthusiastic for going ahead. Could a chain reaction be used in practice? What about limitations of physical space, such as in a submarine? What was known of the availability of uranium? Recognizing that even if the answers to these questions should be discouraging, Jewett opined that it might be wise to push experimentation on a large scale if for no other reason than to disprove optimistic claims: "At the same time it would be foolish to proceed solely on the basis of one-sided enthusiasm and a trust that in an eight-handed poker game the Lord will always enable us to draw the right two cards to complete a royal flush." Millikan responded on the 31st with the opinion that there seemed to be little if any hope in realizing a chain reaction in ordinary (natural) uranium, and that, if this proved so, it would be necessary to enrich U-235, which would be a long and tedious process. While Millikan preferred that efforts be concentrated on problems which would have a good chance of getting into practical use within two or three years, he did suggest that attempting a chain reaction with natural uranium would not be an expensive matter.

Jewett summarized his concerns in a letter to Vannevar Bush on June 6. While having a "lurking fear" that the Academy report might have been over-enthusiastic and not well balanced, he concluded that they should nevertheless proceed with an enlarged program with the proviso that major initial appropriations should be concentrated on the more fundamental aspects of establishing the possibilities of chain reaction. Any approval for other phases of the matter should be reserved for a later time.

Bush's June 7 response to Jewett is worth examining in some detail. He related that Millikan was evidently unaware that "The British have apparently definitely established the possibility of a chain reaction with 238, which entirely changes the complexion of the whole affair." He then proceeded to give Lyman Briggs some uncommon praise: "Briggs has been in a very difficult situation on this matter. I know of no project anywhere where there has been so much need for a balanced, reasoned approach which would, on the one hand, not neglect the possibilities of potential importance but unlikely to develop, and which, on the other hand, would not run wild as the result of unbridled speculation. I think Briggs has done exceedingly well to keep his balance, and to approach the matter on a basis which would seem to me to have good sense. Moreover, I think that Briggs is a grand person to have in the matter, and I have backed him up to the best of my ability, and I intend to do so in the future." Conversely, Bush was concerned with Ernest Lawrence, who was playing the role of a loose cannon: "I finally had to have a very frank talk with him in which I told him flatly that I was running the show, that we had established a procedure for handling it, that he could either conform to that as a member of the NDRC and put in his kicks through the internal mechanism, or he could be utterly on the outside and act as an individual in any way that he saw fit. He got into line and I arranged for him to have with Briggs a series of excellent conferences." Bush praised the Academy report, and added that "[Briggs]… agrees to the enlargement of his section, the adding of a vice-chairman, the adding of a technical side, and in general the gearing up of the affair so as to handle the program to better advantage"; Briggs had been asked for his input on the personnel issue before a meeting scheduled for June 12. Bush also thought that there should be "at least one good sound engineer" on the enlarged Uranium Committee. The last paragraph of his letter revealed growing frustration: "As I have said many times, I wish that the physicist who fished uranium in the first place had waited a few years before he sprung this particular thing upon an unstable world. However, we have the matter in our laps and we have to do the best we can."

Briggs responded on June 11 with an estimate of Uranium Committee expenditures for fiscal year 1942, which would start on July 1. These covered a uranium-carbon experiment at Columbia; a uranium-carbon-beryllium experiment in Chicago; heavy water catalysis at Columbia; an experimental heavy water production plant to be built by Standard Oil in Louisiana; work on centrifuges at Columbia and the University of Virginia; research on diffusion at Columbia; mass spectroscopy under Alfred Nier at the University of Minnesota; and miscellaneous administrative and experimental work at the Bureau of Standards. All of this would run to $583,000 for the first six months. Costs for the balance of the year would depend on the outcomes of various experiments, but perhaps $1 million would be needed for a full-scale chain-reaction experiment and a heavy-water plant. The most immediate need was for $241,000 for materials.

Despite Bush's knowledge of the British opinion that fission bombs were virtually certainly feasible, the NDRC voted the next day to allocate only the $241,000 for materials plus an additional $500 to Nier for preparation of 5 micrograms of U-235. The irony that Briggs was often accused of foot-dragging speaks for itself. Briggs dutifully submitted a revised proposal on July 8 which brought his request down to $357,000, mostly by decreasing requests for the Chicago and Columbia pile experiments; the revised request was approved at a meeting held on July 18. Prospects for the American uranium program were beginning to shift for the better in the early summer of 1941, albeit in fits and starts. The participants could not have been unaware of the world situation: on June 22, Germany invaded Russia, adding a dramatic new dimension to the war.

4.4   July 1941: The Second NAS Report, and the OSRD

At the June 12 NDRC meeting, it was also voted to request to have the NAS again review the proposed program, this time by a committee which included individuals qualified to consider engineering aspects of the situation. Bush put the request to Frank Jewett the next day, and the Academy Committee went back to work, this time under the chairmanship of Coolidge; Compton was traveling at the time. To provide engineering perspective, the committee was augmented by the addition of Oliver Buckley (Bell Labs) and Lawrence Chubb, Director of Westinghouse Electric Research Laboratories. On July 1, the committee met in Washington with Briggs, Gregory Breit, and physicist Samuel Allison of the University of Chicago, and the day after with Pegram and Fermi at Columbia. They submitted their four-page report to Jewett on July 11, who passed it on to Bush on the 15th.

The report did not particularly address engineering aspects as Bush had requested, but, in an appendix written by Ernest Lawrence, related that experiments at Berkeley had verified that element 94 was formed in slow-neutron capture in U-238 and underwent slow-neutron fission. This opened up the prospect of what Lawrence called a "super bomb" if enough 94 could be produced. Given this, the committee considered whether the prospect of military applications was such as to justify allocation of defense monies toward support of an intensified drive on producing atomic fission, concluding that "We are convinced that such support is not only sound but urgently demanded." The report also gave Bush ammunition for reorganizing the project: "The efficient and expeditious conduct of this larger scale attack requires also ... a different pattern of organization from that of the work under the present Uranium Committee ... The project should be under a director able to devote his entire time to it." Costs were projected at over $1 million for salaries and materials for the first year, and the committee also suggested that an isolated laboratory be established to house the relevant work.

Support for the committee's opinion was received from Enrico Fermi, who composed an eight-page report titled "Some Remarks on the Production of Energy by a Chain Reaction in Uranium." Dated June 30, 1941, he described a possible reactor design with lumps of natural-composition uranium metal or oxide distributed in a lattice-like array throughout a moderator, precisely the arrangement he would use in his University of Chicago reactor 18 months later. He estimated that a pile generating one megawatt of fission energy would produce about one gram of element 94 per day. This, however, was for an *uncooled* reactor; by using active cooling with fluid or gas piped through channels, the power level could be raised to tens of megawatts, which would correspondingly increase plutonium production. An intriguing possibility pointed out by Fermi was cooling by liquid bismuth, which would have the advantage of breeding radioactive polonium through neutron bombardment via the reaction

$$\,_{0}^{1}n + \,_{83}^{209}\text{Bi} \rightarrow \,_{83}^{210}\text{Bi} \rightarrow \,_{84}^{210}\text{Po}.$$

The half-life for the bismuth-to-polonium decay is about five days. Slugs of bismuth would be introduced into a pilot-scale reactor at Oak Ridge, Tennessee, and into the production reactors at Hanford, Washington, to breed polonium for use in neutron-generating "triggers" for the Hiroshima and Nagasaki bombs by precisely this reaction. In a July 21 memo to Conant, Bush praised Fermi's report as the first time he had seen anything that approximated engineering data, and that it looked "to be good stuff."

Bush was also rearranging the administration of the NDRC. The NDRC could undertake to issue contracts for research, but lacked authority to underwrite engineering development. To address this, he conceived of a higher-level umbrella organization, the Office of Scientific Research and Development (OSRD). The NDRC would continue, but as a sub-component of OSRD;

Bush would Direct the OSRD, while Conant took on Chairmanship of the NDRC and with it responsibility for the uranium project. The OSRD was established by Executive Order 8807 signed by Roosevelt on June 28, 1941.

4.5　The MAUD Report

Beyond the National Academy reports and the growing restlessness of individual scientists, the single most important stimulus to the American fission project in the summer of 1941 came from Britain: the MAUD report.

The spectacular success of the Manhattan Project under U.S. Army leadership and the fact that the bulk of its facilities were located on American soil have tended to cast the Project as an almost exclusively American affair. But this view trivializes very important British contributions to the effort. Even General Groves, who has been quoted as characterizing the British contribution as "helpful but not vital," observed that "I cannot escape the feeling that without active and continuing British interest there probably would have been no atomic bomb to drop on Hiroshima. The British realized from the start what the implications of the work would be. They realized that they must be in a position to capitalize upon it if they were to survive … and they must also have realized that by themselves they were unable to do the job. They saw in the United States a means of accomplishing their purpose."

American authorities were not unaware of progress in Britain; exchanges between the two countries on scientific matters were well-established before America entered the war. In late August, 1940, a mission headed by Henry Tizard left for a two-month visit to America, where they demonstrated progress that had been made with radar and proximity fuses. A result of this visit was the establishment in Washington of a formal organization to facilitate information exchange, the British Commonwealth Scientific Office. In the spring of 1941, Charles G. Darwin, a grandson of *the* Charles Darwin, was appointed as its Director. Reciprocally, in February 1941 James Conant traveled to London to set up an office of the NDRC; he also met with Churchill on three occasions. Surprisingly, Churchill's personal science advisor, Frederick Lindemann, spoke openly with Conant regarding the Frisch-Peierls memorandum, likely a very serious breach of security. On July 1, Caltech physicist Charles Lauritsen attended a meeting of the MAUD committee at which the main conclusions of its report were discussed.

The final MUAD report was largely the work of Chadwick, who toward the end was working on the manuscript 20 h per day. Lauritsen returned to the United States and briefed Bush in Washington on July 10, just a few days before he received a draft copy of the report which had been transmitted to the NDRC office in London; this was just before the second NAS report landed on his desk.

There were actually two MAUD reports, both authorized by George Thomson on July 15. The first, and the one of interest here, was titled "Use of Uranium for a Bomb"; the second was "Use of Uranium as a Source of Power." The first part of the bomb report summarizes the situation in non-technical language for government officials in a few pages. It opened with a description of why a critical mass exists for a fissile isotope, how a bomb could be triggered by bringing together two subcritical masses, the probable effects of the explosion (estimated as equivalent to 1800 tons of TNT for 25 lb of U-235), and a discussion of materials and costs. A lengthy technical appendix describes how a fast-neutron chain reaction cannot be sustained in U-238 due to the presence of inelastic scattering and neutron capture, how the efficiency of a bomb could be estimated, factors that affect the determination of critical mass, estimates of damage, and the characteristics of a diffusion plant. Depending on values adopted for cross-sections, secondary neutron numbers, and whether or not a bomb was tamped, the report estimated the critical mass to be anywhere from about 2–43 kg; the latter figure is remarkably close to the presently accepted value for untamped uranium-235.

The overall conclusion of this remarkably clear-headed report was that a uranium bomb was possible and likely to lead to decisive results in the war. The government was urged to give the project high priority, predicting that it could be done in about two and a half years. Chadwick dodged the question of whether Britain should undertake the project alone or jointly with the United States, an issue which he may have felt lay outside the committee's purview of providing purely technical advice. Tizard felt that Britain should collaborate with the United States, while Chadwick and Lindemann were in favor of a purely British effort.

George Thomson personally handed Bush and Conant copies of the MAUD report on October 3, but under terms which did not permit its disclosure to the Compton Committee. Despite that prohibition, Thomson had met with both the Uranium Committee and the Compton Committee to apprise them of the situation. The MAUD bomb report would have a significant if officially unacknowledged impact on the preparation of a third National Academy report later in the year.

4.6   "If You Tell Me This Is My Job, I'll Do It"

As the drama of the MAUD report was unfolding, Conant was reorganizing the Uranium Committee. Briggs would remain as Chair, while George Pegram would serve as Vice Chair. The other members were to be Gregory Breit, Harold Urey, Sam Allison, Henry Smyth of Princeton University, and Edward Condon of Westinghouse Electric. Briggs also added four consultant subcommittees, to deal with Separation (enrichment), Power Production, Heavy Water, and Theoretical Aspects. These were respectively chaired by Urey, Pegram, Urey, and Fermi. Merle Tuve, Alexander Sachs, and Albert Einstein had disappeared from the mid-1940 makeup of the committee; so had Jesse Beams, although he would continue as a member of the Separation Group. Henceforth, the Uranium Committee would be known as Section S-1 of the OSRD.

British physicists continued to pressure their American counterparts to push ahead with a bomb project. During August and September, 1941, Marcus Oliphant traveled around the United States, speaking with various physicists about the project. George Thomson had instructed Oliphant to make discrete inquiries as to why nothing seemed to be happening in response to the MAUD report, but discretion was not Oliphant's style. He was particularly outraged that Lyman Briggs had locked away minutes of MAUD meetings in his office safe without passing them on to colleagues. In a 1982 memoir, Oliphant would characterize Briggs as "inarticulate and unimpressive". On September 11, W. D. Coolidge wrote to Frank Jewett to describe a visit Oliphant had made to General Electric in Schenectady, during which he revealed that only 10 kg of U-235 would be needed to produce a blast equivalent to 1000 tons of high explosive. James Conant felt that Oliphant's talking to Coolidge might have been a breach of secrecy, but many American scientists have credited Oliphant for spurring the S-1 program forward.

Oliphant also visited Berkeley and met with Ernest Lawrence, who was so impressed with British progress that he began thinking of how he might turn his 37-in. cyclotron into a large-scale mass spectrometer for separating isotopes. In September, Lawrence related Oliphant's story to Conant and Compton during a visit to Chicago, apparently not realizing that they already knew of it. Lawrence stressed the importance of element 94 to making a bomb, and again expressed his dissatisfaction at the slow pace of work in the United States. In his memoirs, Compton relates how he met with Conant and Lawrence in the living room of his house. After Lawrence had given his description of the prospect for fission bombs, Conant asked him: "Ernest, you say you are convinced of the importance of these fission bombs. Are you ready to devote the next several years of your life to getting them made?" After a brief hesitation, Lawrence's answer was "If you tell me this is my job, I'll do it."

Upon returning to Britain, Oliphant was horrified to learn that the government had decided to turn over operation of the MAUD Committee and the British program, now code-named "Tube Alloys", to Imperial Chemical Industries. ICI's effort would be headed by the company's research director, Wallace Akers, a very competent and diplomatic industrial chemist, but Oliphant's felt that ICI was completely ignorant of the nuclear physics involved and was more motivated by the prospect of securing lucrative possibilities in postwar energy generation. This response would be paralleled in America when scientists learned that the Army would be taking over their work and that industrial contractors would be involved. Oliphant resigned from the committee in protest, although he did later concede Akers' competence.

4.7   "A Fission Bomb of Superlatively Destructive Power..."

October 9, 1941, was a pivotal day in the history of the American atomic bomb program. That morning, Vannevar Bush met with, among others, President Roosevelt and Vice-President Henry Wallace to inform them of developments. Bush summarized the meeting in a memo sent to James Conant later the same day. The President had made it clear that considerations of policy were to be restricted to a group comprising himself, the Vice-President, Secretary of War Henry Stimson, Army Chief of Staff General George C. Marshall, and Bush and Conant. This group would come to be known as the Top Policy Group. From this point forward, American scientists would have to funnel their thoughts on policy issues regarding the fission weapons that they would create through Bush and Conant. That the President had charged a group to consider nuclear weapons policies indicated that the highest levels of leadership of the country were beginning to understand the implications of a successful full-scale commitment to the uranium project.

General George C. Marshall (1880–1959) and Secretary of War Henry Stimson (1867–1950), ca. 1942. Source: http://commons.wikimedia.org/wiki/File:George_marshall%26henry_stimson.jpg

During the meeting, Bush described British conclusions regarding critical mass, the size of isotope-separation plants, costs, time schedules, and raw materials. The meeting endorsed exchange of information with the British on technical issues, and also considered post-war control of nuclear materials. Bush advocated that a broader program ought to be handled independently of the then-present organization, a notion with which the President agreed. Roosevelt instructed Bush to not proceed with any definite steps on the expanded plan until receiving further instructions, but Bush essentially emerged from the meeting with the authority to determine if a bomb could be made, and at what cost. A further key result of this meeting was that Roosevelt wrote to Winston Churchill to offer that Britain and America work jointly as essentially equal partners to develop the bomb. His offer would be badly mishandled in London.

The same day, Bush also requested a third National Academy Report. This time he gave the committee very clear direction as to what he was after. He wrote to Arthur Compton, referring to having received a "communication from Britain" which dealt with the technical aspects of "the matter under consideration by the committee." The British report was available only to himself and Conant, but this would have the advantage that the Compton committee's work would provide an independent check on things. While Bush acknowledged that the way in which the committee wished to conduct its study was up to itself, he offered some topics for consideration: critical mass; the mutual velocity of approach of the subcritical masses during bomb core assembly; efficiency; premature explosions; and isotope separation methods.

Bush copied his instructions to Briggs, adding that he was not able to pass on the British report (which Briggs evidently had anyway). Despite Briggs' position as chair of the S-1 Committee, lines of authority were shifting toward Bush and Conant.

For its third report, the committee was expanded to include MIT chemical engineer Warren K. Lewis, Harvard explosives expert George Kistiakowsky, and future (1966) chemistry Nobel Laureate Robert Mulliken of the University of Chicago. Compton's group went back to work, meeting with Fermi, Urey, Wigner, Seaborg, and others. On October 21, they held a meeting at the General Electric laboratories in Schenectady, which Robert Oppenheimer attended.

Left: George Kistiakowsky (1900–1982); Right: Warren K. Lewis (1882–1975). Sources: AIP Emilio Segre Visual Archives; http://en.citizendium.org/wiki/Warren_K. _Lewis

The Committee transmitted its report, dated November 6, to Jewett on November 17. In a brief cover letter, Compton reported that the committee was "unanimously of the opinion that the prosecution of this program is a matter of urgent importance." The full report ran to 60 pages, comprising four sections. First came a six-page cover letter which summarized the conditions needed for a fission bomb, the expected effects of such bombs, estimates of how long it might take to produce them, and the costs involved. Then came a 20-page appendix, evidently written by Compton, which forms the technical core of the report. Calculations dealt with critical radius, the effects of a surrounding tamper, and the efficiency of the anticipated explosion. This appendix was effectively the parent document of another prepared in 1943 which would come to be known as the *Los Alamos Primer*. Then came another appendix, prepared by George Kistiakowsky, which describes the probable destructive action of fission bombs. Lastly appeared an 18-page report prepared by Robert Mulliken which discusses the feasibility of various isotope separation methods.

The summary letter got right to the essence on its first page: "A fission bomb of superlatively destructive power will result from bringing quickly together a sufficient mass of element U235. This seems to be as sure as any untried prediction based upon theory and experiment can be." The critical mass of U-235 was estimated as hardly less than 2 kg nor greater than 100 kg, and the expected efficiency at between 1 and 5%. It was difficult to assess the destructive capabilities of such a weapon because the theory for describing high-pressure shock waves was not then well-advanced, but the committee conservatively estimated an equivalent of about 30 tons of TNT per kg of U-235; this would prove to be a drastic underestimate. As for obtaining fissile material, centrifugal and diffusion separation methods were approaching the stage of practical tests. It was estimated that if the program was pursued with all possible support, fission bombs might be available in significant quantity within three or four years; the bombing of Hiroshima would occur three years and nine months to the day from the date of the report. As to finances, the committee estimated a rough cost of $80–$130 million, not including the cost of electrical power for operating the enrichment plants. Ultimately, the electromagnetic separation plant alone at Oak Ridge would consume nearly four times this level of funding.

Compton closed with a series of recommendations. The immediate needs were to build and test trial units of centrifugal and diffusion separators, to secure samples of separated U-235 and U-238 for physical-constants measurements, and to begin work on the engineering aspects of enrichment plants. Finally, he suggested that it may be necessary to reorganize the entire program, an idea which must have pleased Bush. In a separate private letter to Bush on the same day, Compton suggested assigning key men responsibility for solving certain problems and giving them adequate funds to "get the answers in their own way." Urey, Lawrence, Beams, and Allison were suggested as appropriate "key men."

The difference between the third Academy report and its two predecessors is stunning. In a draft history of the project prepared by James Conant in May, 1943, he remarked that (paraphrased): "A historian of science two generations hence who might come across the three National Academy reports might well be bewildered by the change. In July 1941 the Committee was speaking of the need for a successful demonstration of a controlled chain reaction. On November 6 the Committee concludes that the availability of bombs may determine military superiority." Conant attributed this shift in emphasis to a general feeling that war was felt to be much nearer and more inevitable in November than in May, and that advocates of a head-on attack on the uranium issue had become more vocal and determined. In a comment that presaged the postwar perception that the Manhattan Project was essentially an exclusively American affair, Conant wrote, somewhat disingenuously, that "It must be remembered that the British report … had not been seen by any member of the National Academy Committee even by November." This is true, but was a selective truth.

4.8   November 1941: Bush, FDR, Reorganization, and the Planning Board

Vannevar Bush wasted no time in using the third Academy report to bolster what he had reported to President Roosevelt on October 9. On Thursday, November 27, he transmitted the report to the President and the Top Policy Group; ironically, this was about the time that a Japanese task force set sail on its mission to Pearl Harbor. While advising that the cost of and time to produce bombs would be greater than the MAUD report had suggested, Bush felt that the matter called for serious attention. He offered that he would wait to be instructed by the President before taking any steps toward any specific program, but in the meantime he was forming an engineering group to study plans for possible production and accelerating relevant research.

By presenting the MAUD and Academy reports as independent but concurrent in their conclusions, Bush played them brilliantly as political cards. A handwritten note from Roosevelt accompanying return of the report to Bush on January 19, 1942, has been taken by many historians to be essentially the initiating Presidential approval for the American atomic bomb program.

Franklin Roosevelt to Vannevar Bush, January(?) 19, 1942. The note reads "OK—returned—I think you had best keep this in your own safe FDR"

Bush and Conant proceeded with their reorganization. On November 26, Bush offered the position of Director of a Planning Board to Eger V. Murphree, a highly-regarded chemical engineer and Vice-President of Research and Development for the Standard Oil Development Company. The Board would be charged with presenting Bush with recommendations for production and contracts for engineering studies. Murphree accepted, subject to his being free to select a group of consultants to serve as an advisory committee. His appointment was formalized in a letter of the 29th in which Bush made clear that while the Board was free to consult with Briggs and the Academy Committee, Murphree was to report directly to Bush, who was sharing overall responsibility for the program with Conant. The new organizational structure, which effectively orphaned Lyman Briggs's S-1 Section, was laid out by Bush at an NDRC meeting on November 28 and was included in a March, 1942, report to Roosevelt.

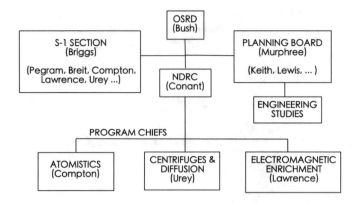

Simplified Manhattan Project organizational chart, ca. March, 1942. From Reed, Atomic Bomb: The Story of the Manhattan Project (2015) Fig. 3.3

The reorganization was not the only step in Briggs' marginalization. On December 1, Bush received a report from Harold Urey and a letter from Henry Smyth. Urey had just returned from a visit to London, and reported that Chadwick believed there to be a 90% chance of a practical fission bomb. Urey felt that "nothing else in the entire war effort should be placed ahead of it. If the Germans get this bomb the war will be over in a few weeks." Smyth was blunter, asking whether Briggs was in charge of the whole of the uranium work and free to call on the Uranium Section or to ignore it at his discretion, or were recommendations to the NDRC supposed to represent the judgment of the members of the S-1 Section and simply transmitted by Briggs as chairman of the Section? Smyth understood the situation to be the latter, but the infrequency of meetings and their informal character left him feeling that the situation was functionally more nearly the former.

Bush sent Smyth's letter to Briggs the next day, along with suggestions as to how to present the reorganization, essentially an accomplished fact, at a Section meeting scheduled for December 6. Physics research would continue, but S-1 members had to understand that policy issues were not their affair. Briggs should announce plans that he was to discuss with Compton regarding splitting the research into parts and putting key individuals in charge of each, while Bush would ask eminent chemical engineers to advise him directly in regard to engineering aspects. Bush wanted a clear-cut division between scientific and engineering studies, but with the latter informing the former.

Eger Murphree got to work, meeting with Harold Urey on December 2 to review isotope enrichment methods. Based on analyses carried out at Columbia and experiments conducted by Jesse Beams at Virginia, Murphree estimated that some 20,000 centrifuges would be needed to produce one kilogram 235 per day, and advocated setting up a pilot plant with 10 such machines. John Dunning at Columbia was working on developing diffusion membranes by an etching process; for this method, Murphree estimated that a full-scale plant would require some 4,000 successive enrichment stages. The centrifuge method would eventually be abandoned, but the diffusion plant would be built.

4.9   December 1941—January 1942: The Pile Program Rescued and Centralized

The one major aspect of the program left unaddressed in the third Academy Committee report was the possibility of developing reactors to synthesize plutonium. This work was salvaged by the personal intervention of Arthur Compton at the December 6 meeting of the S-1 Section held in Washington—the day before Pearl Harbor. Details of the meeting were related in a letter from Bush to Murphree on December 10, and an account was published by Compton in his postwar memoir *Atomic Quest.*

The first main outcome of the meeting was the creation of three development programs, with each being led by a Program Chief. Harold Urey would be responsible for research on separating uranium isotopes by diffusion and centrifugation, as well as work on heavy water. Ernest Lawrence, who now gained an official position and assigned duties, was to investigate electromagnetic methods of isotope separation. As Bush described Compton's role, it was to be concerned with fundamental atomic physics; in particular, measurements of physical parameters of materials. Compton put it more immodestly as "My job was to be the design of the bomb itself." At this point the meeting adjourned with the understanding that there would be another gathering in two weeks to shape plans more firmly.

After the meeting, Bush, Conant, and Compton went to lunch at the Cosmos Club on LaFayette Square, where, as Compton relates it, he argued that further thought should be given to producing plutonium as an alternative to enriching uranium. Conant expressed concern as to the time that would be needed to perfect the chemical extraction of plutonium from bombarded uranium, which would be complicated by intense radioactivity. Compton claims to have responded with "[Glenn] Seaborg tells me that within six months from the time plutonium is formed he can have it available for use in the bomb." Conant's reply was that "Glenn Seaborg is a very competent young chemist, but he isn't that good." Despite Conant's skepticism, Compton was authorized to see what could be done towards producing plutonium via a chain-reaction; just how capable Seaborg would prove to be would be borne out by subsequent events. Henry Smyth refers to Compton's charge as "an afterthought."

Compton's advocacy of the pile program was well-founded. At Columbia, Fermi and Szilard had been continuing their experiments to determine how neutrons slowed down and diffused through graphite, with particular attention to measuring the net number of neutrons produced per each consumed in a fission. This is known in nuclear physics circles as the reproduction factor, and is designated with the letter k. A k value of unity or greater is necessary to maintain a chain reaction. The experiments involved a four-ton graphite column of dimensions 3 by 3 by 8 feet, with a source of neutrons placed inside. Strategically placed detectors mapped the number of neutrons present and their energy distribution. These investigations revealed that the high-speed neutrons emitted in fissions were practically all reduced to thermal velocities after traveling through about 40 cm of graphite.

Around July, 1941, the first "lattice" experiment was set up at Columbia. This comprised a graphite cube about eight feet on each side, with some seven tons of uranium oxide enclosed in iron cans distributed throughout. Larger such structures followed. In September, funded with $40,000 from Briggs's Uranium Committee, the group set up their largest pile yet, comprising thousands of graphite bricks four inches square by twelve inches long. The completed pile measured eight feet square in footprint by eleven feet high. Embedded within were 288 cubical cans filled with uranium oxide, each measuring eight inches on each side. In a report written in late March, 1942, Fermi's group reported that extrapolating the results to a pile of infinite size indicated a k value of about 0.87. The last pile to be built at Columbia, in the spring of 1942, did away with any potential neutron-capture by the cans by replacing them with cylindrical slugs of uranium oxide about three inches in diameter by the same height, formed by compressing powdered oxide into solid form. Each slug weighed about 4 lb, and over 2,100 went into a pile which was again eight feet square in footprint but now 11 feet, four-inches high. To reduce the effect of neutron capture by air, the entire pile was coccooned in a tin enclosure and equipped with an air-handling system so that it could be operated in a carbon-dioxide atmosphere and heated to evaporate any neutron-capturing water. By May, 1942, a k-value of 0.98 was achieved.

America's entry into the war galvanized the uranium project. On December 10, Murphree wrote to Bush regarding progress with centrifugation and to reiterate his argument for developing a pilot plant of 10–25 machines at an estimated cost of $75,000–$150,000. On the 13th, Bush authorized Murphree to go ahead. The same day, Bush sent letters to Compton, Lawrence, and Urey, outlining the new organizational structure and their individual responsibilities. On December 16, Bush met with Vice-President Wallace and Secretary of War Stimson to discuss the third Academy report and to apprise them of the ongoing reorganization of the project. In a summary of the meeting in a memo to Conant, Bush indicated that the group felt that work should proceed as fast as possible on fundamental physics, engineering planning, and pilot plants. The cost estimate had escalated to four to five million dollars. The most important point of the discussion that day, however, was that Bush had made clear that he felt that the Army should take over when full-scale construction was started, and, to that end, felt it would be appropriate to have an officer become familiar with the project. This seems to be the first time that this suggestion, which would have such far-reaching consequences, was made at such a high level. Another advantage of bringing in the Army would be that the necessary budget could be hidden among that organization's enormous wartime appropriations under innocuous-sounding items such as "Expediting Production" or "Engineer Service—Army." While some key Congressmen and Senators were briefed on the Manhattan Project from time to time, the vast majority of legislators had no idea of its existence.

Arthur Compton wasted no time in undertaking his new responsibilities. In a December 20 "Urgent" letter to Bush, Conant, and Briggs, he laid out an ambitious plan for work at Columbia, Princeton, Chicago, and Berkeley. There were to be three major components: theoretical problems regarding nuclear explosions; production of a chain reaction; and determination of relevant physical constants. His goals were to obtain a chain reaction by October 1, 1942; to have a pilot plant for the production of plutonium in operation by October 1, 1943; and to be producing useable quantities of plutonium by December 31, 1944. His projected budget for the first six months of 1942 came to $1.2 million, nearly half of which was for purchase of uranium alloys, graphite, and beryllium. He also proposed bringing Robert Oppenheimer into the theoretical work; Oppenheimer had prepared part of the criticality analysis in the third Academy report.

Ernest Lawrence was busy as well. On December 20, Conant endorsed his recommendation to enter into a contract with the University of California for $305,000 for the coming six months. Lawrence's objective was to determine, within that time, whether or not an electromagnetic method could be feasible as a large-scale uranium separation technique. His group had just prepared their first substantial sample of U-235, some 50 micrograms mixed with about 200 micrograms of U-238.

Also left unaddressed in the late-1941 reorganization of the uranium project was the question of centralizing any or all parts of the effort. Compton had ideas on this as well. Over January 3–5, 1942, he held a planning session in Chicago with representatives from Berkeley, Columbia, and Princeton, and then another at Columbia on January 18. On the 22nd he wrote to Conant to indicate that he and Lawrence proposed to concentrate "both programs"—presumably enrichment and pile studies—at Berkeley, with the approximate date of the transfer being February 10. The advantage to this would be the availability of cyclotrons at Berkeley, although there was some concern with the west coast being a target for Japanese bombs. However, Campton apparently underwent a change of heart, for on the afternoon of January 24, Lawrence wired Conant and Bush to relate that "The latest of Compton's several successive proposals namely maintain status quo excepting moving Princeton to Chicago is in my judgment acceptable only as temporary arrangement. I sympathize with his difficulty in decision as there are numerous conflicting factors however we do need above all vigorous action." Later that evening came another wire from Lawrence: "Just learned Compton's decision to move Columbia [and] Princeton to Chicago which is much better than moving Princeton only." With this action, all pile and physical-constants work would move to Chicago, but electromagnetic research would sensibly remain in Berkeley.

On February 20, Conant summarized progress on various fronts in a report to Bush. Contracts totaling over one million dollars had been authorized to 12 different institutions, mostly for periods of six months. A full re-evaluation of the whole program should be planned for July, 1942, but, as things stood, four approaches to acquiring fissile material were still in play: the electromagnetic, centrifuge, and diffusion methods of uranium enrichment, and synthesizing plutonium via reactors. In just a few short weeks, Lawrence's electromagnetic method had shot to the top of the list of enrichment options. Reconfiguration of his 184-in. magnet as a mass spectrometer was expected to be ready by July 1, and was predicted to yield 0.1–10 g of U-235 per day by September. Planning a series of giant spectrographs, Lawrence was looking to produce perhaps a kilogram of U-235 per day by the summer of 1943. The centrifuge pilot plant was expected to be in operation by August 1, 1942. If this could be put into large-scale operation, it could produce one kilogram of U-235 every 10 days by July, 1943. The properties of plutonium were completely unknown, but a few micrograms were expected from Lawrence's cyclotrons by June, 1942. Expenditures were expected to amount to about $3 million by August 1, but the costs of large-scale construction were, as Conant put it, "anybody's guess." If all methods continued to be pursued and a decision on which to retain was postponed to January, 1943, some $10 million would be called for.

Lawrence began making steady advances with electromagnetic separation in early 1942. In January, he produced 18 micrograms of material enriched to 25% U-235; in February, three 75-microgram samples enriched to 30% became available. On March 7, an ion source designed to give a 10 mA (milliamperes) beam was delivering somewhat more than that to the collecting anode, and Lawrence felt ready to design and construct a 100-mA source and to begin planning for a 1-amp setup based on ten such units. Six days later he reported to Bush and Conant that the 10-mA source was yielding 25 mA, that he was proceeding with design of the 100-milliamp unit, and that he was considering design and construction of a multiple mass spectrograph system using the 184-in cyclotron magnet which would involve a dozen separate ion sources, each rated for 0.1–0.5 A. With a one-amp beam corresponding to one gram of U-235 per day, quantity production would then be underway. Lawrence hoped to have four sources in operation by July, with the entire plant in operation by autumn at a cost of perhaps a half-million dollars. In time, his cyclotrons, reconfigured as calutrons and after going through seemingly interminable teething problems, would succeed in producing U-235 on a large scale.

4.10 Spring 1942: Time Is Very Much of the Essence

On March 9, 1942, Vannevar Bush sent an expanded version of Conant's February 20 report to Roosevelt, Stimson, Marshall, and Wallace. This report would prove as pivotal as the one delivered the previous October.

Work was underway at full speed. The amount of fissile material necessary for a bomb appeared to be less than previously thought, the anticipated effects were predicted to be more powerful (the equivalent of 2,000 tons of TNT for a single bomb), and the possibility of actual production seemed more certain. The amount of necessary fissile material was estimated to be 5–10 lb, to which would be added a heavy tamper. While Bush felt that America might be engaged in a race with Germany toward realization of such weapons, he had no indication of the status of any German program; ironically, physicist

Werner Heisenberg had reported on possible uses of atomic energy to a meeting of the Reich Research Council in Berlin only 11 days earlier. The program was rapidly approaching the pilot plant stage; by the summer of 1942, the most promising enrichment methods could be selected and plant construction could be started. Bush urged that at that time, the whole matter should be turned over to the War Department. A twenty-unit centrifuge pilot plant was under construction, and it was estimated that a full-scale plant could be completed by December, 1943. A pilot-scale diffusion plant was being constructed by the British, and Lawrence's electromagnetic method, "a relatively recent development," might offer a shortcut in both time and plant requirements in offering the possibility of practicable quantities of material by the summer of 1943. The report mentioned power production by reactors only briefly as such developments were expected to be some years off, and did not mention plutonium at all. To help maintain security, the project was subdivided; full information was not being given to every worker. Despite this, Bush felt that the whole enterprise was more vulnerable to espionage than was desirable, an additional reason for the work to be placed under Army control as soon as production was embarked upon. Roosevelt's March 11 response was a clear go-ahead.

THE WHITE HOUSE
WASHINGTON

March 11, 1942.

MEMORANDUM FOR DR. VANNEVAR BUSH:

I am greatly interested in your report of March ninth and I am returning it herewith for your confidential file. I think the whole thing should be *pushed* not only in regard to development, but also with due regard to time. This is very much of the essence. I have no objection to turning over future progress to the War Department on condition that you yourself are certain that the War Department has made all adequate provision for absolute secrecy.

F.D.R.

President Roosevelt to Vannevar Bush, March 11, 1942

4.11 Enter the Army

The project was now on a fast track, with cost being no object. Within days of FDR's OK, the wheels of Army bureaucracy began to turn. On March 14, Harvey Bundy, a Special Assistant to Secretary of War Stimson, wrote to Bush to confirm that General Marshall had authorized Brigadier General Wilhelm Styer as the Army's contact for S-1. Styer was Chief of Staff to Lieutenant General Brehon Somervell, who commanded the Army's Services of Supply.

April 1, 1942 saw another report from Conant to Bush, the result of a conference he and Briggs had had that day with Compton regarding bomb theory and plutonium. Compton's schedule for achieving a chain reaction had been moved forward to November 1, and the expected date for production of experimental quantities of plutonium from a 5-MW pilot reactor had advanced from early 1944 to May 15, 1943, a projection which would prove too optimistic. The projected date for quantity production, one kilogram per day, had been pushed back to July, 1945, presuming the development of 100-MW reactors; this prediction would be beaten. To support all of this, Conant requested an additional $422,000 in funding to July 1. He also raised the issue of sites for production plants, suggesting that one be chosen within the next month. It should be "located in a wilderness," such that reactors or factories for any of the enrichment methods could be built. The site would need to be secure from espionage, take into account the safety of the workers and the surrounding population, have considerable electrical power available, and would require an adequate supply of cooling water if reactors were to be constructed. This would all require construction not only of the plants, but of living quarters, machines shops, laboratories, and other support facilities.

Conant called a meeting of Murphree, Briggs, and the Program Chiefs for Saturday, May 23; General Styer was also invited. On the 14th, Conant wrote to Bush to relate that significant issues were approaching decision points, and that, unless the ultimate decision was a green light on everybody's hopes and ambitions, there would be some "disgruntled and disheartened" people who might "take the case to the court of public opinion, or at least the top physicists of the country." Decisions were approaching gargantuan scales. Fissile material preparation by centrifugation, diffusion, and electromagnetism were still in the running on about equal footings, as were reactors with or without heavy water moderators. All would be entering pilot development within next six months, and production plants should be under design and construction even before the pilot plants were finished. What Conant called this "Napoleonic approach" could run to $500 million. Despite this astronomical cost, Conant felt that an all-out program might be justified: it seemed fairly certain that all methods would yield a weapon, which meant that the probability of the Germans developing such devices was also high. In view of Roosevelt's dictum that time was of the essence, Conant also felt that if they discarded some of the fissile-material production methods now, they may unconsciously end up betting on the "slower horse"; a delay of even only three months could be fatal if within such time Germany could employ a dozen such bombs against England. Finally, he offered some thoughts as to the Army's eventual role, suggesting that while that organization might be willing to take on production or even pilot plants, he felt that it ought not take over research. As to administration: "I do not believe Briggs should be brought back into the picture with any more authority. I am quite sure that Beams, Lawrence, and Murphree should go full time into the Army, probably as officers". He closed with a sense of exasperation: "If the whole matter were out of our hands, it would be a relief, but I am inclined to think a good deal would be lost and eventually it might come back again!" Bush replied on May 21, indicating that he wanted Briggs, Murphree, and the Program Chiefs to report through Conant, giving outlines of the program for each method for next six months and the ensuing year, judgments on how many programs should be continued, and suggestions as to what parts of the program should be eliminated if there were limitations on people, money, and material.

Even as his influence within the project was diminishing, Lyman Briggs still had to deal with his share of headaches. On the day of the May 23 meeting, he received a letter from Gregory Breit in which he announced that he was leaving Compton's project and his position as coordinator of fast-neutron research, to which he had been appointed only in February. Breit was furious over what he saw as lax security at Chicago, alleging that there were individuals at Chicago who were strongly opposed to secrecy and that Compton was being influenced by satisfying personal desires and ambitions. In anticipating that the bomb would exceed the power of ordinary weapons by orders of magnitude, Breit felt that it would be necessary to have adequate security not only during the war but also for decades afterwards, and urged government control of the whole matter, with bomb design under the control of one of the armed services. In a sense, Breit jumped the gun: many of his ideas would come to be reality within a year. His resignation did, however, open the door for Robert Oppenheimer to head the fast-neutron work.

On Monday, May 25, Conant reported to Bush on the May 23 meeting, which had run from 9:30 a.m. to 4:45 p.m. The bottom line was that if the urgency of securing fissile material justified an all-out program, then the group recommended an extensive program to run from mid-1942 to mid-1943 in which every method would receive support: $38 million for a 100 grams-per-day centrifuge plant to be ready by January, 1944; $2 million for pilot-plant and engineering work on a one kilogram per day diffusion plant; $17 million for a 100 grams-per-day electromagnetic separation plant to be ready by September, 1943; $15 million for reactors to produce 100 g per day of element 94; $3 million for a half-ton per month heavy water plant, and miscellaneous research valued at just over $2 million. Throwing in $5 million for contingencies brought the total to $85 million. If cuts were necessary, the reactor and electromagnetic programs could be decreased by $10 million and $2.5 million; delaying centrifuge construction to 1943 would save $18 million, but cost six months in lost time. The group did not choose between cut options, but predicted that bombs could be expected to be ready by July 1944 with the full program.

As a July 1 funding cutoff-date loomed and momentum gathered within the Army toward formal establishment of the Manhattan Engineer District (MED), large-scale decisions began to get made. On June 10, Bush conferred with Generals Marshall and Styer; they decided to proceed with the electromagnetic method and the "boiling project" (reactors), as they would cause the least disruption to critical materials. Styer was to study the impact of the proposed centrifuge and diffusion programs on other essential programs. In a summary memo to Conant the next day, Bush indicated that it was understood that Styer would inform other officers in the Services of Supply, and would plan to take over all production aspects of the project on July 1. The Planning Board would also be turned over to Styer at that time, who would be at liberty to modify it as he saw fit. On June 13, a few days prior to the formal establishment of the MED, Bush and Conant sent a 6-page status report to Wallace, Stimson, and Marshall; from them the document would go to the President for final approval. The estimated yield of fission bombs had been raised to several thousand tons of TNT, and it was estimated that with a suitably ambitious program, a small supply of such bombs could be ready by mid-1944, plus-or-minus a few months. To avoid the danger of concentrating on any one method, Bush echoed the full slate of recommendations of Conant's May 25 report. He also suggested that it was time to arrange for a committee to consider the military uses of the material produced. The uranium project was about to enter a significant new phase of its life.

4.12   The S-1 Executive Committee, the Manhattan Engineer District, and Bohemian Grove

Bush presented the June 13 report to Roosevelt on Wednesday, June 17, 1942. His one-page cover letter, the archived copy of which bears an iconic "V.B. OK FDR", states that it was contemplated that all financing for the project would be handled by the War Department's Chief of Engineers of the Army. On the same day, General Styer telegraphed orders to Colonel James C. Marshall of the Syracuse (New York) Engineer District to report to Washington to take command of what was being called the DSM Project: Development of Substitute Materials.

In many histories of the Manhattan Project, Colonel Marshall suffers much the same fate as Lyman Briggs in being thrust into relative obscurity against the presence of more forceful personalities. Only three months after his appointment to the project, Marshall was replaced as commander by a much more aggressive officer, Leslie Richard Groves. Under Groves, Marshall retained the title of District Engineer until July, 1943, at which time Groves eased him out of that position in favor of Marshall's own deputy, Colonel Kenneth D. Nichols. This terminated Marshall's association with the Manhattan District, but he was by no means inactive during his tenure with the project.

The wartime organization of the Army was immensely complex. In March, 1942, a reorganization of the Army bureaucracy saw the designation of three overall commands: Army Ground Forces (AGF), Army Air Forces (AAF) , and Army Services of Supply, which later became the Army Service Forces (ASF). The latter, the one of interest here, was under the command of Lieutenant General Brehon Somervell. A Lieutenant General carried three stars; beneath that rank came Major General (2 stars) and Brigadier General (1 star). Above Lieutenant General was General (4 stars, such as George C. Marshall, no relation to James Marshall). The Army Corps of Engineers (CE) was one of the operating divisions of the ASF. General Styer was Somervell's Chief of Staff, and the Chief of Engineers was Lieutenant General Eugene Reybold, who was appointed to that position on October 1, 1941. Within the Corps of Engineers lay the Construction Division, which was headed by Major General Thomas Robins. On March 3, 1942, Leslie Groves, then a Colonel, was appointed Deputy Chief of Construction.

Left to right: General Groves (1896–1970) and Robert Oppenheimer; a formal portrait of Groves; Kenneth D. Nichols (1907–2000). Sources: http://commons.wikimedia.org/wiki/File:Groves_Oppenheimer.jpg, http://commons.wikimedia.org/wiki/File:Leslie_Groves.jpg, http:// commons.wikimedia.org/wiki/File:Kenneth_D._Nichols.jpg

One of the most valuable sources of information on initial Army involvement in the Manhattan Project is a diary kept by Colonel Marshall, *Chronology of District X,* which runs from June 18, 1942, to January, 1943, with a few sporadic entries thereafter.

Marshall's assignment was unusual. Normally, the Chief of Engineers oversaw projects through an "Engineer District." An individual designated as District Engineer reported to a Division Engineer, who headed one of eleven geographical divisions of the United States. But Marshall's new District had no geographical restrictions; in effect, he was to have all of the authority of a Division Engineer. While the terms Manhattan Project and Manhattan Engineer District are often used interchangeably, they do not mean the same thing. Marshall initially located his headquarters in New York City. When Groves was assigned to be Commanding General, he became senior to Marshall, and set up his headquarters in Washington. The District office itself remained in New York until Marshall's departure in 1943, at which time Nichols moved it to Oak Ridge. The term Manhattan Project never was an official one, and only came into general use after the war.

Styer briefed Marshall on his new assignment on the afternoon of June 18. On returning to the Office of the Chief of Engineers, Marshall informed a number of other officers, including Groves, of his new command. While Groves claims in his memoirs that he was "familiar" with the Project in its initial stages as a part of his overall responsibilities but knew little of its details, Marshall's diary makes it clear that he was a very active participant from the outset. Groves often advised Marshall as to contractors and procedures, and was involved in suggesting the "Manhattan District" name. Marshall and

Styer met with Vannevar Bush the next morning, at which time they saw Conant's May 25 report and Bush's June 17 letter to FDR. A meeting with Bush, Conant, and the Program Chiefs was set up for June 25. Groves was asked to undertake a survey of sites around the country that would have suitable power available to run the anticipated uranium enrichment plants and plutonium-producing reactors. Marshall also decided that day that he wished to have Nichols, his deputy at Syracuse, accompany him to his new District. Nichols held a Ph.D. in civil engineering, and would become one of the driving forces of the Manhattan Project.

While the Army was coming up to speed on the Project, Vannevar Bush moved to effect yet another rearrangement of the S-1 organization. Following a meeting with Marshall on June 19, he wrote to Briggs and Conant to apprise them of the appointment of an "S-1 Executive Committee" within the OSRD, which would replace the somewhat largish S-1 Section committe. Conant would chair the group; the other members would be Briggs, Lawrence, Urey, Compton, and Murphree. Allison, Beams, Breit, Condon, and Smyth would continue to serve as consultants. The Planning Board would remain in existence, but would report in an advisory capacity to the Chief of Engineers. To set the stage for the planned June 25 meeting, responsibility for the various projects discussed in the May 25 report were divided between the new S-1 Executive Committee and the War Department. The Committee was to recommend contracts for centrifuge and diffusion pilot plants, research and development, a five grams-per-day plant for the electromagnetic method, the heavy water project, and miscellaneous research. The War Department was to take on the 100 g/day centrifuge production plant, engineering and construction of a 1 kg/day diffusion plant, a 100 g/day electromagnetic plant, and a pilot-scale reactor to produce 100 g of plutonium per day.

Work on bomb physics also progressed during the summer of 1942. On May 19, Robert Oppenheimer wrote to Ernest Lawrence with the optimistic prediction that with a total of two or three experienced men and perhaps an equal number of younger ones, it should be possible to solve the theoretical problems of building a fast-fission bomb. Beginning in the second week of July, Oppenheimer gathered a group of theoretical physicists at Berkeley to consider the detailed physics of bomb design. The participants included some of the most outstanding physicists of the time, including Hans Bethe, Edward Teller, and Robert Serber, all of whom would work at Los Alamos.

Left: Hans Bethe's (1906–2005) Los Alamos identity badge photo. Right: Robert Serber's (1909–1997) Los Alamos identity badge photo. Sources: http://commons.wikimedia.org/wiki/File:Hans_Bethe_ID_badge.png, http://commons.wikimedia.org/wiki/File:Robert_Serber_ID_badge.png

The discussions at Berkeley covered the entire spectrum of design issues, and even the possibility of hydrogen bombs. A particularly important issue was the danger that impurities in plutonium could cause a low-efficiency explosion. The problem was not the presence of impurities per se, but an indirect effect that harked back to the discovery of the neutron. Reactor-produced plutonium is a prolific alpha-emitter, and, as Bothe and Becker, the Joliot-Curies, and Chadwick had found, alpha particles striking nuclei of light elements tend to create neutrons: so-called (α, n) reactions. If chemical processing of plutonium left behind light-element impurities, this effect could give rise to a premature detonation. This issue would almost prove the undoing of the plutonium project.

The June 25 Army/S-1 meeting was held at the OSRD, and saw a number of crucial decisions made. All of the major players were present: Marshall, Nichols, Styer, Bush, Conant, Lawrence, Compton, Urey, Murphree, and Briggs. Styer felt that the manufacturing plants should be set up somewhere between the Allegheny and the Rocky mountains to protect them from enemy coastal bombardment. Up to 150,000 kW of power would be needed by the end of 1943 to operate all of the electromagnetic, centrifuge, pile, and diffusion plants. The necessary site size for all of this was estimated to be some 200 square miles, preferably in the shape of a 10 by 20-mile rectangle. The construction and engineering firm of Stone and Webster (S&W) of Boston was suggested for site development, housing construction, engineering and construction of the centrifuge plant, and, if they were agreeable to the idea, to start work on a plant for Lawrence's electromagnetic method.

Stone and Webster was already involved with the diffusion project through Eger Murphree, and Groves had contracted with them on a number of Army construction projects; it was apparently he who suggested the firm to Marshall. It was also decided to enter into a contract with University of Chicago to operate a pilot-scale reactor to be built by S&W in the Argonne Forest Preserve outside Chicago. That reactor, the X-10 pile, would ultimately be built in Tennessee. Other suggested contractors were E. B. Badger and Sons, also of Boston, for a heavy-water plant to be set up in Trail, British Columbia, and, for the diffusion plant, the M. W. Kellogg Company of New Jersey, a firm with extensive experience in design and construction of petroleum refineries and chemical facilities.

On June 27, Nichols met with John Lotz, the President of Stone and Webster, who was enthusiastic that his firm was being considered for construction of the plants, and also expressed interest in contracts to operate them. A formal meeting with Lotz and S&W engineers and managers was held in Groves' office on June 29 to hammer out details of the contract; Groves was also to see to approval of land purchases in Chicago and Tennessee. On the same day, Marshall decided to establish his District Headquarters on the 18th floor of the Corps of Engineers North Atlantic Division building at 270 Broadway in New York City, hence the designation "Manhattan Engineer District". Stone and Webster had offices in the same building.

By August, Compton was urging Marshall to select an operating contractor for the various plants. Since it was anticipated that the operator of the Argonne plutonium-extraction facility would also operate the works for the production plants, that organization should observe construction of the plant at Argonne. Compton suggested the DuPont corporation, SODC, or Union Carbide and Carbon as operator. Marshall, however, was reluctant to bring in any more firms for security reasons, and proposed instead that S&W add operations to their responsibilities, with the provision that they could secure technical assistance from other organizations.

Groves' survey identified eastern Tennessee as a likely site for production plants. The area was supplied with ample power by the Tennessee Valley Authority (TVA), which, by May, 1942, boasted an installed capacity of 1.3 million kW. This would grow to 2.5 million kW by the end of the war, at which time the TVA was supplying some 8% of the nation's electricity. Over July 1–3, Marshall and Nichols visited the Knoxville area to inspect a site of roughly 17 by 7 miles near the town of Clinton. The area was promising not only for its good power and railroad accessibility, but also for possessing several parallel northeast-to-southwest valleys separated by 200–300 foot ridges, which could be used to segregate different production areas and would provide protection in case of a catastrophe within any one of them. The Clinch river provided natural boundaries on three sides of the area, and Tennessee State Highway 61 defined the north side.

One story surrounding the selection of the Clinton site—perhaps fictional—is that Roosevelt asked Tennessee Senator Kenneth McKellar, an influential member of the Appropriations Committee, if he could find a means of hiding the funding for a supersecret national defense project. McKellar's response is said to have been "Well, Mr. President, of course I can. And where in Tennessee do you want me to hide it?"

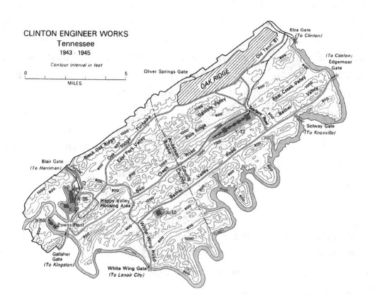

The Clinton Engineer Works site, east-central Tennessee. Knoxville is located about 15 miles east of Solway gate. The locations of the Y-12, K-25, X-10, and S-50 facilities are marked. Tennessee highway 61 became the Oak Ridge Turnpike. Source: http://commons. wikimedia.org/wiki/File:Oak_ridge_large.gif

During the early days of the Manhattan District, Marshall and Nichols were on the road almost constantly in what must have been a brutal pace. Immediately after the Knoxville trip, Nichols visited Chicago on July 6-7, where, along with Enrico Fermi and S&W representatives, he inspected Compton's Metallurgical Laboratory and the proposed 1,000-acre Argonne Forest site located about 10 miles west of the University. On the ninth, Marshall and Nichols were back in Washington for another meeting with S&W representatives to discuss uranium supplies, priorities, the Trail plant, the

Tennessee site, liaison with the British, use of silver as a substitute for high-priority copper in the magnets for Lawrence's cyclotrons, and funding issues. Marshall's diary for that day recorded that Groves was perturbed with what he viewed as indefinite dates for when various parts of the project would get underway.

On July 13, Nichols and Groves prepared a memorandum requesting that the Corps' Real Estate Division secure a lease on the Argonne site. July 14 saw another conference with S&W to deal with contract legalities, purchasing procedures, and road relocations in Tennessee. It was estimated that the site would require housing for 5,000 people, a number which would prove to be a drastic underestimate. Stone and Webster wanted the site obtained by August 10 so that an administration area and 200 associated housing units could be built in October. On July 20, Marshall was in San Francisco to meet with local S&W engineers and Ernest Lawrence. Marshall and the engineers were concerned with a general lack of organization of the work at Berkeley, and encouraged the group there to begin construction of a pilot plant and design of a full-scale plant. Despite these misgivings, Marshall was of the opinion that Lawrence's electromagnetic approach was ahead of the other three fissile-material methods, and that it should be exploited to the fullest extent without delay. The next day, Nichols traveled to Boston to meet with Badger and Sons regarding timing and priorities for the Trail plant, a thorny issue by virtue of its own need for considerable quantities of copper. The following day, he was back in Washington to confer with Groves regarding uranium ore and approval for the formal organization of the new District, which was to be named as soon as the site in Tennessee was chosen. Marshall favored "Knoxville District" as that would be their postal address, but Groves preferred something less revealing.

On July 29, the Real Estate Division got back to Nichols with an estimate of the cost of the Tennessee site. This would involve not only the direct cost of the land, but also relocation of cemeteries and utilities, road closures, and compensation for crop values. The total was estimated at $4.25 million for 83,000 acres, of which 3,000 were owned by the TVA. Some 400 families would have to be relocated. The OSRD approved the acquisition the next day, but Marshall told General Robins that he was unwilling to proceed with acquisition of the site or to begin any construction until Compton's pile process had proven itself. On August 6, Nichols was back in Boston to confer with S&W on supplier contracts with the Metal Hydrides Company, Mallinckrodt Chemical, the Consolidated Mining and Smelting Company of Canada, and DuPont. On the 11th, Marshall conferred again with Groves regarding drafting a General Order which would establish the new District.

Groves still objected to "DSM"; they decided that "Manhattan" was the best place name they could use, and so the "Manhattan Engineer District" was born. The name began to appear in Marshall's diary the next day. On August 13, the day the General Order establishing the District was issued, Marshall traveled to St. Louis to visit the Mallinckrodt Chemical Works to discuss a contract for purification of 300 tons of uranium oxide. Back in Washington on the 14th, he dealt with a contract for the Tennessee Eastman Corporation as an operating contractor for the Clinton site. On the 18th, Nichols was in California, where he learned that Lawrence was willing to proceed immediately with work on the design for the full-scale plant to be sited in Tennessee. While Marshall continued to hold off on acquiring the Tennessee site through August, much other groundwork was accomplished. On the 24th, he and Nichols conferred with Eger Murphree to discuss the idea of contracting with the SODC to operate the reactor pilot plant. The next day, Nichols visited Westinghouse in Pittsburgh to consider centrifuge design; he witnessed a meter-long centrifuge in operation, at least until its motor burned out at 25,000 rpm. Back in Washington for the 26th, Nichols and Marshall conferred again with Conant, Murphree, Urey, Compton, Lawrence, Briggs, and S&W representatives to review all production methods under consideration. A target date of August 1, 1943, was set for the electromagnetic pilot plant to be operation.

Of all of the conferences held between the S-1 Executive Committee and the Army, one of the most important occurred over September 13–14, 1942, at Bohemian Grove, an exclusive campground located within the Muir Woods National Monument just outside San Francisco. As Compton wrote, decisions made at that meeting were destined to shape the entire Project.

The committee's first recommendation was to complete construction of the Argonne Forest site, and to locate Fermi's first critical pile there. A second pile was to be built there later for purposes of producing some plutonium, with the understanding that chemical processing plants to handle the separation of plutonium would be erected in Tennessee. Second, it was recommended that the Army and Stone and Webster enter into a subcontract with a chemical company to develop the separation facilities. Dow Chemical, Monsanto Chemical, and the Tennessee Eastman Corporation were suggested, but those facilities would ultimately be designed, constructed, and operated by DuPont. Third came a recommendation for the Army to enter into a commitment, estimated to cost $30 million, to build a 100 gram-per-day U-235 electromagnetic separation plant of 100–400 vacuum tanks in Tennessee. A sub-recommendation to this was that the OSRD sponsor construction of a pilot electromagnetic plant comprising five vacuum tanks, also to be located in Tennessee. Finally, it was voted to recommend to the Army that construction of the heavy-water plant in British Columbia should be completed by May 1, 1943. The diffusion and centrifuge methods were not considered at this meeting, at least as far as the minutes reflect the discussion.

Within a week of the Bohemian Grove meeting, Leslie Groves would be placed in command of the Manhattan District. What had been a demanding pace was about to become frenetic.

4.13   "Oh, that Thing"

Despite the historical significance of Groves' appointment to take on overall command of the Manhattan District, the record of that event is murky. The decision to place Groves in command was apparently made on September 16 by Somervell and Styer. When Groves later asked Styer about the circumstances, his reply was that General George Marshall wanted Styer to take on the job, but Somervell objected to the prospect of losing Styer. Somervell discussed the matter with Marshall, who instructed him to come up with someone suitable, and Somervell and Styer decided that Groves would be appropriate. Styer may not have wanted to take on the job in any event, as apparently both he and Somervell were skeptical of the idea of a weapon based on atomic energy.

In his memoirs, Groves claims that he learned of his new assignment on the next morning, Thursday, September 17, when Somervell caught up with him just after he had finished testifying before a congressional committee on a military housing bill. Groves claims that he had been offered an overseas assignment, and was disappointed when Somervell told him he could not leave Washington because "The Secretary of War has selected you for a very important assignment, and the President has approved the selection." When Groves realized what Somervell had in mind, he claims that his response was "Oh, that thing." On meeting with Styer later that morning, Groves was also informed that he was to be promoted to Brigadier General. His response to this was to ask that he not be placed in official charge until the promotion had gone through, believing that this would put him in a stronger position to deal with the academic scientists involved in the project: it would be better if he were thought of as a General instead of as a Colonel. The promotion would become official on September 23. Colonel Marshall was on the west coast on September 17, and the diary entry for that day was made by Nichols, who refers to himself and Groves visiting Styer to learn of the new arrangement. Marshall returned on the 19th; subsequent entries make no comment regarding his new, subordinate position, although he continued as District Engineer.

The text of Somervell's one-page directive placing Groves in charge of the project read as:

September 17, 1942

MEMORANDUM FOR THE CHIEF OF ENGINEERS
SUBJECT: Release of Colonel L. R. Groves, C.E., for Special Assignment

1. It is directed that Colonel L. R. Groves be relieved from his present assignment in the Office of Engineers for special duty in connection with the DSM Project. You should, therefore, make the necessary arrangements in the Construction Division of your office so that Colonel Groves may be released for full time duty on this special work. He will report to the Commanding General, Services of Supply, for necessary instructions, but will operate in close conjunction with the Construction Division of your office and other facilities of the Corps of Engineers.

2. Colonel Groves' duty will be to take complete charge of the entire DSM project as outlined to Colonel Groves this morning by General Styer.

    a. He will take steps immediately to arrange for the necessary priorities.
    b. Arrange for a working committee on the application of the product.
    c. Arrange for the immediate procurement of the site of the TVA and the transfer of activities to that area.
    d. Initiate the preparation of bills of materials needed for construction and their earmarking for use when required.
    e. Draw up plans for the organization, construction, operation and security of the project, and after approval, take the necessary steps to put it into effect.

BREHON SOMERVERLL

Lieutenant General

Commanding

Groves graduated fourth in his West Point class of November 1918, and also trained at the Army Engineer School, the Command and General Staff School, and the Army War College. His career in the Corps of Engineers was marked by steady advancement, and by 1942 his workload was enormous. Under Robins' supervision, he was responsible for overseeing all Army construction within the United States, as well as at off-shore bases. Camps, airfields, huge ordnance and chemical manufacturing plants, depots, ports, and even internment camps for Japanese-Americans all came under his purview. At the time the Army became involved in the Manhattan Project, the Corps of Engineers was engaging almost one million people under contracts consuming some $600 million per month; Manhattan was a drop in the bucket in comparison. This experience gave Groves intimate knowledge of how the War Department and Washington bureaucracies functioned, and of which contractors could be depended upon to competently undertake the design, construction, and operation of large plants and housing projects. Over the course of the war, the Corps of Engineers would place more than $12 billion worth of construction within the United States, including over 3,000 command installations and nearly 300 major industrial projects. In the spring of 1942, one of Groves' projects was the construction of the Pentagon, which was completed within sixteen months of ground being broken.

Groves ran a remarkably tight headquarters. He and a staff of just a couple dozen administered the Manhattan Project from a small suite of offices on the fifth floor of the New War Building at the intersection of Twenty-First street and Virginia Avenue NW in Washington. The building is now part of the State Department; because of subsequent renovations, the original offices no longer exist.

It is not uncommon to read of Groves as being arrogant, arbitrary, insensitive, overbearing, and high-handed. More appropriate adjectives might be mission-focused, supremely competent, and able to get things done. His ability to juggle the multiple responsibilities of the Manhattan Project was remarkable. Colonel John Lansdale, Groves' head of security for the Project, offered this assessment: "General Groves was a man of extraordinary ability and capacity to get things done. Unfortunately, it took more contact with him than most people had to overcome a bad first impression. He was in fact the only person I have known who was every bit as good as he thought he was. He had intelligence, he had good judgment of people, he had extraordinary perceptiveness and an intuitive instinct for the right answer. In addition to this, he had a sort of catalytic effect on people. Most of us working with him performed better than our intrinsic abilities indicated."

The relationship between Groves and Kenneth Nichols was apparently somewhat strained, despite its productivity. After the war, Nichols offered this reflection:

First, General Groves is the biggest S.O.B. I have ever worked for. He is most demanding. He is most critical. He is always a driver, never a praiser. He is abrasive and sarcastic. He disregards all normal organizational channels. He is extremely intelligent. He has the guts to make timely, difficult decisions. He is the most egotistical man I know. He knows he is right and so sticks by his decision. He abounds with energy and expects everyone to work as hard or even harder than he does … if I had to do my part of the atomic bomb project over again and had the privilege of picking my boss I would pick General Groves.

Groves' first meeting with Vannevar Bush was not auspicious. Styer had not had time to inform Bush of Groves' appointment, and Bush was reluctant to answer questions. After the meeting, Bush sent a note to Stimson's assistant Harvey Bundy, expressing doubt that Groves had sufficient tact for the job: "I fear we are in the soup." Another meeting between Groves and Bush two days later went much more smoothly; Groves later claimed that they became fast friends.

Groves got to work promptly in his new command. On September 18, his first full day in charge, he dispatched Nichols to New York to confer with Union Minière President Edgar Sengier to reach an agreement to purchase that firm's 1,200-ton stock of uranium ore being held in storage in the United States. Nichols also made arrangements to ship to and store in the United States ores then being held in the Belgian Congo, and to assign those ores a prior right of purchase for the United States. Nichols and Marshall then visited the offices of Stone and Webster, where a $66-million estimate for engineering development for the four alternate production methods, construction of electromagnetic and reactor pilot plants, materials procurement, and town site development was hammered out. On the 19th, Groves issued a directive to purchase the Tennessee site. He also resolved in one step the issue of priority assigned to the project that had been a holdup for months. With a letter in hand addressed to himself which granted the project AAA priority— the highest possible—Groves appeared at the office of Donald Nelson, head of the War Production Board. Nelson initially refused to sign, but reversed himself when Groves said he would have to recommend to the President that the project be abandoned because the WPB was unwilling to cooperate. On the 21st, Groves and Marshall met again with Bush, where they learned that the Navy had been left out of the project at the explicit direction of the President. Despite that injunction, Groves and Nichols visited the NRL later the same day to see a 14-stage liquid thermal diffusion facility that was under construction. Ross Gunn was desirous of coordinating Navy efforts with the Army, but Groves' impression was that the Navy effort lacked urgency.

One of the directives in Somervell's memo of September 17 was that Groves should arrange for a "working committee on the application of the product." This occurred on September 23, the day of Groves' promotion. At a meeting with Stimson, General Marshall, Conant, Bush, Styer, and Somervell, it was decided to appoint a Military Policy Committee (MPC) comprising Bush (as Chair; with Conant as his alternate), Styer, and Rear

Admiral William Purnell of the Navy. The charge of the MPC was to determine general policies for the entire Project. Formally, Groves was to sit with the committee and act as an Executive Officer to carry out policies that it determined, but in practice the committee usually ended up reacting to what he had already done. Groves cut short his attendance at the meeting to undertake a tour of the Tennessee site.

After returning to Washington from Tennessee, Groves and Nichols met with Stone and Webster officials on September 26, at which time it was decided to approach DuPont to develop and operate the plutonium-extraction plants. Founded in 1802, the E. I. DuPont de Nemours and Company was considered to be "the colossus of American explosives and propellant production." The firm had vast experience with designing, constructing, and operating a wide variety of chemical processing facilities; by the end of the war, DuPont had built 65% of total United States Ordnance Department powder production facilities. After some arm-twisting, DuPont accepted, on October 3, a contract to design and build the plutonium separation plants. Groves, impressed by the security advantages of DuPont's practice of building its own plants, soon began envisioning a much bigger role for the company in the Manhattan Project.

On October 2, Arthur Compton presented Groves with a proposal for development of four reactors. There were to be the first experimental pile, to be in operation at the Argonne site by December 1; a 10-MW water-cooled pilot reactor in Tennessee to be in operation by March 15, 1943, for the purpose of generating small amounts of plutonium for testing development of the separation process; a 100-MW liquid-cooled unit in Tennessee to be in operation by June 15, 1943; and a helium-cooled 100-MW unit, also to be located in Tennessee, to be in operation by September 1, 1943. By planning for two 100-MW plants, Compton wanted to insure adequate production given the uncertainties and problems that would inevitably arise. It was anticipated that the liquid-cooled plant could be constructed more quickly, but the form of cooling was not specified; both ordinary water and heavy water were still in the running. Groves assured Compton that a decision on an operating contractor would be made in about three weeks.

Three days later, Groves paid his first of many visits to Compton's Metallurgical Laboratory at the University of Chicago. Despite the high opinion he gained of the scientific competence of the Chicago group, he was horrified to learn that they casually considered their estimate of the amount of fissile material needed for a bomb to be correct within a factor of ten, an enormous margin of uncertainty for an engineer. Groves later said that he felt like a caterer who was being asked to prepare for a dinner for which anywhere between ten and a thousand guests might show up. They also discussed pile-cooling methods, eventually settling on (gaseous) helium, but that decision would be changed within three months.

Groves began pressuring DuPont to take a leading role in the plutonium program. He decided to relieve Stone and Webster of responsibility for that program, having come to the conclusion that every aspect of it, from design through operation of both piles and the separation facilities, should be overseen by a single firm. On October 31, Groves and Conant met with two DuPont Vice Presidents, Willis Harrington and Charles Stine. Groves pressed them to take on the pile program, stating that he felt that DuPont could handle all aspects of the project better than any other company in the country. Harrington and Stine were skeptical; chemistry, not nuclear physics, was DuPont's forte. On November 10, Groves, Nichols, and Compton visited DuPont's headquarters in Wilmington, Delaware, to put the case directly to the company's President, Walter Carpenter. Groves played to Carpenter's patriotic sympathies, emphasizing the importance that Roosevelt, Stimson, and General Marshall attached to the plutonium work. In response to Carpenter's concern that the background knowledge necessary to design and build piles was not yet sufficient, Groves emphasized that the paramount importance of the project to the war effort required proceeding directly. Carpenter concluded that the company could not refuse, but the issue would have to be put before the firm's Executive Committee; the company also insisted that a full review of the project be undertaken before deciding what role it would play.

Groves returned to Wilmington on November 27 to meet with the Executive Committee, before which he reiterated the arguments made to Carpenter. The Committee concluded that the pile project would probably be feasible, but insisted that the government be willing to indemnify DuPont against any losses or future liability claims due to the unusual hazards that would be involved. Concerned that liability claims for radiation-induced illnesses could begin cropping up twenty or thirty years in the future, DuPont insisted that a trust fund be set up to cover such claims; it would be funded to the extent of $20 million. Groves agreed, and a contract was signed on December 21. Since the company had no desire to produce plutonium after the war, it insisted that any patents revert to the Government, waived all profits, and accepted only payment for expenses plus a fixed fee of $1.00.

The company established a separate corporate structure to organize its pluto-nium activities, the so-called TNX Division. Described as a "task force within a matrix organization," TNX would have two subdivisions: a Technical Division to carry out design, and a Manufacturing Division to advise DuPont's Engineering Division on construction of facilities and their operation.

Before Groves could get a review committee together, another problem arose on two fronts almost simultaneously. On November 3, Glenn Seaborg reported to Robert Oppenheimer his concern that even very minor light-element impurities in plutonium could lead to an uncontrolled predetona-tion via (α, n) reactions. Seaborg estimated that plutonium purity would have to be controlled to one part in a hundred billion, else the entire pile project could be at risk. On November 14, Wallace Akers informed James Conant that British scientists were concerned about exactly the same issue. Groves asked Lawrence, Compton, Oppenheimer, and Edwin McMillan to investigate the situation. They reported back on the 18th that the problem would perhaps not be quite as drastic as Seaborg feared, but DuPont engi-neers remained very concerned with the desired plutonium purity, which would have to be better than 99%. Ultimately, the purity issue would be eclipsed by another problem, spontaneous fissioning of pile-produced pluto-nium.

On November 18, Groves appointed a five-person review committee, heav-ily populated with DuPont representatives. The group was headed by Warren Lewis, the MIT chemical engineer who had been involved with the third National Academy report of a year earlier. The other members were Craw-ford Greenewalt, a DuPont chemical engineer and former student of Lewis; Thomas Gary, a manager in the company's Engineering Department design division; and Roger Williams, an expert on plant operations in the company's Ammonia Department. The fifth member was to be Eger Murphree, but he had to withdraw on account of illness. Williams would be assigned over-all responsibility for the TNX Division, and Greenewalt would be assigned to head the Technical Division. Greenewalt's job would come to involve almost continuous commuting between Wilmington and Chicago; Groves and Compton considered that he did a superb job. Greenewalt would go on serve as DuPont's President from 1948 to 1962.

Crawford   Greenewalt (1902–1993) in the late 1970's. Source: AIP Emilio Segre Visual Archives, John Irwin Slide Collection

The committee assembled in New York on Sunday evening, November 22, and began their work the next day with a visit to Columbia to review Harold Urey's gaseous diffusion research. On Thanksgiving day, November 26, they arrived in Chicago, where Compton presented them with a 150-page document titled "Report on the Feasibility of the 49 Project." This report explored all aspects of the proposed pile process: uranium-graphite designs utilizing helium, liquid bismuth, or water-cooling; a uranium-heavy water system where the heavy water would serve as both coolant and moderator; problems of extracting plutonium from irradiated uranium; health and safety issues; radioactive by-products; and proposed time and cost schedules. That evening, the committee left Chicago to see Ernest Lawrence at Berkeley, where they witnessed calutrons in operation. On their way back East they stopped again in Chicago on December 2, where Greenewalt, serving as the group's representative, witnessed the first criticality of Enrico Fermi's CP-1 pile.

The Lewis committee finished its report on Friday, December 4; Groves received it on the following Monday. Their main conclusions were somewhat surprising: Although the electromagnetic method was probably the most immediately feasible approach for producing U-235, they felt that the diffusion process probably had the best chance for ultimately producing it at the desired rate of 25 kg per month, whereas the pile process (plutonium) might provide "the possibility of earliest achievement of the desired result."

The report offered five main recommendations: proceed immediately with the design and construction of a 4,600-stage diffusion plant to produce one kilogram of U-235 per day (anticipated cost $150 million); expedite design and construction of a pilot-scale pile and full-scale helium-cooled piles to produce 600 g of Pu-239 per day ($100 million); expedite development work on the electromagnetic method; install a small electromagnetic plant to produce 100 g of U-235 for experimental purposes ($10 million); and construct a heavy water plant capable of distilling two tons of that material per month ($15 million). In total, $315 million should be available early in 1943, in addition to $85 million that was already available from funds under the control of the Chief of Engineers. The explosive power of a fission bomb was estimated at 12.5 kt TNT equivalent, a figure which would prove close to *Little Boy's* yield at Hiroshima. As to predicted availability, it was estimated that there was a small chance of production prior to June 1, 1944, a "somewhat better" chance beginning before January 1945, and a "good chance" during the first half of 1945. There was still fear that Germany may be six months or a year ahead of America.

The Military Policy Committee met on December 10 to review the committee's report. This meeting would prove as crucial to the development of the Project as had the Bohemian Grove conference of three months earlier. The MPC endorsed all of the review committee's major recommendations, deciding to proceed with the kilogram-per-day diffusion plant and a 500-tank electromagnetic plant to obtain some early production of U-235 even though it would be in small quantities (100 grams per day). The Committee proposed that no intermediate-size piles be constructed, favoring instead that a full-scale pile program be undertaken directly at a site other than where the uranium plants would be located. With its proximity to Knoxville, the Clinton site appealed to neither Groves nor DuPont from a safety perspective; to site the piles there would require acquiring some 75,000 acres of land beyond what had already been taken. The X-10 pilot-scale reactor and separation facilities, which DuPont accepted a contract to design and construct on January 4, 1943, would, however, remain in Tennessee; DuPont referred to these facilities a "semi-works." The same day, Groves contracted with Westinghouse Electric, General Electric, and the Allis-Chalmers Manufacturing Company to produce components for the electromagnetic plant.

With removal of the production piles from Tennessee, Arthur Compton attempted to argue that the pilot-scale pile should be built at the Argonne site near Chicago so that personnel from the University would not have to relocate. This idea was vetoed by DuPont engineers, who feared that the scientists would interfere by insisting on endless design changes. Undeterred,

Compton came back with yet another proposal: that his group should be allowed to build its own pile at the Argonne site in order to create enough plutonium for research purposes. Groves traveled to Chicago on January 11 to press for the Tennessee location. Compton acquiesced, but the Chicago group did receive an unanticipated and initially unwanted consolation prize. Left unspecified in DuPont's January 4 contract was the question of who would operate the pilot plant. Despite their commitment to build and operate the main production plants, DuPont officials were reluctant to also agree to operate the pilot plant, preferring, as was corporate norm, to assign that task to research staff. DuPont proposed that the University operate the pilot plant, a suggestion which shocked Compton: universities do not normally operate industrial plants. In March, 1943, the University agreed to take on the operating responsibility, essentially doubling the size of its campus. The University remained as the operating contractor until July 1, 1945, when the task was taken over by the Monsanto Chemical Company.

4.14 December 1942: A Report to the President

On December 16, the MPC decisions were transmitted to President Roosevelt in a 29-page report under Bush's name; it had also gone through the Top Policy Group. This report was remarkable for both its summary of the situation at the time and its analysis of pending issues regarding international relations and postwar possibilities. In OSRD records, both the report itself and Bush's cover letter bear "OK FDR" scrawls.

Bush opened with a summary of work on the project to date. A site was being acquired near Knoxville to locate plants for the electromagnetic and diffusion methods. Another was being procured in New Mexico for a secret bomb-design laboratory. Centrifuge work was being limited to research only as that method now looked less promising then it had a few months earlier. A ten-stage pilot diffusion plant was under construction, and a 4,600-stage full-scale plant was being planned. An experimental pile had been constructed and operated. Because of possible hazards with full-scale piles, the MPC considered it essential that the President authorize the War Department to enter into contracts where United States would assume all risks.

The latter part of the report was devoted to speculations on the possibilities of atomic power. These anticipated many issues still relevant today: "There remains, however, little doubt that man has available a new and exceedingly potent source of energy … It is decidedly unfortunate, however … that the operation of such a power-plant pile inevitably involves the incidental production of a material, which is, to a high degree of probability, a super explosive … Certainly if, in the future nations are to construct and use power plants utilizing atomic power, and especially if a super-explosive is a possible by-product, the United States must be one of those nations."

The report also addressed implications in the areas of control of atomic energy and post-war international relations of what it called "a turning point in the technical history of civilization." Issues included the status of heavy water, uranium ores, accrual of patents to the Government, and relations with the British and Canadians. As to the latter, Bush reported that there had been complete scientific interchange between the British and American groups, but the subject was now entering a new phase with the involvement of the Army in developing production plants. Since the line between research and development was nebulous, Bush felt that the situation demanded a "new and clear" (i.e., Presidential-level) directive on future U.S.—British relations in the atomic field. Since the British had no intention of engaging in production of U-235 or Pu-239, Bush felt that passing any American knowledge to the British in those areas would be of no use to them during the war. British research in the area of heavy water had been transferred to a group in Montreal operating under the auspices of the National Research Council of Canada, but Bush felt that not having that group available to the American program would "not hamper the effort at all fatally." The British were well-along in diffusion research, but here again Bush felt that a complete cessation of interchange in that field might somewhat hinder, but not seriously "embarrass" the United States' effort.

Having mustered his arguments, Bush summarized with a statement that hinted at postwar American isolationism: "it appears (a) that there would be no unduly serious hindrance to the whole project if all further interchange between the United States and Britain in this matter were to cease, and (b) there would be no unfairness to the British in this procedure." Bush closed his discussion of the interchange issue by offering suggestions as to possible policy approaches. In a time-honored bureaucratic tactic, he presented three possibilities, with the politically unpalatable extremes presented first and second in order to pave the way for the third, which he evidently preferred. These were to cease all interchange; have complete interchange in both research and development; or restrict interchange only to information that the British could use directly. Within the latter option there would be no interchange on the purely American electromagnetic program, unrestricted interchange on the design and construction of the diffusion plant, research-only interchange (no plant design information) on the manufacture of plutonium and heavy water, and no interchange on the bomb-design laboratory that would be located at Los Alamos. Some historians have suggested that Bush viewed British interests as oriented more toward advantage in postwar commercial development of nuclear power than any wartime application, and so saw no justification for having an American-funded effort aid such development.

Bush's assessment of the implications of nuclear energy was sobering: "The whole development of atomic power, if it arrives as a new complication in an already complicated civilization, as now appears to be very probable as an event certainly of the next decade, may be an exceedingly difficult matter with which to deal wisely as between nations. On the other hand, it may be capable of maintaining the peace of the world."

Roosevelt approved the recommendations on December 28, including the choice of the third interchange option, although the issue of relations with the British would prove to be far from settled. With the President's approval, work that had begun three years earlier with a commitment of $1,500 was approaching a cost anticipated to be on the order of $400 million.

By the spring of 1943, Groves and the Military Policy Committee were firmly in charge of the Manhattan Project. All OSRD research and development contracts were transferred to the Manhattan District as of May 1, 1943. Contracts had been let for the design, construction, and operation of enrichment facilities and plutonium-producing reactors, and the intricacies of bomb physics were being explored at Los Alamos. The Planning Board and the S-1 Executive Committee essentially disappear from the history at this point, although Groves did retain James Conant and Richard Tolman as personal scientific advisors.

A sense of how the things had changed was captured by Conant in his May, 1943, draft history of the project: "For eighteen months this highly secret war effort has moved at a giddy pace. New results, new ideas, new decisions and new organization have kept all concerned in a state of healthy turmoil. The time for "freezing design" and construction arrived a few weeks past; now, we must await the slower task of plant construction and large-scale experimentation. The new results when they arrive will henceforth be no laboratory affair, their import may well be world shattering. But as in the animal world, so in industry: the period of gestation is commensurate with the magnitude to be achieved."

5

Piles and Secret Cities

When Arthur Compton decided to centralize nuclear-pile research at the University of Chicago's Metallurgical Laboratory in early 1942, the first goal on the path to large-scale plutonium production would be to show that a self-sustaining chain reaction could be created and controlled. Enrico Fermi began moving his Columbia pile group to Chicago in early 1942 to join forces with Samuel Allison's group. Fermi himself commuted between New York and Chicago through the winter and early spring of 1942, moving permanently in April. Five months were required to move Columbia uranium, graphite, and personnel to Chicago.

The first serious issue was supply of critical materials. Without tons of uranium metal and very pure graphite, there would be no point in attempting a large-scale pile project. While the purity requirements for graphite were stringent, there was at least an established graphite industry in America, and the Speer Carbon Company and the National Carbon Company were able to produce the necessary material. On the other hand, commercial production of uranium at the time was a relatively small-scale enterprise: the element was used only as a coloring agent in glasses and ceramics, as a source of radium, and in some specialty lamp filaments produced by the Westinghouse company. Westinghouse produced its uranium via a photochemical process which involved exposing large vats of solutions to sunlight, but this was far too slow for large-scale production. After considerable work, the problem of reducing uranium salts to a readily-handled metallic form metal was solved by Frank Spedding of Iowa State College and Clement Rodden of the National Bureau of Standards, who devised a chemical process by which pure uranium metal

© Springer Nature Switzerland AG 2020
B. C. Reed, *Manhattan Project*,
https://doi.org/10.1007/978-3-030-45734-1_5

could be produced by the ton. Large-scale production was contracted to the Mallinckrodt Chemical Company in Saint Louis, Missouri.

As uranium and graphite began to become available, Fermi and his group built a succession of sub-critical "exponential" piles, so named because mathematical analyses indicated that the density of neutrons within an operating pile would fall off exponentially toward the edges. Between September 15 and November 15, 1942, sixteen piles were constructed at Chicago (in addition to thirteen built at Columbia) to help inform the decision of optimal lattice size and to test batches of graphite and uranium. These piles had used radium/beryllium neutron sources; in a pile large enough to sustain a chain-reaction, spontaneous and cosmic-ray-induced fissions would be sufficient to ensure self-start-up, which would have to be quenched with control rods while the pile was under construction.

By October, enough material was on hand to begin planning for a critical pile. The original intent had been to build the first chain-reacting pile at the Argonne Forest site outside Chicago, but a labor disruption threatened postponement. In early November, Fermi approached Compton with the idea of performing the experiment at the University itself. Building an experimental nuclear reactor in the heart of a metropolitan area may sound like lunacy, but Fermi had done his calculations carefully and was confident that the reaction could be safely controlled; he planned on multiple redundant safety systems. Compton, fearing that University administrators would veto the plan, decided to authorize it on his own responsibility. He described the plan at a meeting of the S-1 Executive Committee held on November 14, and wrote in his memoirs that James Conant's face went white. The site chosen was a 30 by 60-foot squash court under the west stands of the University's Stagg Athletic Field. According to some sources, mistranslations in Soviet reports had the reactor being located in a "pumpkin patch."

5.1   "It was an Awesome Silence..."

Unlike its predecessors, Critical Pile number one (CP-1; also known as Chicago Pile 1) was built in the shape of a somewhat flattened ellipsoid with an equatorial radius of 388 cm (about 13 feet) and a polar radius of 309 cm (10 feet). This configuration helped minimize the ratio of the surface area to the volume of the pile, thereby cutting down on neutron loss. The original design had called for a spherical shape, but the quality of materials, particularly the availability of pure uranium metal, permitted getting away with a somewhat smaller structure than was originally envisioned. Layers of solid graphite bricks alternated with ones within which slugs of uranium were embedded, with the slugs configured to form a cubical lattice of side length 21 cm (about eight inches) as the pile was built up. This length was the average displacement over which neutrons would become thermalized after successive strikes against carbon nuclei; there would be no use in making the lattice size any larger. The bottom layer of graphite lay directly on the floor of the squash court, with the assembly supported by a wooden framework. Herbert Anderson scoured Chicago lumberyards for what he called an "awesome number" of four-by-six-inch timbers. In case it would prove necessary to enclose and evacuate the pile to improve the reproduction constant, Anderson also arranged, with the Goodyear Rubber Company, for a cubical rubber balloon 25 feet on a side. In practice, the balloon enclosure was not used.

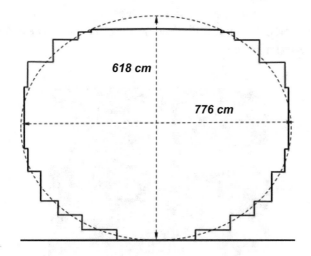

Side-view sketch of the shape of CP-1 and its equivalent ellipsoid. The dimensions are from side-to-side and bottom-to-top of the ellipsoid. Adapted from Fermi (1952)

Construction began on November 16, the day the balloon arrived, with physicists and hired laborers working twenty-four hour days in two twelve-hour shifts under the supervision of Anderson on the night shift and Walter Zinn during days. Two special crews machined graphite and pressed uranium oxide powder into solid slugs using a die and hydraulic press. Albert Wattenberg, who had joined Fermi's group while a student at Columbia, recalled that between mid-October and early December, 90-hour work weeks were not uncommon, with crews often smoking on the job to skip meals and save time. Two layers per shift was the normal rate of construction. Graphite was received from manufactures in the form of bricks of square 4 1/4-inch cross-section and lengths varying from 17 to 50 inches. With planers and wood-working tools, the bricks were cut to 16.5-inch lengths and milled to smooth 4 1/8-inch cross-sections; surfaces were held to tolerances of 0.005 inches, and lengths to 0.02 inches. About 14 tons of bricks could be processed per eight hours of work. In all, CP-1 incorporated 385.5 tons of graphite, some 40,000 bricks averaging about 20 lb each.

The uranium was in the form of pure uranium metal (just over 6 tons) and uranium oxide (about 40 tons); the slugs of pure metal were placed in the center of the pile. Holes of diameter 3-1/4 inches were drilled into bricks on 21-cm centers to receive the slugs, some of which were cylindrical and some pseudo-spherical. A total of 19,480 slugs were pressed, with about 1,200 being produced every 24 h. Fully one-quarter of the bricks needed to have holes drilled in them. Sixty to 100 holes could be drilled per hour, but the drill bits would become dull after doing only about 60 holes; some 30 bits had to be resharpened every day.

Walter Zinn, left, stands atop the partially reconstructed CP-1/CP-2 reactor at the Argonne site. Photo Courtesy of Argonne National Laboratory; http://www.flickr.com/photos/argonne/5963919079/

The pile was arranged with ten horizontal slots into which cadmium-sheathed wooden rods could be inserted. Cadmium is a voracious neutron capturer, and the rate of reactivity could be controlled by inserting and withdrawing rods as necessary. When construction was underway, all rods would be fully inserted and locked in place, but once per day they would be temporarily removed and the neutron activity level measured. As each layer was completed, Fermi computed an effective pile radius. A graph of the square of the effective radius (a measure of the surface area of the pile, through which neutrons could escape) divided by the number of neutron counts per minute (an indirect measure of the volume of the pile) versus the number of layers was essentially a descending straight line. As the neutron flux became closer and closer to exponentially diverging—the signature of a self-sustaining reaction—the surface-to-flux ratio would decline. By extrapolating the line to zero, Fermi could predict the layer at which criticality would occur.

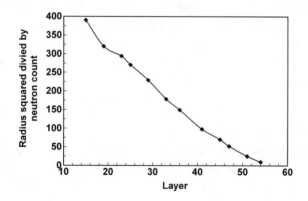

Square of pile radius divided by neutron count versus number of completed layers. Data from Fermi (1952)

Any one cadmium rod was sufficient to bring the reaction below criticality at any time. In addition, two vertical safety "zip" rods and one automatic control rod were also incorporated into the pile. During normal operation, all but one of the cadmium rods would be withdrawn. If neutron detectors signaled too great a level of activity, the zip rods would be automatically released, accelerated by 100-pound weights. The automatic control rod could be operated manually, but was also normally under the control of a circuit which would drive it into the pile if the level of reactivity exceeded a designated level, but could also withdraw it if the intensity fell below a desired minimal level. As a last-resort measure, a three-man crew would stand atop the pile ready to dump buckets of liquid cadmium solution. This group has been called the "suicide squad"; a better term might have been "Chicago pile hit squad". By late November, it was clear that the pile would become critical on completion of its 56th layer. Fermi decided to add a 57th layer, which would be laid during the night of December 1–2. He instructed Anderson not to start the reaction that night in order that Laboratory staff and a representative of the visiting Lewis Committee could be present for the historic event on the next day.

The witnesses assembled at 8:30 on the morning of December 2 on a balcony at the north end of the squash court, about 10 feet above the floor. Including Fermi, 49 people were present to witness the dawn of the nuclear age. On the floor below, George Weil would handle the final cadmium rod, which was at layer 21. Fermi had calculated in advance the expected neutron intensity for each position of Weil's rod, and had a pocket slide rule on hand with which he checked readings against his predictions throughout the day. At 9:45, he ordered the electrically-driven safety rods removed. The neutron count grew and steadied out. One of the safety rods was tied off to the balcony railing, with Norman Hilberry standing by with an axe in order to cut the rope in case the automatic shutdown system failed. According to some sources, the phrase "to scram" a reactor—execute an emergency shutdown— is an acronym for "safety control rod axe man."

Shortly after 10:00, Fermi ordered the automatic safety rod withdrawn. This was done, and again the neutron count grew and leveled off. At 10:37, he instructed Weil to pull the last rod out to 13 feet; the count again leveled within a few minutes. Fermi ordered the rod withdrawn another foot, and then, at 11:00, another six inches. Additional withdrawals at 11:15 and 11:25 were not enough to achieve criticality, as Fermi had anticipated. Proceeding cautiously, Fermi ordered the automatic control rod reinserted as a check; the intensity dropped accordingly. At about 11:35 the automatic rod was withdrawn and the cadmium rod adjusted outwards. Suddenly a loud crash occurred. The threshold safety intensity had been set too low, and one of the zip rods had deployed itself. Fermi decreed that a lunch break was in order, and directed that all control rods be reinserted.

The group reassembled at 2:00 p.m. To check that the neutron flux returned to its pre-lunch reading, Fermi ordered all rods withdrawn except for Weil's. Satisfied with the neutron count, he then directed that one of the zip rods be inserted; the neutron count obediently declined. He then ordered Weil to withdraw the cadmium rod by one foot. On then directing the zip rod to be removed, Fermi said to Arthur Compton: "This is going to do it. Now it will become self-sustaining. The trace will climb and continue to climb; it will not level off."

Herbert Anderson recorded the time as 3:36 p.m. In his words:

At first you could hear the sound of the neutron counter …. Then the clicks came more and more rapidly, and after a while they began to merge into a roar; the counter couldn't follow any more. That was the moment to switch the chart recorder [to a less-sensitive setting]. But when the switch was made, everyone watched in the sudden silence the mounting deflection of the recorder's pen. It was an awesome silence. Everyone recognized the significance of that switch; we were in the high-intensity regime. … Again and again, the scale of the recorder had to be changed to accommodate the neutron intensity which was increasing more and more rapidly. Suddenly Fermi raised his hand. "The pile has gone critical," he announced. No one present had any doubt about it.

Fermi allowed the pile to operate for 28 min before calling for a zip rod to be inserted. He estimated that at that point the pile was operating at a power of about one-half of a Watt, less than a flashlight battery. Crawford Greenewalt recorded in his diary that "Fermi was as cool as a cucumber." Because of security regulations, no photographs of the completed pile were ever taken; we now have only an artist's rendering of the startup.

Artist's conception of the startup of CP-1. Source: http://commons.wikimedia.org/wiki/File:Chicagopile.gif

A strip-chart recording of the neutron flux, "The Birth Certificate of the Nuclear Age", clearly shows the exponential growth characteristic of a self-sustaining reaction.

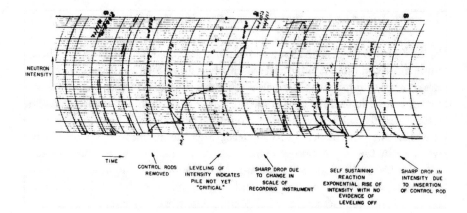

Galvanometer tracing of CP-1 neutron intensity. Source: Courtesy of Argonne National Laboratory; http://www.flickr.com/photos/argonne/7550395714/

Eugene Wigner presented Fermi with a bottle of Chianti, and a paper-cup toast was raised. Many of those present signed the wicker wrapping of the bottle. Arthur Compton excused himself to phone James Conant in Washington with the news. As Compton related their conversation:

> Jim, you'll be interested to know that the Italian navigator has just landed in the new world. The earth was not as large as he had estimated, and he arrived in the new world sooner than he had expected.

> "Is that so? Were the natives friendly?" asked Conant.
> "Everyone landed safe and happy," reported Compton.

In a report written on December 15, Fermi gave a laconic description of the historic event: "The chain reacting structure has been completed on December 2 and has been in operation since then in a satisfactory way."

On the twentieth anniversary of the achievement, Wigner offered a reflection:

> Nothing very spectacular had happened. Nothing had moved and the pile itself had given no sound. Nevertheless, when the rods were pushed back and the clicking died down, we suddenly experienced a let-down feeling, for all of us understood the language of the counters. Even though we had anticipated the success of the experiment, its accomplishment had a deep impact on us. For some time we had known that we were about to unlock a giant; still, we could not escape an eerie feeling when we knew we had actually done it. We

felt as, I presume, everyone feels who has done something that he knows will have very far-reaching consequences which he cannot foresee.... Do we then exaggerate the importance of Fermi's famous experiment? I may have thought so some time in the past, but do not believe it now. The experiment *was* the culmination of the efforts to prove the chain reaction. The elimination of the last doubts in the information on which our further work had to depend had a decisive influence on our effectiveness in tackling the second problem of the Chicago project: the design and realization of a large-scale reactor to produce the nuclear explosive plutonium. This objective could now be pursued with all the energy and imagination which the project could muster.

Fermi computed the reproduction factor to be $k = 1.0006$. Because CP-1 was always operated at very low power, the level of radioactivity created was harmless. Radiation doses are commonly measured in "rems", an abbreviation for "roentgen equivalent in man". A lethal single-shot radiation dose for a human being is about 500 rems; the typical background dose for the entire population of America is about 0.3 rems per year. During CP-1 operation, the exposure level near the pile was about 0.05 rems per minute; at the sidewalk outside the building it was about one-thousandth as much.

Fermi's prediction that the pile could be safely controlled proved correct. When the pile was in steady-state operation under the control of a single cadmium rod, some four hours were required for the reactivity to rise by a factor of two if the rod was pulled out by one centimeter. Even if all rods were removed, the neutron intensity within the pile would have had a characteristic exponential rise time of about two and a half minutes, plenty of time to re-insert a rod. Control could be so finely maintained that it was occasionally necessary to adjust a rod by a centimeter or two in response to the pile's reaction to changing atmospheric pressure, and the temperature sensitivity of the reproduction factor, a key engineering consideration, could be measured by simply opening a window to allow outside air to cool the pile. Fermi described controlling the pile as being as easy as the minute steering adjustments made while driving a car on a straight road.

Because CP-1 was uncooled and unshielded, it was operated most of the time at half-Watt power, although it did operate briefly at 200 W on December 12. But before engineers could extrapolate to production-scale reactors, much more research on control and shielding systems was necessary. After about three months, CP-1 was disassembled and moved to the Argonne site, where it was reassembled as CP-2. CP-2 was essentially cubical in shape, about 30 feet square in footprint by 25 feet high, and incorporated some 52 tons of uranium and 472 tons of graphite. CP-2 was also uncooled, but was shielded on all sides by a concrete wall five feet thick and on its top by a six-inch layer of lead and 50 inches of wood; this permitted operation at a power of a few kilowatts. The rebuilt pile first went critical in May, 1943, and was used for studies of neutron capture cross-sections, shielding, instrumentation, and as a training facility for later production operations. Also built at the Argonne site was the world's first heavy-water cooled and moderated reactor, CP-3, which began operation in May, 1944. This reactor consisted of an upright aluminum tank six feet in diameter, which was filled with about 6.5 tons of heavy water, a design not unlike what physicists in Germany were developing. The tank was surrounded by a graphite neutron "reflector," which was further surrounded by a lead shield and then a biological shield of concrete. The top of the structure was pierced with holes for experimental ports and control and fuel rods, and was shielded with removable bricks of alternating layers of iron and masonite. CP-3 reached its full operating power of 300 kW in July, 1944.

5.2   Scaling Up: X-10

With a self-sustaining reaction having been demonstrated, attention turned to the construction of the X-10 pilot-scale pile. X-10 had multiple missions: to produce plutonium for research at Los Alamos, to test chemical separation, to train operating personnel for the eventual production-scale reactors at Hanford, to serve as a platform for instrument development and cross-section research, and to conduct radiation-damage and biological radiation-effects studies.

Extrapolating from the operation of CP-1, Fermi estimated that a pure-uranium/graphite system could develop a reproduction constant of $k \sim 1.07$, a value great enough to keep open the possibilities of water-cooling the production piles and air-cooling the pilot-scale unit. Crawford Greenewalt had been thinking of helium-cooling for the pilot unit, but air-cooling would be much simpler from an engineering perspective. By January, 1943, X-10's basic specifications had been developed: a 1,000-kilowatt (one megawatt) air-cooled, graphite-moderated pile of cubical shape. The anticipated power level was crucial. Plutonium production in a reactor is directly proportional to its operating power. A reactor fueled with natural uranium produces about three-quarters of a gram of plutonium per day per megawatt of power produced, a mere one-third of the mass of a dime. If X-10 were to run for a full year at its one-megawatt rating, it would theoretically produce about 275 grams of plutonium, assuming perfect chemical separation efficiency. It ultimately achieved better than this.

Formally, the X-10 reactor was under the administration of the University of Chicago's on-site Clinton Laboratories, which was located in a 112-acre site in the Bethel Valley of the Oak Ridge reservation in Tennessee. X-10's core comprised a 73-layer graphite cube, 24 feet square on its base by 24 feet, 4 inches high. Right-angled notches were cut into the sides of the graphite bricks, which, when laid side-by-side, formed 1,248 horizontal diamond-shaped front-to-rear channels into which cylindrical aluminum-jacketed uranium slugs could be fed from the front face of the pile. X-10's 700 tons of graphite was in the form of bricks of cross-section four inches, with lengths varying from eight to 50 inches. The fuel channels were built on eight-inch centers; the pure uranium slugs were just over an inch in diameter by four inches long. A full fuel load would be about 120 tons, but it was anticipated that the pile would go critical with about half that amount. The core was surrounded by a seven-foot thick shield made of a type of concrete that retained some water upon setting; this helped to capture escaping neutrons. With the addition of layers of pitch to prevent the shielding from losing water along with special precast concrete blocks on the front face to align fuel channels, the full outside dimensions of the pile came to some 47 feet long by 38 feet wide by 35 feet high. A removable "graphite thermal column" section of the core could be lifted out to facilitate lattice dimension experiments.

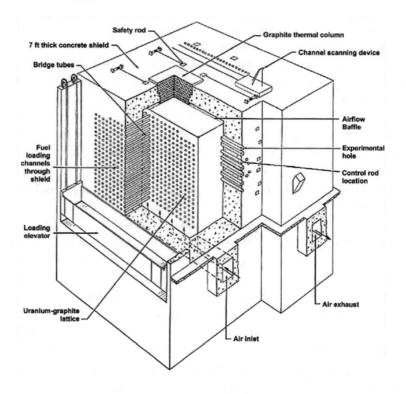

Schematic drawing of X-10 pile. Not all horizontal and vertical channels are shown. Courtesy of Oak RidgeNational Laboratory, U.S. Dept. of Energy. See http://info.ornl.gov/sites/publications/files/Pub20808.pdf

After some period of operation, fuel slugs would be discharged from the back of the pile as new slugs were pushed in. Discharged slugs would fall through a chute into a pit containing 20 feet of water, where their intense short-lived radioactivity would be allowed to die off for a few weeks before they were transported to the chemical separation plant. To fuel the pile, workers rode in an elevator which spanned its front face. While X-10 was not a model for the larger production reactors built at Hanford, some of its design features would find their way into those piles, particularly the procedures for fueling and handling discharged slugs. 183 DuPont employees would train at Clinton before moving to Washington.

Front face of the X-10 pile. Source: http://commons.
wikimedia.org/wiki/File:X10_Reactor_Face.jpg

As with CP-1, the control system for X-10 was deliberately over-designed. Three sets of control rods were incorporated: regulating rods, shim rods, and safety rods. The latter were four eight-foot-long boron-steel rods suspended above the pile. They could be operated manually, but were held in place with electric brakes; in the event of a power failure they would passively fall into the pile. As an emergency backup system, hoppers above the pile could release small boron-steel balls into two vertical channels. During normal operation the pile would be controlled by two horizontal regulating rods which entered from its right side. Four horizontal shim rods provided a means to compensate for variations too large to be handled by the regulating rods, and could effect a complete shutdown by themselves in the event of power failure. Fuel and experimental channels were equipped with plugs which were removed only when the power output was low enough to prevent a dangerous amount of radiation from escaping. The limiting factor in X-10's operation was the capacity of its forced-air cooling system. This initially consisted of two fans each capable of moving 30,000 cubic feet per minute, plus a stand-by steam-driven 5,000 cubic foot unit which would come on-line in the event of a power failure. Before being exhausted from a 200-foot stack, the heated air would be filtered and sprayed with water to remove any radioactivity that it might have picked up.

DuPont began excavation for the pile building on April 27, 1943. Concrete pouring began in June, and the Aluminum Company of America began "canning" uranium slugs, sheathing them in thin aluminum jackets. Graphite stacking began on September 1. By late October, construction and final mechanical testing were complete. Loading of fuel into the central portion of the pile began on the afternoon of November 3, with Enrico Fermi inserting the first slug to give the pile the "Blessing of the Pope". X-10 went critical at 5:07 on the morning of November 4, with only about 30 tons of uranium inserted. After a week the fuel load was increased to 36 tons, bringing the power level to 500 kW. Before November was out, five tons of fuel containing some 500 milligrams of plutonium had been discharged and sent off for chemical processing. In December, empty channels were blocked off with graphite plugs to force the airflow to be concentrated around the installed fuel; this permitted higher-temperature operation and raising the power level to about 800 kW. By February, 1944, the pile was producing irradiated uranium at a rate of about one-third of a ton per day; the efficiency of chemical separation of plutonium from uranium eventually exceeded 90%.

In early 1944, X-10's fuel distribution was reconfigured to further enhance plutonium production. The standard configuration had been 459 channels loaded with 65 slugs each (about 40 tons); this was changed to 709 channels with 44 slugs each. This did not significantly increase the amount of fuel in the pile, but reducing the amount of fuel in the center of the pile relative to that further out permitted operating at a higher power level without attaining too great a central temperature. Improved slug-canning techniques allowed for even higher-temperature operation; by May, 1944, the power level could be increased to 1.8 megawatts. In June and July of that year, installation of two 70,000-cubic-foot fans allowed operation at an impressive four megawatts. X-10 operated with remarkable reliability; the only real problem encountered was a bearing failure in one of the new fans, which necessitated temporary re-installation of one of the 30,000-cubic foot units during the summer of 1944.

Plutonium production began in December, 1943, with a mere 1.5 milligrams being isolated. By mid-1944, tens of grams were being turned out per month, and by the time production ceased when the Hanford reactors were coming on-line in January, 1945, over three hundred grams had been extracted from 299 batches of slugs. It was X-10 plutonium that would lead to the discovery of the spontaneous-fission crisis at Los Alamos in the summer of 1944; had this discovery had to wait for Hanford-produced material, the Nagasaki *Fat Man* bomb would have been delayed by the better part of a year. A bonus of X-10 operation was the production of quantities of radioactive lanthanum, which could be extracted from decaying barium, a direct fission product. This radio-lanthanum ("RaLa") would prove crucial at Los Alamos in developing a diagnostic test of the plutonium implosion bomb. X-10 more than fulfilled its wartime mission.

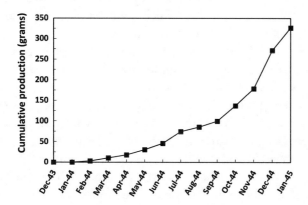

Cumulative production of plutonium from the X-10 pile. Data from National Archives and Records Administration microfilm set A1218 (Manhattan Engineer District History), Roll 6 (Book IV—Pile Project X-10, Volume 2—Research, Part II—Clinton laboratories, "Top Secret" Appendix). By January 1945 cumulative production reached 326.4 grams

5.3   Secret Cities East: Oak Ridge

In 1942, rural eastern Tennessee was still very undeveloped. The population was sparse, most roads were unimproved, and services were minimal. If a secret production complex employing thousands of people was to be constructed, those people, many of them highly-educated and used to the amenities of big-city and university living, would need a place to call home. Perhaps no other statistics speak more forcefully to the scale of the Clinton Engineer Works than the growth of the town established to house its workers. In 1942, Oak Ridge, Tennessee, did not exist. By mid-1945 it would be the fifth-largest city in the state, boasting a population of about 75,000. Located in the northeast corner of the Clinton reservation, it appeared on no maps at the time. Once the Clinton site was closed to public access as of April 1, 1943, Oak Ridge literally became a secret city, and was known to its residents by that term. Of every dollar spent on the Manhattan Project, just over 60 cents went into the Clinton Engineer Works, and the responsibility borne by its commander, District Engineer Colonel Kenneth Nichols—who nominally administered production sites for the *entire* Project - was extraordinary. In 1942, the idea of using plutonium for a fission bomb was a tantalizing but wholly speculative prospect; only uranium looked certain as a bomb material. If plutonium proved unworkable (which it almost did), the success or failure of the Project would be decided at the Clinton site.

The roughly 140 lb of enriched uranium that went into the Hiroshima *Little Boy* bomb would fit very comfortably inside a soccer ball. While it might seem that if all one desired to do was to produce one or a few bombs, a small factory should be adequate. But the nature of the processes involved in enrichment are such that it does not work that way. To fulfill the task in any reasonable length of time means building factories that by their nature can turn out material for dozens or hundreds of bombs once they are in operation; it is all-in or stay home. To produce that 140 lb, the scale of CEW operations would grow to be enormous. The number of construction workers alone would peak at 45,000 in the spring of 1944, and by May, 1945, the entire CEW would employ just over 80,000 personnel.

When the Stone and Webster company took on responsibility for constructing the electromagnetic enrichment plants along with site development and townsite design and construction in the fall of 1942, they envisioned a village to house some 5,000 inhabitants. By October 26, 1942, when S&W submitted its first plan for the area, the estimated population had grown to 13,000. With constant design changes and expansions of the electromagnetic enrichment plant, S&W soon found its resources stretched. In late November, Groves decided to relieve S&W of town-design functions, although the firm would retain responsibility for overseeing construction, utility operations, and road maintenance. Design of housing units was contracted to the architectural firm of Skidmore, Owings, and Merrill of Chicago.

Oak Ridge grew up in three phases of development. The first, known as "East Town" from its location just southwest of the Elza Gate entrance to the reservation, was completed in early 1944 and contained over 3,000 family-type housing units, dormitories, 1,000 trailers, an administration building, stores, recreation areas, schools, churches, theatres, laundries, a cafeteria, and a hospital which would be the birth site of 2,910 babies in the first three years of its existence. Oak Ridge would acquire nearly 100 miles of paved streets, and a further 200 miles would be laid to service the production sites.

By the fall of 1943, the population estimate had grown to 42,000, and phase two was begun about two miles west of East Town. By the summer of 1944, this had added some 4,800 family units, a number of barracks, and fifty dormitories which could house some 7,500 single residents. In early 1945, estimates were again revised upward to an ultimate population of 66,000. The third phase of development, built to both the east and west of the original site, saw the addition of 1,300 family units, 20 dormitories, and hundreds of trailers plus associated services. By the time housing construction was finished in 1945, over 7,000 family houses, apartments containing over 9,000 dwelling units, 89 dormitories, 2,000 five-man "hutments," and seven trailer camps with a total capacity of about 4,000 occupants had been put up. Two sewage-treatment plants, 130 miles of sewer mains, a steam plant, ten elementary schools, two junior-high schools, two senior-high schools, five nursery schools, nine shopping areas, and a number of temporary stores were erected. Trees cut down during building operations were turned into 163 miles of boardwalks. For residents, life was a bargain: rents were minimal and services heavily subsidized; household electricity use went unmetered. The cost of constructing Oak Ridge ran to just over $100 million, not including the building of construction camps which temporarily housed a further 14,000 inhabitants near the various enrichment plants.

To speed construction and minimize costs, Skidmore, Owings, and Merrill restricted plans to nine different types of pre-fabricated houses and three different apartment designs, all wood-frame structures. Many units incorporated interior and exterior panels of "cemesto," a sturdy, fireproof building material made of fiber-board with pressed asbestos-cement panels bonded to both sides. One history of the area records that at one point, housing units fully equipped with appliances and furniture were being turned over from the contractors to the Government at a rate of one every thirty minutes. Intended to be only semi-permanent, many of the original cemesto homes still stand, now prized for their historic value and location close to the commercial center of town.

Typical cemesto homes and a trailer-housing area at Oak Ridge. Source: http://www.y12.doe.gov/about/history/getimages

Oak Ridgehad not only to be constructed, but also managed and operated. For this, Groves approached the Turner Construction Company of New York, which he had used on other projects. Turner established a wholly-owned subsidiary, the Roane-Anderson Company, named after the two counties which the Clinton reservation straddled. On a cost-plus-fee basis, Roane-Anderson managed utilities, police and fire departments, medical personnel, trash collection, school maintenance, cemeteries, cafeterias, warehouses, deliveries of coal and ice, and granted concessions to private operators for grocery stores, drug stores, department stores, barber shops, and garages. The company also operated an extensive bus system, a railroad, and a motor pool. To take some 60,000 riders per day to and from the productions sites required 840 buses, making the system for a time the ninth-largest bus network in the United States. By early 1945, Roane-Anderson had over 10,000 employees on its payroll, although the number declined thereafter as services began to be contracted to other organizations.

5.4   Secret Cities West: Hanford

Like the Clinton facility, the Hanford Engineer Works was constructed from scratch. But in many ways, construction at Hanford was far more challenging than at Clinton. With no cities nearby, DuPont had to plan from the outset for large on-site communities. They decided to build two: a construction camp at Hanford itself, and, more distant, a permanent housing area at Richland for employees and their families. The construction camp was located about 6 miles from the nearest process area, and Richland village was about 25 miles from the piles. Planning for both began in early 1943. Nearly 400 miles of highways and over 150 miles of railroad track would also be constructed. Initial estimates called for housing a construction workforce of 25,000–28,000, but plans for the construction camp were deliberately made scalable, an approach which proved its worth when the construction force alone grew to nearly twice the initial estimate.

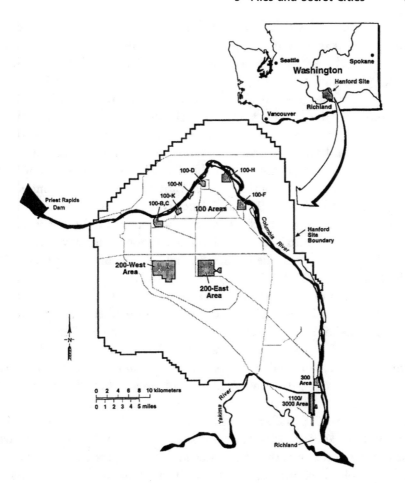

The Hanford Engineer Works site. Piles were built at the 100-B, D, and F sites from west to east along the Columbia river. The 200-North site is not shown on this map; it was about 3 miles north of the 200-East site. The original village of Hanford was on the west bank of the Columbia, due east of the 200 area. Source: HAER, Fig. 1

For Richland, initial estimates projected a population of 6,500–7,500, but this was soon revised to 12,500, then to 16,000, and eventually to 17,500. As at Clinton, living conditions were rough-and-ready, with many families housed in prefabricated and portable residences. Once again, schools, stores, churches, recreational areas, hospitals, utilities, street maintenance, trash pickup, transit services, and fire and police forces had to be provided. Eventually, some 4,300 family dwelling units and 21 dormitories were put up. As at Oak Ridge, an extensive bus system was necessary; during the construction phase alone, some 340 million passenger-miles were driven.

If conditions at Richland were reminiscent of a boomtown, they must have seemed luxurious in comparison to those at the construction camp. Shops to machine graphite, fabricate concrete pipes, and prepare sections of steel plate and masonite panels for reactor shielding were interspersed among houses, heating and water plants, barracks, trailer courts, cafeterias, bars, administrative buildings, theaters, schools, hospitals, and libraries. The first DuPont employee arrived on February 28, 1943, and construction officially began on March 22 with the opening of an employment office in the city of Pasco, about 30 miles from the site. The construction camp began housing workers in April, although some spent their first six months living in tents. Between March, 1943, and August 1944, the local police force, the Hanford Site Patrol, recorded just over 8,000 "incidents," the vast majority of which involved intoxication and burglary, although the tally also included five violent deaths, 19 accidental deaths, and 88 cases of bootlegging. Village police kept a copy of a key to every house in the town.

Construction of the construction camp itself had to come first. Work on the first barracks began on April 6, and by September most people were working nine-hour days six days per week, with some laborers temporarily putting in ten-hour days seven days per week. By August, the cafeteria was serving some 22,000 meals per day, and by November, 5,300 workers were employed in erecting the construction camp alone, which by the end of the year boasted over 100 men's barracks, several dozen women's barracks, seven mess halls, and 1,200 trailers. On December 3, work went to two nine-hour shifts, and on January 1, 1944, a third shift was added. By July, 1944, when construction of the piles themselves was in full swing, the camp was home to 45,000 people. Total project man-hours at Hanford would run to over 126 million, with only about 15,000 lost due to labor disruptions. Walter Simon, DuPont's plant operations manager at Hanford, allegedly said that "Rome wasn't built in a day, but DuPont didn't have that job."

Isolation, sandstorms, and spousally-segregated living conditions made employee turnover an endemic problem. DuPont interviewed over 262,000 applicants and hired over 94,000 to maintain an average workforce of 22,500 over the life of project. Robley Johnson, an official photographer with the project, made some 145,000 ID photos. By the summer of 1944, the turnover rate in construction personnel had reached 21%. To raise morale, DuPont put up recreation halls, taverns, bowling alleys, tennis courts, baseball and softball fields, and brought in nationally-known entertainers. Groves directed that beer be could be sold in whatever quantities were needed. Unskilled laborers were attracted by an average daily pay of $8, twice the $3–$4 rate common in other parts of the country; for skilled laborers, the figures were $15 in comparison to $10. All employees signed a declaration of secrecy, which reminded them that violation of the Espionage Act could result in 10 years in prison and fines of up to $10,000. For every employee of the Manhattan Engineer District, security and surveillance were constant companions; agents would often pose as regular workers.

Oak Ridge and Hanford were never intended to be permanent communities, although that is what they ultimately grew into. During their wartime tenures, Clinton and Hanford each had but a single overriding function: to produce fissile material in sufficient quantity to make the world's first atomic bombs and help end the war. Nothing else mattered.

6

U, Pu, CEW and HEW: Securing Fissile Material

6.1 Y-12: The Lorentz Force Law Goes Big-Time

Ernest Lawrence's electromagnetic method of enriching uranium was the *only* enrichment method discussed at the September, 1942, Bohemian Grove planning meeting. The late-1942 Lewis committee concluded that an electromagnetic plant capable of producing one kilogram of U-235 per day would require at least 22,000 vacuum tanks; to achieve the same output with a gaseous- diffusion plant would require a 4,600-stage installation. At its December 10, 1942, meeting, the Military Policy Committee opted to start with a 500-tank electromagnetic plant. While neither the electromagnetic nor the diffusion approaches to enrichment would be easy, the electromagnetic system had the advantage that its fundamental concepts were proven. Also, since it would operate in a "batch" mode, it could be built in sections, each of which could begin operating independently as soon as it was constructed. In contrast, each section of the "continuous" diffusion plant would have to be operational before it could be connected to preceding and succeeding sections.

The Y-12 electromagnetic enrichment plant represented an enormous scaling-up of using mass spectroscopy to separate streams of ionized atoms of different masses; in the case of uranium, the streams are ions of the 235 and 238 isotopes. Fundamentally, Y-12 used a modified version of Lawrence's cyclotron wherein hundreds of pairs of vacuum tanks were placed back-to-back within a magnetic field. Pairing tanks doubled production over what would be obtained with using just single tanks.

© Springer Nature Switzerland AG 2020
B. C. Reed, *Manhattan Project*,
https://doi.org/10.1007/978-3-030-45734-1_6

Y-12 was located in an 825-acre tract within the Bear Creek Valley of the Clinton Engineer Works site. It would become a mammoth undertaking. The second-most expensive facility of the entire Manhattan Project (about \$478 million in construction and operating costs; \$512 million for the gaseous diffusion plant), Y-12 ranked first as measured by number of personnel: a peak of nearly 22,500 in May, 1945. The complex would come to include nine main processing plants and over 200 auxiliary buildings, totaling some 80 acres of floor space. The entire complex was surrounded by a 5.3-mile perimeter fence with 19 guard towers.

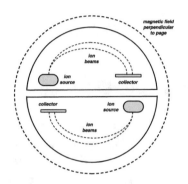 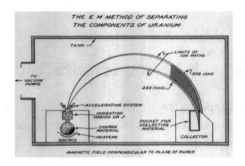

Left: Schematic illustration of two back-to-back calutron "tanks" and a magnet coil, represented by the circular dashed line. The magnetic field is perpendicular to the plane of the page. Right: Sketch of the electromagnetic separation method, reproduced from a Manhattan District History microfilm. Source: A1218(9), image 831

For access to the vacuum tanks for maintenance and to remove accumulated material, it was convenient to place them between adjacent magnet coils instead of within them. By linking together multiple copies of such arrangements with a current-carrying "busbar," dozens of tanks can be operated simultaneously. Since the electrical current must run along a closed circuit, another refinement is to configure the track as a closed loop. Lawrence and his engineers conceived this idea early on, deciding on an oval-shaped arrangement which came to be known as a "racetrack".

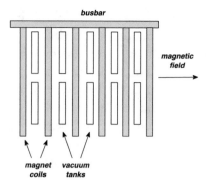

Schematic side-view illustration of part of a calutron "racetrack." In reality, a given track would comprise dozens of tanks and coils

Left: A Y-12 alpha "racetrack." Right: Workers tend to a C-shaped vacuum tank. Sources: http://www.y12.doe.gov/ about/history/getimages; http://commons.wikimedia.org/wiki/ File:Alpha_calutron_tank.jpg

The need for hundreds of tanks was dictated by an effect known to cyclotron engineers as the "space-charge" problem. As the like-charged ion beams travel through the vacuum tank, they repel each other and become disrupted from their ideal circular paths. This sets a practical limit on the strength of the beams, which is usually expressed as an equivalent electrical current. This in turn limits the mass of material that any one vacuum tank can separate per day. In the case of Y-12 calutrons, this capacity was about one-tenth of a gram of U-235 per day in the best of circumstances. To collect 50 kg from one tank (barely enough for a single bomb) would require 500,000 days of operation, or over 1,300 years. It was appreciated from the outset that isolating enough material for a bomb in a year or two would require a facility with at least hundreds of tanks.

By the fall of 1942, experiments with Lawrence's 184-in cyclotron indicated that some 2,000 ion sources would be necessary to separate 100 grams of U-235 per day, the goal set at the Bohemian Grove meeting. Stone and Webster conservatively assumed that no more than one ion source and collector could be fitted into each tank, and began planning on as many as 2,000 tanks. From the capabilities of available electrical power-distribution equipment, it seemed feasible to assume that each production building could house two racetracks, both containing about 100 tanks. If the gap between each pair of coils housed two tanks, each track would require 50 gaps. To provide for two thousand tanks, ten production buildings would be required, as well as fabrication and maintenance shops, laboratories, and generating facilities. A particularly daunting aspect of the system was its vacuum requirements. The tanks would have to be pumped down to and maintained at pressures of about a hundred-millionth of atmospheric pressure. It was estimated that the vacuum volume for the plant would probably exceed by many orders of magnitude the entire evacuated space in the world at the time. Another consideration was facilities for chemical processing. Uranium oxide received from Mallinckrodt Chemical had to be transformed to uranium tetrachloride before being fed into the calutrons, and the processed material, which was washed out of the tanks with acid, had to be collected and purified. Chemical operations alone at Y-12 employed several thousand people.

Through 1942, Lawrence and his engineers concentrated on refining the design of ion sources to incorporate multiple beams. On November 18, he installed a new tank which contained a double source between the poles of his 184-in cyclotron. Both sources were capable of producing two sets of beams, that is, there would be four sets of U-235 and U-238 beams altogether. The system was cantankerous, however; often only two beams could be kept in focus simultaneously. But even two sources per tank would be a major advance, as the total number of tanks could be halved to 1,000. In the meantime, Stone and Webster engineers had to begin designing buildings based on only very rough ideas as to the equipment they would contain; Groves often invested enormous sums in construction before fully-workable enrichment systems had been developed.

Lawrence, Groves, and contractor representatives met at Berkeley on January 14, 1943, to begin planning the Y-12 project in detail. The initial phase of development called for five 96-tank tracks to be housed in three buildings; the tracks themselves would be 122 feet long, 77 feet wide, and 15 feet high; the buildings would require 6-foot foundations to support the weight of the magnets. Groves wanted the first track in operation by July 1, 1943. The first floor of each building (below ground level) would house vacuum pumps. The tracks would reside at ground-level, and above them resided operating galleries from where employees, mostly local young female high-school graduates, continuously monitored and adjusted the ion beams in each tank. The process was labor intensive, requiring some 20 employees per pair of tanks. The magnetic fields had to be kept extremely uniform; a deviation of only 0.6% would result in collecting the wrong isotope. In practice, the fields could be kept stable to about one part in 4,000, and accelerating voltages to one part in 2,000.

Design of the Y-12 facility and its equipment evolved continuously and incrementally. First came a decision to use two-source ion emitters in each tank. In early 1943, Edward Lofgren conceived the idea of building second-stage enrichers which would be fed with slightly-enriched material (about 15% U-235) that emerged from the first-stage tracks, and which would raise the enrichment level to 85–90%. Groves authorized the first two such units on March 17. It was at this point that the original oval racetracks became known as "Alpha" units, and the second-stage enrichers as "Beta" units.

Alpha tracks contained 96 vacuum tanks placed as back-to-back pairs within 48 gaps between the magnet coils, which appear as rib-like structures in the accompanying photo. This number of gaps was chosen because of its large number of even divisors, which provided for greater flexibility in the use of power-supply equipment. The linear structure running across the top of the photo holds the busbar, which was made of square-foot solid silver. The vacuum tanks in these tracks were C-shaped and could be withdrawn on special gantries for material extraction and maintenance. In contrast, Beta units were laid out as two parallel rows of 18 tanks each, sandwiched between D-shaped magnet coils. Beta units incorporated twin-source emitters; their tanks were half the diameter of Alpha units, but operated at twice the magnetic field strength of Alphas. Beam radii were 48 in for Alpha tanks and 24 for Betas; for a 48-in. radius, the U-235 and U-238 beams would be separated by only about six-tenths of an inch at the collector. Both Alpha and Beta units utilized accelerating electrodes operating at 35,000 V. Uranium tetrachloride was used as the feed material as it sublimes directly to a gas when heated, obviating the need to handle liquid feeds. The tetrachloride was load into "charge bottles" containing 5 kg of material for Alpha units and 800 g for Beta units; ovens operating at 475 °C vaporized the tetrachloride, which was then ionized by bombardment with electrons from a tungsten or tantalum filament.

Left: This photograph, reproduced from Manhattan District History microfilms, shows two Beta tracks, one in the foreground and one in the background. Source: A1218(10), image 0231. Right: "Calutron girls" at their operating stations. Each operator monitored the performance of two vacuum tanks, but had no idea what was being produced. Source: http://commons.wikimedia.org/wiki/File: Y12_Calutron_Operators.jpg

Lawrence was relentless in seeking improvements. In July, 1943, he began advocating for multiple-beam sources within tanks, but Groves was reluctant to authorize changes that might delay plant completion. A compromise was struck: the first four Alpha and all Beta tracks would use two-beam sources, but the fifth Alpha track would use four-beam sources. The staff of the Radiation Laboratory was expanded to take on the additional design and engineering tasks; by mid-1944 it would boast 1,200 employees. Research costs for the electromagnetic program ran to about $20 million.

Even as designs evolved, ground was broken for the first Alpha building, 9201-1, on February 18, 1943. Buildings containing Alpha tracks were designated as "9201" structures, while those containing Beta units were numbered "9204". There would ultimately be five Alpha buildings housing nine tracks, plus four Beta buildings housing eight tracks. Altogether, these 17 tracks would contain 1,152 tanks, although not all came online until after the end of the war. The first structure completed at the Y-12 site was building 9731, the "pilot plant" building. Completed in March, 1943, this building housed experimental Alpha and Beta units which were used for training operators. Designated as calutrons XAX and XBX, these units still stand in 9731, now accessible to tourists and identified as Manhattan Project Signature Artifacts by the Department of Energy's Office of History and Heritage Resources.

Left: Building 9731, the light-colored, flat-roofed building at center left, was the first building completed at the Y-12 complex. The large building is a Beta plant. Source: http://www.y12.doe.gov/about/history/getimages. *Right: An Alpha II racetrack. Source: A1218(10), image 0214*

The experimental XAX Alpha unit was first operated on August 17, 1943, by which time Groves was already considering further expansions. After reviewing design improvements at a meeting in Berkeley on September 2, he presented his plan to the Military Policy Committee on September 9. Four additional 96-tank Alpha tracks, called Alpha-II units, would be constructed. They would differ from the original oval configuration in being of rectangular layout like Beta units; tanks at the curved portions of the oval-shaped units, which would be re-named Alpha I units, proved difficult to regulate. In actual operation, the Alpha-II units never performed as satisfactorily as expected, apparently due to complexities introduced in going from two to four ion streams. Two more Beta tracks were also authorized. Production of the vacuum tanks, sources, and collectors was contracted to Westinghouse; General Electric took on responsibility for high-voltage electrical controls, and the magnet coils themselves were fabricated by the Allis-Chalmers company.

One of the most unique aspects of the electromagnetic program was its use of Treasury Department silver to make magnet coils. Normally, copper would have been used, but since that metal was used in shell casings it was a high-priority commodity during the war. Congress had authorized use of up to 86,000 tons of Treasury Department silver for defense purposes; not having to divert large amounts of copper was a boon for the Project's secrecy. Kenneth Nichols met with Undersecretary of the Treasury Daniel Bell on August 3, 1942, to inquire about borrowing 6,000 tons of silver from the Treasury's vaults; Bell informed Nichols that the Treasury's preferred unit of measure was the troy ounce, which is about 10% heavier than a common one-sixteenth-of-a-pound ounce. Secretary of War Henry Stimson formally requested the silver in a letter to Treasury Secretary Henry Morgenthau on August 29. Stimson gave no indication what the silver was to be used for, stating only that the project was "a highly secret matter." His letter stipulated silver of purity 99.9%, and assured Morgenthau that title to the silver would remain with the United States. The deadline for returning the metal was five years from its receipt, or upon written notice from the Treasury that all or any part of it was required for reasons connected with monetary requirements of the United States.

The War Department eventually withdrew more than 400,000 bullion bars of approximately 1,000 troy ounces each from the West Point Bullion Depository in West Point, New York. The first bars were withdrawn on October 30, 1942, and trucked about 70 miles south to a U.S. Metals Refining Company facility in Carteret, New Jersey, where they were cast into cylindrical billets weighing about 400 lb each. By the time casting operations ceased in January, 1944, just over 75,000 billets weighing nearly 31 million pounds had been cast. Remarkably, this weight exceeded the 29.4 million pounds withdrawn from the Treasury. Groves insisted on careful cleanup operations: workers coveralls were vacuumed clean, and machines, tools, furnaces, factory floors, and storage areas that had accumulated years of metal shards were dismantled and scraped clean. Armed guards observed every step in the processing to ensure that all trimmings were recovered. The recovery operation was so successful that more than 1.5 million pounds of silver were gained versus less than 11,000 which were considered lost.

After being cast, the billets were trucked a few miles north to a Phelps Dodge Copper Products Company plant in Bayway, New Jersey. There they were heated and extruded into strips 3 in wide by 5/8 in thick by 40–50 feet long; if all of the Manhattan Project silver was shaped into one strip of that width and thickness, it would reach from Washington to outside Chicago. After being cooled, the strips were rolled to various thicknesses, depending on the particular magnet coils for which they were intended. They were then formed into tight coils (not yet the magnet coils) that were about the size of large automobile tires. Over 74,000 coils were produced, most of which were shipped to Wisconsin for magnet-coil fabrication at Allis-Chalmers in Milwaukee. There they were unwound, joined together with silver solder, and fed into a special machine that wound them around the steel bobbins of the magnet casings. Between February, 1943, and August, 1944, 940 coils were wound, each containing on average about 14 tons of silver. Separately, some

268,000 lb of silver were sent directly to Oak Ridge to be formed into busbar pieces.

This Manhattan District History photo shows magnet coils being wound onto square bobbins, likely Alpha I coils. Note person in lower right foreground for scale. Source: A1218(10), image 0443

Stone and Webster constructed Y-12, but the plant had then to be operated. For that task, Groves contracted with the Tennessee Eastman Corporation, a subsidiary of the Eastman Kodak Company. Eastman's contract was on a cost-plus-fee basis: a basic stipend of $22,500 per month plus $7,500 for each track up to seven, plus an additional $4,000 for each track over that number.

By the summer of 1943, construction was in full swing at Y-12. Stone and Webster's construction payroll hit 10,000 by the first week of September, and would peak at about 20,000. Overall, the company would interview some 400,000 people for construction jobs; building Y-12 would consume 67 million man-hours of labor. Tennessee Eastman began training operators; by November some 4,800 were ready. Ernest Lawrence was awed by the scale of Y-12, relating that "When you see the magnitude of that operation there, it sobers you up and makes you realize that whether we want to or no, that we've got to make things go and come through … Just from the size of the thing, you can see that a thousand people would just be lost in this place, and we've got to make a definite attempt to just hire everybody in sight and somehow use them, because it's going to be an awful job to get those racetracks into operation on schedule. We must do it." Despite the pace of work at Clinton, the site's safety record was remarkable; through December, 1946, only eight fatal accidents occurred: five by electrocution, one by gassing, one by burns, and one fall.

Construction at Y-12, 1944. Source: http://
www.y12.doe.gov/about/history/getimages.php

Problems began to emerge in the fall of 1943, however. Operators had trouble maintaining steady ion beams; electrical failures, insulator burnouts, and vacuum leaks were endemic. Some of the steel vacuum tanks, which weighed about 14 tons, were pulled several inches out of line by magnetic forces, putting stress on vacuum lines; the solution was to secure the tanks to the floor with steel straps. Soon after the first Alpha track was started on November 13, it had to be shut down due to electrical shorts caused by coil windings being too close together and insulating oil being contaminated with rust and sediments. Furious, Groves arrived on December 15 to personally review the situation. The only option was to ship 80 coils back to Milwaukee for rebuilding, while modifying designs to include oil-filtration systems. Refurbishing the coils cost over $470,000.

As the magnets from the first Alpha track were being rebuilt, the second track entered service on January 22, 1944. Despite seemingly endless breakdowns, performance gradually improved as experience was gained by maintenance and operating personnel. By the end of February, it had enriched about 200 g of material to 12% U-235, some of which went to Los Alamos while the remainder was used as feed for Beta calutrons. The rebuilt first alpha track re-entered service on March 3, and the first Beta unit began operation in mid-March. Lawrence began advocating for another expansion, proposing that four new Alpha tracks be added to the nine already authorized. Groves did not authorize any additional Alpha tracks, but did decide to proceed with two more Beta buildings (four tracks), in part to receive partially-enriched material from the gaseous diffusion plant. Construction on the third Beta building began on May 22; coils in these tracks were made with conventional copper windings.

During routine operation, Alpha tracks would be shut down about every tenth day to recover their uranium, and Beta tracks about every third day. After shutdowns, it would take a full day to restore the vacuum to Alpha tanks, and three to four hours for Beta tanks. It took time for productivity to settle into a routine. During the first months of 1944, not more than about 4% of the U-235 in the Alpha sources was making its way to the receivers; for Beta stages the fraction was little better at about 5%. Losses were due mostly to low ionization efficiency of the uranium tetrachloride feed material, and dissociative processes that yielded species other than just singly-ionized molecules. Much of the feed material ended up splattered around the insides of the vacuum tanks, which had to be scraped clean and washed over catchment sinks. Material that adhered to components that were too costly or awkward to pull out and clean was abandoned. More prosaic problems also cropped up. In one case, a mouse became trapped in a vacuum system, preventing proper pump-down. Several days of production were lost, as was the mouse. In another, what Groves described as a "suicidal" bird perched on an insulator outside the building housing Alpha tracks 6 and 7, causing a short. The bird received 13 kV, and the entire building was shut down.

Improvements accumulated through late 1944. Alpha process efficiencies eventually approached about 11%, and Beta 15%. Between October 21 and November 19, U-235 production amounted to 1.5 kg, an amount nearly equal to that of all previous months combined. By December 15, all nine Alpha tracks and Beta tracks 1, 2, and 3 were in operation, Beta tracks 4 and 6 were processing unenriched Alpha feed, and Beta 5 was being used for training. Y-12 operated on an around-the-clock basis.

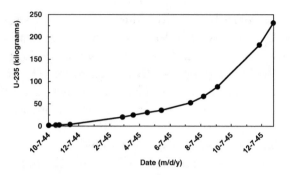

Cumulative production of uranium-235 from Beta stages of the Y-12 plant through early 1946. Data from National Archives and Records Administration microfilm set A1218 (Manhattan Engineer District History), Roll 10 (Book V—Electromagnetic Plant, Volume 6—Operations, Top Secret Appendix, p. 4.)

With all uranium enrichments methods finally coming on-line in early 1945, Groves began to think of harnessing them in series as opposed to treating them as competing in parallel. Following calculations of how to optimize the rate of production of bomb-grade material, the decision was made on February 26 to begin the process by first feeding natural-abundance uranium hexafluoride ("hex") to the S-50 thermal diffusion plant, which would enrich the U-235 content from 0.72 to 0.86%. This product would be fed to Alpha calutrons, but when the gaseous diffusion cascade had advanced to the stage of producing 1.1%-enriched material, the S-50 product would be fed to it to be enriched to that level, after which material would go to Y-12 Alpha units. When enough diffusion stages were on-line to produce 20% enriched material, the Alpha I units would be shut down and K-25's product would be directed to Alpha II tracks, after which it would go to Beta units. Since the various plants at Clinton could achieve different but overlapping levels of enrichment, the sequence of feed steps was adjusted constantly as they came on-line. The S-50 plant could raise the enrichment level from 0.72 to 0.86%, the Alpha stages of the Y-12 plant from 0.72% to about 20%, Beta stages from 20 to 90%, and K-25 from 0.72 to 36%. Y-12's enriched uranium tetrachloride was converted to uranium tetrafluoride for shipment to Los Alamos, with chemical processing carried out in gold trays to minimize contamination. The precious product, accounted for to fractions of a gram, was packed into gold-plated nickel cylinders about the size of coffee mugs, which were then placed into cadmium-lined wooden boxes. The boxes were secured two at a time inside briefcases, which were chained to the wrists of armed Army couriers for a two-day train trip to New Mexico. By April, 1945, Y-12 had produced some 25 kg of bomb-grade U-235; by mid-July the total would reach just over 50 kg. At peak production, Alpha units were yielding a total of about 258 g per day of 10%-enriched material, and Beta units were producing about 200 g per day of material enriched to at least 80%, better than the 100 g per day specified at the Bohemian Grove meeting.

Clinton required an enormous amount of electricity. Each Alpha racetrack consumed 4,580 kW, and each Beta 1,250 kW. By mid-1945, transmission facilities at CEW could provide power at a peak rate of 310,000 kW, of which 200,000 were for Y-12 alone. Peak consumption, nearly 299,000 kW, occurred on September 1, 1945. The total amount of electricity consumed by Y-12 between November, 1943, and the end of July, 1945, totaled to over 1.6 billion kilowatt-hours, about 100 times the energy released by the U-235 *Little Boy* bomb. At its peak of operations in the summer of 1945, Clinton was consuming about one percent of all electrical power produced in the entire United States, much of it flowing through Lawrence's calutrons. Every atom of uranium in the *Little Boy* bomb would pass through at least once through a calutron.

6.2   K-25: The Gaseous Diffusion Program

The K-25 gaseous diffusion complex was the single most expensive facility of the entire Manhattan Project, and also one of the most difficult to design, engineer, and construct. While gaseous diffusion would eventually prove to be the most economical method of enriching uranium, it was nearly stillborn. Its importance is evidenced by the fact that even decades later, the process for fabricating the diffusion membrane is still considered so highly-classified that little information is available regarding its manufacture.

The fundamental principle of gaseous diffusion is that if a gas of mixed isotopic composition is pumped against a porous barrier containing millions of microscopic holes, atoms of lower mass will on average pass through slightly more frequently than those of higher mass. The result is a very minute level of enrichment of the gas in the lighter-isotope component on the other side of the barrier. Since only a small enrichment factor can be achieved in any one step, the slightly-enriched gas has to be pumped on to subsequent enriching stages. By linking together a number of processing "cells" in series in a cascade, bomb-grade material can eventually be isolated. The gas which emerges from each stage slightly "depleted" in the lighter isotope still contains atoms of that isotope, however, and can be sent back "down" the cascade for additional processing.

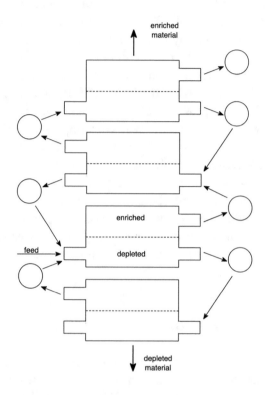

Schematic illustration of a diffusion cascade. Feed material enters the cascade in the second "cell" from the bottom of the diagram. The dashed lines inside each cell represent the diffusion membrane, and the circles represent pumps. Gas enriched in the lighter isotope accumulates toward the top of the diagram, while that depleted in the lighter isotope accumulates toward the bottom. In reality, the cascade is not arranged vertically as this diagram suggests; in the K-25 plant, all cells were at ground level. Sketch by author

The enrichment capability of each cell is dictated by the statistical mechanics of diffusion, and depends on the characteristics of the input gas and the ratio of the masses of the heavy and light isotopes to be separated. In the case of K-25, the feed gas used was uranium hexafluoride, UF_6, for which the maximum possible increase obtainable in the light-to-heavy-isotope abundance ratio in any one diffusion stage is only 0.86%. To achieve a nine-to-one output ratio (90% enrichment) requires over 1,600 consecutive stages. But in practice the process is never as efficient as the theory implies: not all of the material that enters each stage undergoes diffusion, and some gas will naturally diffuse back through the barrier. Detailed calculations indicated that the best arrangement from the point of view of plant size and power requirements would be one in which only half of the gas that is pumped into each stage diffuses through the barrier, with the depleted half being recycled to the preceding stage. In an actual plant, the volume of gas that emerges from the end of the cascade may be only about a hundredth-thousandth that fed in. The number of stages built below the feed point is dictated by a judgment of how economical it is considered to be to continue processing depleted material; in the K-25 plant, the feed point was about one-third of the way along the cascade. It is no wonder that, as Henry Smyth described in his 1945 history of the Project, "many acres" of barrier are needed for a large-scale plant.

The diffusion barrier had to be robust and easy to manufacture, but its most important feature is the size of the diffusion holes. At atmospheric pressure, the distance that a molecule will travel on average before colliding with another is on the order of a ten-thousandth of a millimeter. To achieve true diffusion, the diameter of the holes in the barrier should be no more than about one-tenth of this figure, equivalent to about a ten-thousandth of the diameter of a typical human hair.

Diffusion was well-known to chemical engineers, so it is not surprising that it became the object of attention as a possible enrichment method once understanding of the crucial role of U-235 in the fission process had become appreciated. In late 1940, John Dunning, Eugene Booth, Harold Urey, and mathematician Karl Cohen, all at Columbia University, developed a barrier material from partially fused glass, but it could not stand up to the corrosive effects of uranium hexafluoride. When Vannevar Bush reorganized the project to appoint Program Chiefs in November 1941, Urey was formally designated to lead diffusion work. By that time, Dunning and Booth were experimenting with creating a porous metallic barrier by etching zinc from a sheet of brass, and had succeeded in enriching a small amount of uranium. (Brass is an amalgam of copper and zinc; etching away the zinc rendered the sheet porous). At Oxford University, Franz Simon was developing 10-stage cascade models to test different pumping schemes.

At their May 25, 1942, meeting, S-1 administrators advocated proceeding with a diffusion pilot plant and engineering studies for a kilogram-per-day full-scale plant, recommendations which Vannevar Bush took to President

Roosevelt on June 17. By late October, Booth had a 12-stage demonstration system in operation which achieved a small enrichment of uranium hexafluoride during a five-hour run. The Columbia system involved face-to-face cylinders about four inches in diameter, with dollar-coin-sized barrier samples placed between the faces; the entire assembly fit in a cabinet about eight feet square and three feet deep. When the Columbia work came under Manhattan District auspices, it became christened as the Substitute (or, by some sources, Special) Alloy Materials (SAM) Laboratory. Following the decision to proceed with a kilogram-per-day diffusion plant, research shifted into high gear. By the end of 1943, Urey had over 700 people working on diffusion problems at Columbia alone, plus several hundred more at other universities and industrial laboratories.

When the M. W. Kellogg Company took on the design and engineering of the diffusion plant in late 1942, no suitable barrier material had been developed. Ultimately, development of a useable barrier would prove one of the most difficult aspects of the entire Manhattan Project. The uranium hexafluoride process material had the advantage that it could easily be made into a gas, but it is extremely caustic. The barrier would have to be strong enough to withstand both the corrosive effects of the gas and the high pressures involved. The only element that can withstand the caustic effects of UF_6 is nickel, and in late 1942 a group at Columbia under the direction of Foster Nix of the Bell Telephone Laboratories turned their attention to experiments involving compressed nickel powders. These barriers proved sufficiently rugged, but insufficiently porous. In contrast, fine-enough holes could be realized with an electro-deposited mesh, but the mesh was not particularly strong. The mesh had been developed by Edward Norris, an interior decorator, as part of a paint sprayer that he had invented. Norris joined the Columbia group in late 1941, and by January, 1943, he and chemist Edward Adler had developed a material which looked to have the correct combination of porosity and strength. Construction of a six-stage pilot plant to hand-produce the Norris-Adler barrier was begun at Columbia in February; it would begin operating in July. For a full-scale plant, however, piecework would be impractical; several million square feet of barrier would be required, which meant industrial-scale production. In April, 1943, the Houdaille-Hershey Corporation, a manufacturer of automobile accessories with considerable experience in plating techniques, took on a contract to build and operate a $5-million barrier-production plant to be located in Decatur, Illinois, on the premise that the Norris-Adler barrier would prove amenable to mass production. The diffusion tanks themselves, some as large as 10,000 gallons, were manufactured by the Chrysler Corporation, which had to develop techniques for nickel-plating the insides of the tanks. Over the course of two years, Chrysler would plate some 63 acres of steel surface.

Kellogg's contract with the Army was unusual. The firm was not required to make any guarantee that it could design, build, and get a plant into operation. Financial terms were left unspecified until the work was further developed; the company eventually accepted a fee of about $2.5 million for its efforts. A separate corporate entity, the Kellex Corporation, was set up to carry out the work. As Henry Smyth described it, Kellex was a unique temporary cooperative of scientists, engineers, and administrators drawn from a number of schools and industries for the express purpose of carrying out one job. Percival Keith, a Kellogg Vice President and MIT chemical-engineering graduate, was designated to be in charge of the new corporation, which by 1944 would have some 3,700 employees. The firm also began undertaking its own barrier research, as well as construction of a 10-stage pilot plant in Jersey City, New Jersey, which would eventually be used to test full-size diffusion tanks under simulated operating conditions.

American and British ideas on plant design conflicted. Karl Cohen's 4,600-stage analysis was predicated on a high-pressure, high-temperature single-cascade operation, whereas the British proposed a cascade-of-cascades arrangement which would operate at lower temperatures and pressures. The British approach would be more complex to engineer, but would place less stringent demands on the barrier material and would have the advantage of a shorter equilibrium time. On the other hand, a single-cascade design could be more easily configured to permit the system to be plumbed so that process material could be fed into or drawn from any stage as desired. In his Appendix to Arthur Compton's third National Academy of Sciences report, Robert Mulliken estimated the equilibrium time for a low-pressure plant to be 5–12 days, as opposed to 100 days for a higher-pressure design.

The barrier was not the only issue Kellex faced. Since uranium hexafluoride reacts explosively with grease and moisture, neither could be allowed to mix with the process gas along the miles of pipes that a plant would require. How then to lubricate the thousands of pumps and valves that would be involved? The solution turned out to be a water-resistant chemical, polytetrafluorethane, now known as Teflon. Itself a fluorine compound, Teflon resists attack by fluoride compounds, and has one of the lowest coefficients of friction known of any solid material. Another problem was that the plant would require some 7,000 pumps. But when gases are compressed, they naturally heat up; the pumps had to be cooled as they operated, again with all seals vacuum-tight and non-leaking. K-25's pumps were supplied by the Allis-Chalmers Company, which also manufactured the Y-12 magnet coils.

By mid-1943, work was proceeding on the Decatur plant and surveying was underway for the K-25 plant. The barrier issue was approaching crisis proportions. The Norris-Adler barrier, for which the Decatur plant was being configured, was proving brittle, plagued with pinholes, and difficult to manufacture in uniform quality. A key advance was made in June of that year when Clarence Johnson, a Kellex engineer, developed a new barrier using a method coyly described in official histories as combining the techniques of Norris, Adler, and Nix. A Military Policy Committee meeting held on August 13, 1943, concluded that a suitable barrier would probably be forthcoming, but research would have to continue on both the Columbia (Norris-Adler-Nix) and Kellex (Johnson) processes for the time being.

By the end of 1943, the time to make a decision was approaching. At that point, Groves did something unusual in turning to British scientists to review the situation. On December 22, he met in New York with a 16-strong British contingent which included Franz Simon and Rudolf Peierls. The group was briefed by representatives from Kellex and Columbia, following which they adjourned to visit various laboratories before preparing their report, which was considered at a four-hour meeting at Kellex headquarters on January 5, 1944. The British felt that Johnson's barrier would be easier to manufacture and likely eventually prove superior to the Norris-Adler version, but, if time was the determining factor, the research already accumulated on the latter represented an important advantage. Houdaille-Hershey, however, was becoming pessimistic about being able to produce the Norris-Adler barrier on a large scale. Kellex engineers countered that even with a switch to the Johnson barrier, they could have K-25 in operation by Groves' target date of July 1, 1945. Groves announced his decision at a visit to Decatur on January 16: the plant would be converted to fabricate the Johnson barrier.

As with Y-12, construction and operations were handled by two different contractors. Groves needed an operating contractor for K-25, and two days after his Decatur visit convinced the Carbide and Carbon Chemicals Corporation, a subsidiary of Union Carbide, to take on that task on a cost-plus-$75,000-per-month basis. The company appointed one of its vice-presidents, physical chemist and engineer George Felbeck, to be its K-25 project manager.

K-25 would require enormous amounts of steam to heat the process material and operate pumps. A concern in this regard was that a power interruption would not only delay production, but could set up pressure waves that could reverberate through the cascade and damage equipment. Fearing interruptions or sabotage, Groves did not want to rely on the TVA for electrical power, and decided that K-25 would have its own 238-MW steam-electric generating plant (enough to power a city the size of Boston), which would feed the main plant through protected underground cables. To construct the generating plant and the main K-25 plant itself, Groves chose the J. A. Jones construction company of Charlotte, North Carolina. He was familiar with the firm; Jones had built more Army camps than any other contractor in America. Jones would ultimately engage over sixty subcontractors in what was one of the largest construction projects in the world to its time. Work got underway in May, 1943, with a surveying party laying out a site for the generating plant on the bank of the Clinch river; surveying for the K-25 plant itself got underway later that month. When work on the power plant was begun, design of pumps for K-25 had not been settled; the plant had to be designed to provide power at five separate frequencies. Despite this complication, it came online on Mach 1, 1944, only nine months after construction began.

Groves' original intent had been that K-25 would be capable of producing 90% U-235, using diffusion stages incorporating barriers in the form of annular bundles. However, detailed calculations indicated that available pumps would be most efficient up to an enrichment of 36.6%, beyond which a different cell design and other pumps would be required. This prompted Groves, in early 1943, to consider limiting K-25 to about 36% enrichment, with its product to be fed to the Beta calutrons then being authorized. Groves formally announced the cutback at the August 13 MPC meeting, and asked Kellex to supply estimates on when 5, 15, 36.6, and 90% plants might be expected to come into operation; the 90% figure would involve a later extension plant known as K-27.

K-25 was constructed in a 5,000-acre area in the northwest corner of the Clinton reservation, about 15 miles southwest of Oak Ridge. To provide for a flat working area, almost 3 million cubic yards of earth were moved. Construction on the main building was begun on September 10, 1943, and the first concrete was poured on October 21. K-25's dedicated temporary construction camp was idyllically known as Happy Valley; its population would peak at about 17,000.

The four-floor main process building, laid out in the shape of a giant letter U, was enormous. Each side section was 2,450 feet long (just under a half-mile) by 450 feet wide; the total width exceeded 1000 feet. Some 12,000 construction drawings detailed a facility with a total floor area of about 120 acres, some 80% of that of the Pentagon. Over 500 miles of pipes and a half-million valves would be involved, with the latter varying in size from 1/8 to 36 in. Kellex planned for a total of 2,892 diffusion stages. The construction force peaked at just over 19,600 in April, 1944; by June the plant was

37% complete and the estimated cost had risen to $281 million. Ideally, as increasingly-enriched uranium accumulated toward the upper "end-stages" of the cascade, pumps and cells of steadily decreasing sizes could be used. As this would have involved complex and expensive manufacturing, Kellex decided on five sizes of pumps and four types of cells. The building itself comprised 54 sub-buildings, with the cascade divided into nine "sections," which, although they would normally operate as part of an overall cascade, could be operated individually. The fundamental operating entity was a "cell," a unit of six individual diffusion tanks.

An aerial view of the K-25 plant. Source: http://commons.wikimedia.org/wiki/File:K-25_Aerial.jpg

The basement housed lubricating, cooling, and electrical equipment. The diffusion tanks themselves resided on the ground floor, while the second aboveground floor served as a pipe gallery; the top floor housed operating equipment. A central control room equipped with some 130,000 monitoring instruments was located in the top floor of the base of the U. Kellex divided its construction plan into five steps, designated as "Cases." Case I, to be completed by January 1, 1945, would see through to completion one cell for testing, then a building with a 54-stage pilot-plant, and finally enough functioning plant (402 stages) to produce 0.9% U-235. Cases II, III, and IV would subsequently take the process to 5%, 15%, and 23% enrichment by June 10, August 1, and September 13, respectively. Case V, to achieve 36%, was to follow as soon as possible thereafter.

Cleanliness requirements during construction were practically at a surgical level. Workers wore special clothing and lintless gloves; even a thumbprint could leave enough moisture to be disastrous. Areas where process piping was being installed were equipped with pressurized ventilation and fed with filtered air. Some pieces of equipment required up to ten separate cleaning steps to remove all traces of dirt, grease, oxides, and moisture. Welding, which eventually involved 1,200 machines in simultaneous operation performing 14 specialized techniques, was done inside inflatable balloon enclosures. Since the entire plant would have to be constantly monitored for leaks, inert helium gas was fed into the piping system, and its presence sniffed for by sensitive portable mass spectrometers developed by Alfred Nier. Hundreds of Nier's devices were manufactured by General Electric and deployed throughout the plant; Nier was also involved in developing a system of over 50 fixed devices used to monitor the flow of various chemicals at locations throughout the building. These devices reported data back to the central control room, from where a single person could monitor the entire plant. Another challenge was that projected nickel requirements for piping exceeded the entire world production of that metal. Again drawing on his knowledge of industrial firms, Groves contracted with Bart Laboratories in New Jersey, which specialized in electroplating oddly-shaped objects. Bart engineers were able to develop a method of electroplating the insides of pipes by using the pipe itself as the electroplating tank; rotating the pipe as current was passed through molten nickel ensured a uniform deposit of metal.

Progress was detailed in monthly reports from Nichols to Groves. On April 17, 1944, the first six-stage cell was operated briefly as part of a preliminary mechanical test. By May, barrier of sufficient quality was beginning to become available; quantity production began in June. By August, operators could begin training at the 54-stage pilot plant located at the base of the U, using nitrogen in place of UF_6. On September 22, the first four diffusion tanks were received from Chrysler, but two were returned for tests of the effects of railroad handling. By November 9, the first dozen tanks were installed. A month later, Chrysler had shipped 324 tanks, of which about 200 were installed. By the end of 1944, the plant was 65% complete, and 60 of the 402 stages of Case I were ready to be turned over to Carbide operators. By early January, 1945, all tanks necessary for Case I had been received, and Chrysler was producing 65–70 per week; by the end of the month the total number shipped would near 800.

After a period of leak testing and instrument calibration, the first process gas was introduced into the system on January 20, 1945. On March 10, Nichols reported that 102 of the 402 stages in Case I were in "direct recycle" operation, and that almost 1,100 tanks had been received. By March 12, two more buildings were connected to the system, and on the 24th, all of Case I went on-line. By early April, just over half of the total 2,892 tanks had been received, and Cases I and II were producing 1.1%-enriched U-235, which signaled that the facility could begin receiving its first slightly-enriched feed from the S-50 thermal diffusion plant. This occurred on April 28, by which time over 1,500 tanks were installed or ready for installation. By early June, all tanks had been shipped, nearly 1,500 were in operation, and K-25 was feeding 7%-enriched product to Beta calutrons. On August 7, the day after the Hiroshima bombing, Nichols reported over 2,200 stages in operation. His report for August operations, dated September 6, indicted that all 2,892 stages were in operation by August 15, the day after the Japanese announced their surrender. When the entire plant was operating, enrichment increased to 23%.

Ultimately, K-25's product was not limited to 36% enrichment. In early 1945, Kellex developed plans for a 540-stage "extension" plant, which came to be known as K-27. By mixing waste output from the main K-25 cascade with natural uranium, K-27 produced a slightly enriched product which could be fed to the upper stages of K-25, increasing both its production and enrichment. Groves authorized construction of K-27 on March 31, 1945; it entered full operation in February, 1946, by which time all enrichment operations were being conducted by gaseous diffusion.

K-25 was a remarkable engineering accomplishment. While the plant really came into its own only after the close of the war, Groves' gamble bequeathed America a means of enriching uranium that would operate flawlessly for years thereafter.

6.3   S-50: The Thermal Diffusion Program

The Manhattan Project's liquid-thermal-diffusion program has tended to be regarded as being of second-team status when compared to its much more gargantuan electromagnetic and gaseous-diffusion cousins. The S-50 plant was erected hastily, operated for only a short time, and enriched uranium by only a small degree (from 0.72 to 0.86% U-235), but its contribution was vital in giving the trouble-plagued electromagnetic enrichers a head-start on their efforts.

The prime mover behind the thermal diffusion method was Ernest Lawrence's graduate student, Philip Abelson, who, among others, confirmed the discovery of fission at Berkeley in early 1939. Abelson formally received his Ph.D. in May of that year, just a few months after his fission-confirmation work. He remained in Berkeley over the summer to complete some work on radioactive decays, and in September moved to Washington, D.C., to take up a position that Merle Tuve offered him at the Carnegie Institution of Washington. In the spring of 1940, he took a brief leave to return to Berkeley to complete his neptunium-discovery work with Edwin McMillan, efforts which directly motivated Glenn Seaborg to search for plutonium. After returning to Washington, Abelson began to consider possible approaches to enriching uranium, and after reviewing the research literature decided to explore liquid thermal diffusion.

There was considerable political wrangling between the Army and the Navy over the liquid diffusion method. Development was begun by the Navy, but it was later appropriated essentially wholesale by the Manhattan District. District documents include a brief summary of the Navy work to late 1942, but them jump abruptly to the S-50 project proper in mid-1944. However, there are now available a number of sources that fill in the gaps in this history, in particular two Naval Research Laboratory reports which list Abelson as first author. The first, dated January 4, 1943, describes progress up to that time, at which point the NRL had a small pilot-plant running. The second, dated September 10, 1946, covers the engineering theory of thermal diffusion plants and the full history of the method between 1940 and 1945.

The fundamental principle on which liquid thermal diffusion is based is that if a fluid—which in this context can be a liquid or a gas—comprising two isotopes of an element is subjected to a thermal gradient, the lighter isotope will accumulate toward the hotter region, while the heavier will collect toward the cooler region. As a consequence, fluid containing the lighter isotope will be of lower density and will rise by convection, while that containing the heavier isotope will fall. Competition between this effect and the ordinary diffusion of isotopes through each other leads, after some hours or days, to equilibrium between the two processes. The theory of thermal diffusion was first developed by David Enskog in Sweden (1911) and Sydney Chapman in England (1916); its experimental proof was established by Chapman and F.W. Dootson in 1917. In Germany, Klaus Clusius and Gerhard Dickel first used a "column" approach in 1938 by placing a hot wire along the axis of a vertical tube, and achieved a small enrichment of neon isotopes. Soon thereafter, Arthur Bramley and Keith Brewer of the U.S. Department of Agriculture conceived the idea of using two concentric tubes at different temperatures. Abelson adopted the Bramley and Brewer approach, using steam to heat the inner tube and water to cool the outer one, while injecting the process fluid into a narrow annular space between them.

The time for a column to achieve equilibrium depends on the difference in temperature between the two tubes, their annular separation, and their lengths. The ultimate important characteristic of such a column is its separation factor, which specifies its enrichment capability. If a column has a separation factor of 1.2 (which was the case for the S-50 columns) and natural uranium is used, the percentage of U-235 after processing will be 0.720% times 1.2, or 0.864%.

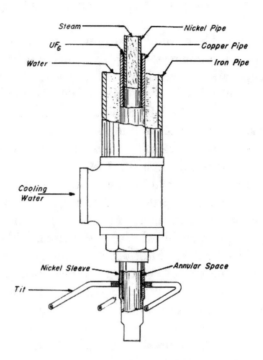

Sectional view of a thermal diffusion process column. Uranium hexafluoride (UF₆) consisting of a mixture of light (U-235) and heavy (U-238) isotopes is driven into the narrow (0.25 mm) annular space between the nickel and copper pipes; the nickel pipes were 1.25 in outside diameter. The desired lighter-isotope material is harvested from the top of the column. At the top and bottom of each tube, three small projecting "tits" provided access to the annular space for supply and withdrawl of material. From Reed (2011)

Abelson's 1946 report indicates that his first column experiments were carried out at the Carnegie Institution in July 1940; his goal was to repeat the German work by exploring diffusion of solutions of various potassium salts. Unfortunately, his attempt to use a solution of uranium salts produced what he called "an insoluble mess" at the bottom of the column. Merle Tuve became concerned that the experiments would produce radioactive contaminants, and began to look for another location for them. Tuve was a member of Lyman Briggs' Uranium Committee, and Briggs made space available to Abelson at the National Bureau of Standards. Abelson moved his experiments to the Bureau in October, 1940, by which time the NRL had entered into a contract with Carnegie to support the work. The NRL furnished Abelson's equipment, Carnegie paid his salary, and the Bureau of Standards provided laboratory space and an assistant chemist.

Between July 1, 1940, and June 1, 1941, Abelson constructed 11 columns of lengths between 2 and 12 feet, diameter 1.5 in, and annular separations between 0.5 and 2 mm. Experiments with water solutions of potassium salts showed that the equilibrium time and separation factor depended sensitively on the annular separation. A run with hex in a 12-foot column in April, 1941, yielded a small enrichment, but the measured value was only roughly equal to the probable error of measurement. On June 1, 1941, Abelson formally became an employee of the NRL, where a decision had been made to pursue study of liquid diffusion using 36-foot columns. These first Navy columns were collectively called the "experimental plant," to distinguish them from a later pilot plant. Abelson achieved enrichment of chlorine isotopes with his first column, but in November of that year it was ruined by decomposition products of carbon tetrachloride.

Between January and September of 1942, Abelson constructed five more experimental columns using a hot-tube temperature of 286 °C. These were built with annular spacings of 0.53, 0.65, 0.38, 0.2, and 0.14 mm, and yielded separation factors of 2% (January 1942), 1.4% (March 1), 9.6% (June 22), 21% (July), and 12.6% (September). Abelson regarded the 9.6% result as the first indisputably successful application of the method with uranium. Particularly encouraging was that the "pseudo-equilibrium time," the time for the column to produce a separation of one-half of the equilibrium value, was only eight hours. The optimal annular spacing appeared to be around 0.2 mm; a spacing of 0.25 mm would be used in the S-50 units. In July, the Navy authorized the construction of a pilot plant with fourteen 48-foot columns with annular spacings of 0.25 mm to be built at the Anacostia Naval Station in Washington.

As the Navy group gained experience, their fortunes within formal Project administration were declining. According to Hewlett and Anderson, President Roosevelt made it clear to Vannevar Bush around March, 1942, that the Navy was to be excluded from S-1 affairs. Bush had evidently had a bumpy relationship with the Navy in any event. Admiral Harold Bowen, who had been on the original Uranium Committee and was Director of the Navy's

Bureau of Engineering, had criticized the OSRD for supplanting military-service laboratories and thus diverting funding away from the NRL. Admiral Alexander Van Keuren, who became Director of the NRL in 1942, was outraged by the Army's expenditure of what he described as "astronomical sums" of money on the uranium project.

Fortunately, efforts to shut the Navy out of the work of the S-1 Committee were not entirely successful. In a letter to Conant on July 27, 1942, Harold Urey brought up Abelson's experiments, remarking that "This work has not been correlated with the other work of the Committee, for reasons that I do not understand, but efforts should be made … to be sure that the work of that laboratory ties in with the general purpose of this committee." Bush asked Briggs to get more information from Ross Gunn. In September, Briggs reported that Abelson was experimenting with 36-foot columns, and estimated that seven such columns in series would produce a doubling of the U-235 percentage. The catch was that the equilibrium time for such an arrangement, which was not specified, would be impracticably long.

By November 15, the Anacostia pilot plant was essentially complete, and by December 1 (the day before CP-1 went critical), five columns had been charged with material. The timing was propitious: in late November, the S-1 Executive Committee again reassessed enrichment methods, and decided to include the work at the NRL in its review despite its being officially orphaned. Groves, Warren Lewis, and three DuPont employees visited the Anacostia plant on December 10, which for two full weeks between December 3 and 17 ran continuously with no shutdowns and a minimum of human intervention. On the 12th, Lewis wrote to Conant that the NRL work "is certainly of such interest that the development work ought to be continued intensively." He went on to report that the NRL workers expressed a desire for help by suitable experts, and told them that he would do anything he could to make "such men available through the NDRC." Conant replied on December 14, indicating that he would see if anything could be "done along these lines."

To work around Roosevelt's injunction to exclude the Navy, Bush wrote to Rear Admiral William Purnell (Military Policy Committee) on December 31 to express "the hope that the work of the Naval Research Laboratory can be expedited so that a comparison can be made with other processes, and that … the S-1 Executive Committee will do all it can to help." Noting that the Lewis committee felt that the NRL needed further facilities and manpower, Bush declared that "I would feel much gratified if you found it possible in some way to aid the [NRL]," and added that "Dr. Briggs has already undertaken to assure that any information that we have that can be of service to

NRL ... is made available to them." Purnell sent Abelson's reports to Conant, who had them reviewed by Briggs, Urey, and Eger Murphree, which group he also asked to visit the Anacostia facility. Accompanied by Lewis and chemical engineer William I. Thompson of Standard Oil, they did so, and submitted a report on January 23, 1943. Their assessment was that the NRL had made excellent progress, but they had concerns over a lack of solid production data: no appreciable amount of material had as yet been withdrawn from the columns. The Navy group envisioned as most realistic a plant of 21,800 columns of length 36 feet, which would produce one kilogram per day of 90% U-235. Individual columns would have a separation factor of 1.307, and their equilibrium time would be 625 days. Construction and operation costs for 625 days were estimated at $72 million. As with K-25, an important requirement was that the heated inner tubes of the columns would have to be made of nickel. But even with appropriate strategic-materials priorities, product could not be expected until some 38 months following a decision to proceed, which would mean early 1946.

On January 25, Murphree wrote to Briggs to emphasize that thermal diffusion could serve as an alternative to the initial stages of the K-25 plant. Briggs forwarded this idea to Conant, who on the 27th recommended that the NRL group should obtain more data and that an engineering group should study the process. Groves forwarded the documents to another review committee comprising Lewis and several DuPont executives, including Crawford Greenewalt. They did not concur that thermal diffusion should become a substitute for gaseous diffusion, but did recommend continued research and preliminary engineering studies. The S-1 Executive Committee endorsed this conclusion on February 10. On the 19th, Murphree and Urey proposed a program of experiments which would include testing the reproducibility of results for different tubes and drawing samples in order to quantify the approach to equilibrium. Briggs sent a copy of the proposal to Conant, and on 23rd followed up by suggesting to Conant that the S-1 Committee hoped that he would transmit the proposal to the Director of the NRL. Conant relayed this request to Groves the next day, but he took no immediate action—his hands were likely full at the time with getting construction of the Y-12 and K-25 plants underway and with finding a site for the Los Alamos Laboratory. Also, his thinking at the time was directed to enrichment methods that would turn natural uranium into bomb-grade product in essentially one step as quickly as possible; the idea of using different enrichment methods in tandem had not yet emerged.

By the time Abelson, Gunn, and Van Keuren prepared their January, 1943 report, nine columns had been constructed at the Anacostia facility. Six were already operating, some for up to 500 h. The earlier 36-foot "experimental" columns had been dismantled and checked for corrosion; none was found. Between February and July, the NRL group constructed 18 columns, which operated for a cumulative total of 1,000 days. By September, they had produced some 236 lb of slightly enriched hex, which they sent to the Metallurgical Laboratory in Chicago.

Groves kept himself informed of progress at Anacostia. On July 10, 1943, he wrote Conant that "progress at the Naval Research Laboratory ... has reached a point where it will be desirable to have this situation reviewed by the S-1 committee," and asked Conant "to take charge of this review and render a report." Conant notified Admiral Purnell that he proposed to appoint a committee consisting of Lewis, Urey, Murphree, and Briggs to again review the NRL work, expressing his hope that the NRL "would not regard such a visitation as an intrusion but rather as one more indication of the desire of the S-1 Committee to be of any assistance ... to the group which is doing such interesting and excellent work." On September 8, the committee conveyed to Conant the same concerns regarding cost, steam requirements, and long equilibrium times as they had in January, but did favor the S-1 Committee and the Manhattan District supporting work on improving the efficiency of the process. Such support never materialized.

Abelson and his group pressed on, proposing the development of a larger pilot plant or small production plant for the explicit purpose of "providing insurance against the complete failure of the Manhattan Project." Such a plant would require far more steam at higher pressures than was available at the Anacostia station, so they undertook a survey of other naval establishments. This quickly focused in on the Naval Boiler and Turbine Laboratory at the Philadelphia Naval Yard. Van Keuren, Abelson, and Gunn visited the site on July 24, 1943, and on November 17 formal orders were signed to authorize construction of a 300-column plant. They decided to first proceed with a 100-column installation (strictly, a "rack" of 102 columns) on the rationale that such a basic unit could be duplicated as desired if expansion was warranted. Construction at Philadelphia began about January 1, 1944, with completion scheduled for July. The plant's 48-foot columns were to be operated as a cascade of seven stages, which was expected to deliver about 100 g of product per day at a concentration of 6% U-235. The inner nickel tubes of the columns were formed from four 12-foot columns welded together, with nickel spacer buttons spot-welded to the tubes at 90° intervals at 6-in spacings. Hung from steel racks, the tubes were heated by feeding condensing steam at the tops of their interiors; condensate was removed from the bottom for recirculation. The outer copper tubes were cooled by water flowing upward between them and external 4-in iron tubes. When operating at a hot-wall temperature of 286 °C, about 1.6 kg of material resided within a single 48-foot column at any time. The power consumption for producing the steam was substantial, about 11.6 megawatts for one 102-column rack.

The circumstances of the resurrection of Manhattan District interest in thermal diffusion had an almost comedic flavor. In Abelson's telling, it began when Briggs obtained Gregory Breit as an advisor on nuclear matters. Breit evidently knew that a high-ranking naval officer had been assigned to work at Los Alamos, and one day in early 1944 Abelson received instructions to prepare a brief summary of the NRL work and to appear at 8 p.m. on the balcony of the Warner Theatre in Washington, where he would encounter a naval officer who would whisper a code word. That officer was William S. Parsons, who was in charge of ordnance engineering for the uranium bomb at Los Alamos. In another version of the story, Hewlett and Anderson have it that Parsons visited the Philadelphia Navy Yard in the spring of 1944, and "discovered" that Abelson was building a thermal diffusion plant. Richard Rhodes has depicted the situation as more of a conspiracy between Abelson, Oppenheimer, and Parsons, with Abelson first making an effort to get information through to Los Alamos, and Oppenheimer and Parsons protecting the Navy by concocting the cover story that Parsons happened to learn of the NRL work on a visit to Philadelphia.

Left to Right: Commander William Parsons, Rear Admiral William R. Purnell, and Brigadier General Thomas Farrell on Tinian island, August, 1945. Source: http://commons.wikimedia.org/wiki/File:080125-f-3927s-040.jpg

However covered, Oppenheimer wrote to Conant on March 4 to indicate that it seemed probable that some of the isotope-separation work being carried out at the NRL might be relevant to the purification of plutonium, and asked that Abelson's reports be sent to him. Conant cleared the request with Admiral Purnell, commenting "that the chances that they will find anything of use is slight, but I hesitate to turn down the request from that hard-pressed area." Conant forwarded the reports to Oppenheimer, who alerted Groves on April 28. Oppenheimer indicated that if the 100-column NRL plant were operated in parallel, it could theoretically produce 12 kg of material per day enriched to 1% U-235, and that the method might increase the electromagnetic-plant production by some 30–40%. Groves waited until May 31 before appointing Lewis, Murphree, and Richard Tolman to investigate the situation once again. The group visited Philadelphia on May 31 and June 1, and turned in their report to Groves on June 3, just three days before the D-Day invasion of Europe. Work on the 100-column plant was well-advanced; it was expected to begin operation about July 15. The committee considered Oppenheimer's estimate of 12 kg per day of 1% U-235 to be optimistic, but felt that 10 kg per day of 0.95% U-235 was feasible.

Groves now moved with his typical dispatch. On June 5 he sent Lewis and Conant to Oak Ridge to confer with Colonel Nichols to discuss the feasibility of constructing a thermal diffusion plant there. They decided that the 238-MW powerhouse being constructed for the K-25 plant could provide sufficient steam for such a plant, at least until K-25 went into operation. At 11.6 megawatts per rack of columns, 238 megawatts could provide power for between 20 and 21 racks; 21 were built. Groves decided to proceed with construction of the plant on June 24. On the 26th, Groves, Tolman, and Lieutenant Colonel Mark Fox, who had been appointed chief of the thermal diffusion project at Oak Ridge, visited the Philadelphia installation to inspect it and collect blueprints. The next day, Groves contracted with the H.K. Ferguson Company of Cleveland, Ohio, to construct the plant in 90 days; a second contract with Ferguson would follow for its operation. Groves initially demanded that the plant be in full operation in four months, with its first production unit operating 75 days after the beginning of construction. In a July 4 letter to Fox, he revised the schedule to demand that all units be in operation in 90 days. The 75-day requirement would be met, but not the 90-day one.

The main S-50 process building was 522 feet long, 82 feet wide, and 75 feet high. The most pressing initial problem was to find contractors to produce the large numbers of columns; twenty-one manufacturers were consulted before the Grinnell Company of Providence, Rhode Island, and the

Mehring and Hanson Company of Washington agreed to attempt the job. The outside diameter of the inner nickel tubes had to be maintained to tolerances of ±0.0003 in, and the clearance between the nickel and copper tubes to ±0.002 in. Since neither nickel nor copper tubes could be drawn in 48-foot lengths, shorter tubes had to be welded together. The first order for columns was placed with Mehring and Hanson on July 5.

The S-50 facility. The main process plant is the long, dark building to the left of center. The K-25 powerhouse (three smokestacks) is to its right, and a tank farm for supplying oil for the "new boiler plant" is to the left. The new boiler plant itself is between the main process building and the river. Source: http://commons.wikimedia. org/wiki/File:S50plant.jpg

S-50 was designed as twenty-one copies of the 102-column Philadelphia installation, with 2,142 columns operated in parallel to provide a large quantity of slightly enriched U-235 as feed for the Y-12 and K-25 plants. Each rack was arranged as two rows of 51 columns, which for purposes of steam supply were divided into three groups of seven sections. Columns could be isolated from each other for maintenance or product removal by freeze-off water directed through intercolumn connectors. Erected adjacent to the K-25 powerhouse on the bank of the Clinch River, the pace of construction of the S-50 plant was phenomenal. Ground was broken in early July, and foundations laid less than three weeks later. Installation of process equipment began on August 17, and the first columns were received from Grinnell on August 27. By September 16, sixty-nine days after the start of construction, 320 columns were on hand, one-third of the plant was complete, and preliminary operation of the first rack had begun. Of the twenty-one racks, number 21 was completed first, and was used for training operators. The first process

material was introduced into that rack on October 18, and the first product was drawn off on October 30. Operation of S-50 was carried out on a
cost plus $11,000 per month fee basis by the Fercleve Corporation (from
*Fer*guson of *Cleve*land), a wholly-owned subsidiary of Ferguson established
to avoid the possibility of labor trouble when employing non-union laborers;
Ferguson normally operated on a unionized basis. When power demand for
the K-25 plant began to increase in early 1945, plans were made to construct
a new boiler plant to service S-50. Construction began with site clearing on
March 16; the boilers arrived on April 26, and steady operation was underway by July 13. Ironically, the plant was completed on August 15, the day
after the Japanese surrender was announced.

Production from S-50 began in October 1944, with 10.5 lb of output.
During routine operation, enriched product was removed at two-to-four-
hour intervals from the tops of columns. By mid-January, 1945, large-scale
production was underway, with ten of the 21 racks producing, and construction of all racks nearly complete. By March 15, all 21 racks were yielding
product, and in April, S-50 output began to go directly to the K-25 plant.
Cumulative production amounted to nearly 45,000 lb by the end of July, and
just over 56,500 lb by the end of September. If all of this was of 0.86% U-235
concentration, this would represent some 220 kg of U-235, enough for
almost four *Little Boy* bombs. This productivity was less than the 10 kg per
day of 90% U-235 that the Lewis committee had estimated in June, 1944,
because the S-50 columns were operated in parallel, not series. S-50's mission
was to produce a large quantity of slightly-enriched material, as opposed to a
small quantity of greatly-enriched material.

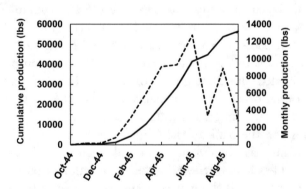

*S-50 production: monthly (dashed line, right scale);
cumulative (solid line, left scale). Source: MDH, Book
VI—Liquid Thermal Diffusion (S-50) Project, Top
Secret Supplement*

The cost of the S-50 plant was about $20 million, plus about $2 million in research costs borne by the Navy. This represented a mere 1% of the total cost of the entire Manhattan Project, and less than one-twentieth the cost of either Y-12 or K-25. Kenneth Nichols estimated that S-50 contributed to shortening the war by about nine days.

6.4 Hanford

Leslie Groves rightly characterized the project to synthesize plutonium on a large scale as truly pioneering. The Hanford Engineer Works represented even more of a gamble than its counterpart in Tennessee. The uranium enrichment facilities at Clinton were complex, plagued with difficulties, and subject to constant design changes, but at least involved processes that were familiar to engineers. But in 1943 there was no established nuclear industry, or even a recognized discipline of nuclear engineering. Nobody had ever before designed or constructed a large-scale reactor, and there was no cadre of experienced operators ready to walk into a control room. The dangers of large-scale nuclear technology could be immense. An electrical short in a calutron or an explosion in a diffusion tank at Oak Ridge might set back production temporarily and endanger a small number of workers, but an explosion in a reactor could potentially spread radioactive fission products over hundreds of square miles and endanger tens of thousands of lives.

A reactor fueled with natural uranium produces about three-quarters of a gram of plutonium per day per megawatt of operating power; to produce 10 kg consequently requires some 13,000 MW-days of trouble-free operation. At Hanford, three 250-MW reactors were built, implying about 17 days between 10-kg lots once steady operating conditions had been reached. But with this 10 kg comes an immense waste problem. If a fuel slug remains in a reactor for an average of 100 days, then about 1% of the full fuel load (about 250 tons per reactor) must be ejected every day. The fission products trapped in each day's output corresponds to over four million curies, the equivalent of over four tons of radium. And that is just the beginning. Only about one atom per 13,000 in a fuel slug will be transmuted into plutonium, and those transmuted atoms have to be extracted by first dissolving the slug in acid and then processing the resulting fluid with a complicated series of chemical reactions. The result is a handful of plutonium and a very daunting waste-disposal problem. It is no wonder that Groves insisted on an isolated location for the Manhattan Project's reactors.

6.5   The Hanford Site

Hanford and Oak Ridge developed in parallel following the Military Policy Committee meeting of December 10, 1942, which determined that plutonium production piles should be removed from the Clinton location. A completely separate site had to be procured for them. On December 14, Colonel Nichols, Arthur Compton, and Lieutenant Colonel Franklin Matthias, Groves' Area Engineer for the plutonium project, traveled to Wilmington to discuss pile design and site selection with DuPont officials; Matthias had served as Groves' Deputy Manager of construction for the Pentagon. Four helium-cooled 250-MW piles and two chemical separation plants were planned for, with a goal of producing 600 g of plutonium per day. The piles, which would each require some 250 tons of uranium fuel, were to be separated from each other by at least one mile, and the chemical separation plants from each other by four miles. Each pile was to be a self-contained unit, independent of the others in case of a disaster at any one of them. Laboratories would have to be at least eight miles from the separation plants, and a village for housing workers was to be at least 10 miles upwind from the nearest pile or separation plant; the village would eventually be located some 30 miles from the piles. To allow for the possibility of up to six piles, the site would require an area of about 15 by 15 miles square.

Matthias reported back to Groves, who directed him to make inquiries as to locations where suitable electric power would be available. On December 16, Matthias, Groves, and DuPont representatives met to draw up more specific criteria. Some 100,000 kW of power would have to be continuously available. While cooling the piles with circulating helium gas was the preferred design at that time, water-cooling was being given serious consideration, and Groves wanted to cover all bases: the site would require a water supply of 25,000 gallons per minute in case water-cooling was chosen, which proved to be the case. Level terrain with conditions suitable for heavy construction was desirable, with plenty of sand and gravel available for producing large quantities of concrete. Hanford would ultimately require over 780,000 cubic yards of concrete, enough for a 390-mile highway 20 feet wide by 6 in thick. Overall, an area of close to 700 square miles was required, preferably in the form of a rectangle of about 24 by 28 miles which would completely enclose a 12 by 16-mile plant area. The setting should be remote, with no settlement of population greater than 1,000 within 20 miles. Consideration was given to locating the site within a 44 by 48-mile buffer area from which all residents would be removed, but this idea was eventually dropped.

Matthias and a group of DuPont representatives spent two weeks scouring the western United States in search of possible sites, looking at eleven altogether. Two sites in each of California and Washington looked promising. In California, these were near the Shasta Dam and the Hoover Dam on the California-Arizona border. In Washington, one site lay near the Grand Coulee Dam in the central-northeast part of the state, and the other, which had the advantage of access to the Bonneville Power Authority, was near the town of Hanford on the Columbia river in the south-central part of the state. Matthias reported back to Groves on December 31 that the group was unanimously enthusiastic about the Hanford location. Groves inspected the site personally on January 16, 1943, and gave his approval. On February 9, Undersecretary of War Robert Patterson approved acquisition of more than 400,000 acres for the site of the Hanford Engineer Works (HEW). Hanford was the last site selected for the Manhattan Project.

Overall, the Hanford reservation comprised 670 square miles—about half the area of the state of Rhode Island—over a roughly circular area which extended 37 miles at its greatest north-south extent by 26 miles in maximum east-west breadth. The site was distinctly unappealing. Flat, semi-arid, and covered in grayish sand and gravel, the area could be swept by blinding sandstorms that lasted two or three days and which left everything coated in dust. Despite this, land acquisition proved to be a chronic bane for Groves and Matthias. About 88% of the land was being used for grazing, 11% was farmland, and less than 1% was occupied by three small towns: Richland (population about 200–250), Hanford (about 100), and White Bluffs, which was about the same size as Richland. Acquisition was complicated by the presence of a number of interests: some 157,000 acres were owned by federal, state, or local governments; 225,000 by private individuals; 46,000 by railroads; and 6,000 by irrigation districts.

The first tract of land was acquired on March 10, but resistance soon arose among individuals and irrigation districts over what they thought to be low property valuations, inadequate advance-notice allowances—sometimes as little as 48 h to vacate—and insufficient compensation for crops. The issue reached a Military Policy Committee meeting on March 30, and then a cabinet meeting on June 17. At the latter, President Roosevelt, concerned with possible wartime food shortages, wondered if another site could be chosen. Groves had to explain to Henry Stimson that both DuPont and the Manhattan District had concluded that the Hanford site was the only one in the country where the work could be done. In the late spring of 1943, the Corps of Engineers agreed to reappraise all tracts that had not yet been acquired. A number of cases went to trial, and settlements on over 1,200 tracts were averaging no more than seven per month until Groves requested more judges from the Department of Justice and an end to the habit of juries inspecting the areas in question. Despite more expeditious proceedings, a number of landowners had to be evicted by court orders, and the acquisition program was still ongoing as of late 1946 when the Manhattan District History was being prepared. By that time, some $5 million had been spent acquisitions; Groves thought many of the settlements to be exorbitant.

6.6   Pile Design and Construction

Well before Groves took command of the Project or CP-1 had demonstrated the feasibility of a self-sustaining reaction, scientists and engineers at Arthur Compton's Metallurgical Laboratory were exploring possible configurations for production piles. The complexities were legion: design of reactors and separation plants; cooling and control systems; determining relevant chemical and metallurgical properties of uranium and plutonium; effects of reactor materials on the efficiency of the chain reaction; and ensuring human and environmental safety. In early 1942, one of Compton's first actions upon centralizing the pile program was to establish an Engineering Council to consider suggestions for pile design. For chief engineer, he chose Thomas V. Moore, a veteran of many years experience in the petroleum industry. A younger member of the group was John Wheeler of the Bohr and Wheeler fission theory. Initially, much of their effort focused on investigating helium-cooled, uranium-graphite configurations.

On June 18, the Council gathered to consider designs that might be suitable for production-scale piles. One plan was to use an actively-cooled Fermi-type lattice, even though it would be necessary to dismantle the pile to retrieve the irradiated uranium. Walter Zinn suggested an arrangement of uranium in graphite cartridges that would move through a graphite block at about three feet per second, which would be fast enough to obviate the need for cooling the pile itself. John Wheeler proposed alternating layers of uranium and graphite, with the uranium-bearing layers connected to shafts to draw them out of the pile. Another concept, which was ultimately adopted, was to use cooled rods of uranium that extended through a large graphite block.

The most pressing issue was to decide how the piles should be cooled. If they were to be gas-cooled, two alternatives looked feasible. Air cooling was familiar to engineers, but would involve some neutron loss. Helium cooling was attractive in view of its chemical inertness and that element's low neutron-capture cross-section. But all gases have relatively poor heat-transfer properties, which would mean that large volumes of gas would have to be pumped under high pressures, an issue which would complicate design of compressors and pumps. As for liquids, water-cooling was also familiar territory for engineers, but water captures neutrons and corrodes unprotected uranium metal. Some Met Lab scientists favored heavy water as it could serve as both a coolant and a moderator, but that material was scarce. A drawback of any form of liquid-cooling was that a leak could render the pile inoperative or cause an explosion if the coolant became vaporized under high pressure. During the summer of 1942, Moore and his team concentrated on designing a helium-cooled pile, while John Wheeler and Eugene Wigner continued to research the possibility of water cooling. At the same time as pile design went forward, there was also the issue of plutonium separation chemistry to consider. At one point, 12 different separation methods were under consideration; in May, 1943, DuPont officials decided on a bismuth-phosphate process for units at both Clinton and Hanford.

The successful operation of CP-1 indicated that water cooling would be feasible for large-scale piles. Within five weeks of CP-1's first criticality, Eugene Wigner and his group had designed a 500-MW pile wherein a thin film of water would flow over aluminum-sheathed uranium slugs which would be contained within aluminum tubes which ran through a graphite moderating structure. After being irradiated, slugs would be ejected from the back of the pile and collected in a pool of water to let their radioactivity die off before being transported away for chemical separation. Curiously, Wigner anticipated an operational lifetime for the reactor of only 100 days. This is the system that came to be used in the Hanford piles, although they ended up operating for 25 years.

In mid-February 1943, DuPont decided to terminate research on the helium-cooled design in favor of Wigner's water-cooled design. The decision to shift to water cooling was a significant one, and involved a number of competing factors. Wigner had objected to helium on the grounds that the reactor would have to run at a very high temperature, perhaps 400–500 °C, which would mean serious material stress problems. Helium cooling would also require handling and purifying large volumes of gas, and maintaining a leakproof pressure enclosure for the pile. While it was expected that water cooling would reduce the reproduction factor by perhaps 3%, DuPont engineers had become impressed by Wigner's design, and were confident that the chain reaction could be maintained. The decision in favor of water-cooling came *after* the Hanford site had been chosen, but the Columbia river was more than able to supply the requisite amount of water. Helium was not totally abandoned, however: it would be used to provide an inert operating atmosphere for the pile, which still meant providing for a pressure enclosure. Eugene Wigner personally reviewed all blueprints, and accrued 37 patents on various kinds of reactors.

Three major types of working areas were laid out over the Hanford reservation. The piles themselves would be located in "100" areas: 100-B, 100-D, and 100-F, each about one mile square. The separation facilities were located about 10 miles south of the piles in "200" areas: 200-E, 200-W, and 200-N, for East, West, and North, respectively, with the 200-N area used as a storage area for irradiated fuel slugs. The 300 area, located just a few miles from Richland, was where uranium slug fabrication and testing took place. Each pile also required a plethora of support facilities: retention basins to hold spent cooling water until its radioactivity had declined to the point where it could be safely returned to the Columbia; water pumping and treatment plants; refrigeration and helium-purification facilities; fuel-storage areas; steam and electricity substations; and fire and first-aid stations. Equally monumental would be the three chemical separation plants. Colloquially known as Queen Marys after the famous ocean liner, each would be 800 feet long by 65 feet wide by 80 feet high. Irradiated fuel from the piles traveled to the Queen Marys in lead-lined sealed casks aboard railroad flatcars; one of the locomotives used in this process is now on display at the B-reactor site.

Left: The 100-B area, looking northwest, January, 1945. The Columbia river is in the background. The pile building itself is adjacent to the more distant water tower. Source: http://commons.wikimedia.org/wiki/File:Hanford_B_site_40s.jpg. *Right: Queen Mary separation building. Source:* http://commons.wikimedia.org/wiki/File:QueenMarysLarge.jpg

Initial plans called for eight 100-MW piles laid out along the banks of the Columbia, designated as 100-A through 100-H. When the 250-MW water-cooled design was chosen, it was decided to cut the number of piles to three, to be located at the B, D, and F sites. The A and H sites were left vacant as safety areas; B-pile was about 7 miles southwest of D, and F about 9.5 miles southeast of D. Various other reactors were built at Hanford after the war, but there never was an A-pile.

The B-pile building under construction. (HAER, Photo 3)

The first pile built was the B-pile, and a particularly rich record on its construction and operation is available in a Department of Energy Historic American Engineering Record document. Survey work for the B area was completed on April 15, 1943; ground was broken for a retention basin on August 27, and layout of the reactor building itself, the 105-B building, began on October 9. 105-B had a footprint of 120 feet by 150 feet, and was 120 feet high. Including shielding, the outer dimensions of the pile itself were 37 feet from front to rear (roughly west to east), 46 feet from side to side (north-south), and 41 feet high. The graphite core measured 36 feet wide by 36 feet tall by 28 feet from front-to-rear.

At the front face of each pile was the charging area, where slugs of uranium metal were loaded into 2,004 aluminum process tubes, each 44 feet long. The charging area was large enough to permit removal of fuel tubes for repairs if necessary. At the back of the pile was the discharge face, from which irradiated slugs would fall into a 20-foot deep pool for storage and transfer. The control room was situated on the left side of front face of the pile on the ground floor. Above the control room was a "rod room" from where nine 75-foot long control rods could be electrically or manually deployed. Exclusive of the pile itself, each pile building used 390 tons of structural steel; 17,400 cubic yards of reinforced concrete; 50,000 concrete blocks; and 71,000 bricks. The piles themselves were welded to be gas-tight, and contained 2.5 million cubic feet of masonite; 4,415 tons of steel plate; 1,093 tons of cast iron; 2,200 tons of graphite; 221,000 feet of copper tubing; 176,700 feet of plastic tubing; and some 86,000 feet of aluminum tubing.

The bottom-most layer of the pile structure was a 23-foot thick concrete footing, cast to accommodate instrument and gas-transfer ducts. Atop the footing lay a 1.5-in steel baseplate. Each pile was surrounded on all sides by water-cooled cast-iron blocks which formed a thermal shield wall approximately 10 in thick. The bottom layer of this shield served as a base for the graphite bricks of the pile, and absorbed over 99% of the heat generated by fission reactions. The cast-iron blocks were machined to accuracies of 0.003 in, and were interlocked to provide a radiation barrier; holes bored through the shield for fuel-channel tubes had to match corresponding holes in moderator bricks to 1/64 of an inch. Working outward, the thermal shield was surrounded by a 4-foot thick biological shield comprised of over 350,000 blocks of alternating layers of steel and masonite, known as B-blocks. This layer reduced the ambient radiation by a factor of 10 billion; to achieve the same effect with concrete would have required a wall 15 feet thick. The entire assembly was then surrounded by a steel outer shell, which served as a containment structure for the pile's helium atmosphere.

The total volume of land excavated at Hanford amounted to some 25 million cubic yards, about 10% of that moved during construction of the Panama Canal. Material was brought to the site in rail cars, some 40,000 in all over the course of the project. Organizing the construction was a mammoth task; over two years, DuPont placed over 47,000 purchase orders and engaged 74 subcontracts with firms in 47 states. Remarkably, the project was brought on-line a year ahead of schedule at a cost only about 10% above that estimated in mid-1943.

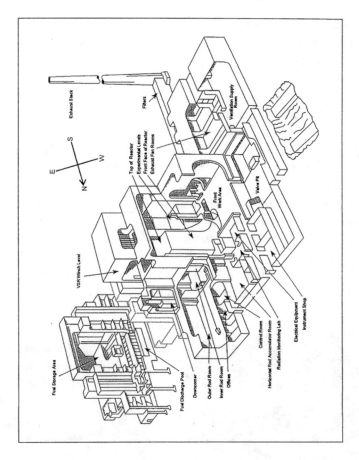

Cutaway view of the Hanford B pile. (HAER, p. 133)

As with K-25, a particular issue in the construction of the piles was the quality of welding joints. Once a pile had been activated, it would be next to impossible to correct any internal problems; all joints had to be done properly the first time. Each pile required over 50,000 linear feet of welds, which had to be smooth to a tolerance of 0.015 in. This task was assigned to the highest-quality welders, who received a special pay grade and had to submit to background checks and periodic tests; only about 18% of applicants qualified. Welds were inspected by use of X-rays or penetrating dyes; each weld was stamped with a welder's identification number.

Each pile comprised some 75,000 graphite moderating bricks, most just over four inches square by 48 in long. About one in five were bored length-wise to accommodate fuel tubes spaced 8 and 3/8 in apart. The squareness tolerance of the bricks was held to 0.004 in to ensure snug fits, and their corners were bevel-cut to provide passages for the helium atmosphere. Each brick weighed about 50–60 lb, and their neutron-capturing boron content was held to 0.5 parts per million. Bricks were milled in a restricted-access building; each was stamped with a quality code so that the best ones would be placed in the centers of the piles. A small test pile was built in the 300-area to check the fit of each brick, with the location of each recorded in order that layers could be correctly reconstructed in the real pile. After each layer of bricks was stacked, it was vacuumed to remove any contaminants. Milling of bricks for the B-pile began on December 10, 1943, and laying was finished on June 1, 1944, just a few days before the invasion of Europe. Cleanliness was so critical that DuPont even had a laundry procedure which specified what soaps and detergents could be used to clean worker's clothes.

Left: Laying the graphite core of B-reactor. The rear face of the reactor is toward the lower left, and the inside of the front face to the upper right. (HAER, Photo 6). Right: Front face of F-pile, February, 1945 (HAER, Photo 21)

6.7   Fueling the Beasts

At the heart of each pile was its assembly of graphite moderator bricks, aluminum channels, and fuel slugs. Eugene Wigner's early-1943 design for a 500-MW pile called for 1,500 tubes piercing a graphite cylinder 28 feet in diameter by 28 feet deep. DuPont engineers modified the design by adding 500 fuel channels to make a roughly square-faced arrangement.

The record as to who suggested the overdesign is unclear. Some sources indicate that Hood Worthington, the head of DuPont's design effort, followed what was normal chemical engineering practice at the time and invoked a one-third overcapacity margin. Other sources suggest that the idea was proposed by John Wheeler and Enrico Fermi, who were concerned about possible neutron-capturing fission products poisoning the chain-reaction. While many physicists thought that the overdesign would make the piles more expensive than necessary to construct and operate, the conservatism would pay off: the additional tubes beyond Wigner's 1,500 contributed only about 10% of the reactivity of the central ones, but would prove to be crucial to achieving the piles' design power ratings. One DuPont engineer estimated that had the additional tubes not been provided for from the start, eight to ten months would have been necessary to revise pile design and construction in response to a neutron-capturing crisis that emerged on the startup of B-pile.

As constructed, each pile comprised a square central area of 42 tubes on a side, for a total of 1,764. To those were added 240 tubes arranged as two rows of 30 tubes each, centered on each of the four sides of the square. This gave a total of 2,004 tubes, each of which was uniquely numbered so that operators at the front and back faces of the pile could open the same tube simultaneously for refueling. Piles were shut down during re-fueling operations, during which tons of irradiated slugs would be discharged from their back faces. During normal operation, each tube contained 32 active fuel slugs of outside diameter 1.44 in by 8.7 in long; short slugs were used to minimize warpage due to thermal expansion. Each slug contained about ten pounds of natural uranium; with some 64,000 slugs inside the pile (2,004 tubes times 32 active slugs per tube), the usual fuel load was about 250 tons. Fuel slugs were supported inside the tubes by two ribs which ran along the bottom of each tube, an arrangement which left an annular gap of less than a tenth of an inch for the flow of cooling water; about 14 gallons worth would pass through each tube per minute. The piping was arranged such that if water flow to a tube was stopped for refueling or maintenance, the tube would remain full of water.

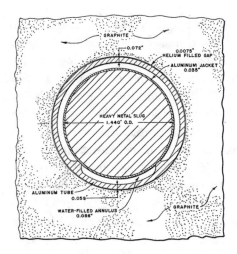

*Cross-section of a fuel tube assembly
(HAER, Fig. 9, p. 143)*

Early IBM computers were used to track the irradiation history of each fuel slug in order that the amount of plutonium production could be predicted. Dummy slugs (inert spacers and neutron-capturers) were used to help control the neutron flux within the pile; in routine operation a pile would contain almost as many dummy slugs as active ones. Dummy slugs could be reused, but since they too would become slightly radioactive they also had to undergo a period of post-use thermal and radiological cooling before being recycled into a pile.

Fueling was accomplished by operators riding in a loading elevator using a charging machine to push fuel slugs into process tubes, which simultaneously caused irradiated ones to emerge from the back of the pile. In operation, tubes typically averaged 59 fuel and dummy slugs. The rear face of the pile was surrounded by a 5-foot thick concrete wall, and workers would normally vacate the area after they had opened the discharge tubes but before pushing began. A discharge elevator on the rear face carried a cab which was shielded with seven inches of lead and was equipped with a periscope and power tools. After free-falling into the collecting pool, slugs would be sorted into buckets of active and dummy units. After an hour or two, their radioactivity would drop by a factor of ten, and then by another factor of ten after 60 days.

Vital to the safe operation of each pile was its once-through cooling system. For all three piles, the total cooling water consumption was equivalent to that of a city of about 1.3 million inhabitants. Some 30,000 gallons of water was pumped through each pile per minute, but only a small fraction of that would be inside the core at any moment. By using a single-pass arrangement, outlet temperatures could be kept low enough to ensure rapid heat dilution of the effluent water in the Columbia, which has a flow rate at Hanford on the order of 54 million gallons per minute. The cooling water would nevertheless become slightly radioactive from its pass through the pile; to allow short-lived fission products to decay, effluent was held in a 7-million gallon retention basin for three to four hours before being returned to the river. Intake and discharge lines were guarded by grates to prevent fish from swimming up them.

The cooling system contained multiple backups. Primary circulation was provided by electric pumps, with steam-driven pumps idling on the same lines in case of a power failure; the primary pumps were fitted with 4,600-lb flywheels so that they would keep running for 20–30 s until the steam pumps came up to full power. Each pile was also equipped with two elevated 300,000-gallon water tanks which could dump their contents into the piles by gravity feed.

In addition to fuel and cooling management, another concern was the operating environment of the piles. Ordinary air would not do: nitrogen captures neutrons, and air also contains a small amount of argon, which becomes radioactive upon capturing neutrons. Each pile was enclosed in a steel casing through which helium was pumped at a rate of about 2,600 cubic feet per minute; helium also had the advantage of being fairly thermally conductive for a gas. Pressure-tests of B-pile began on July 20, 1944, the same day as an unsuccessful assassination attempt against Adolf Hitler.

6.8   Controlling the Beasts

Control of the Hanford piles was effected by a system of boron-steel control and backup rods similar to those used in the X-10 pile. At Hanford, nine 75-foot long, water-cooled horizontal control rods entered from the left side of the pile as seen from its front face. Hydraulically and electrically driven, these were arranged in three rows of three rods each, set five feet apart both vertically and horizontally. Seven were shim rods which controlled the bulk of the pile's reactivity. These could be moved at speeds of up to 30 in per second, and could effect a complete shutdown of a pile unless a complete loss of cooling water occurred. The other two were regulating rods, which were used to handle finer minute-to-minute adjustments and could be moved at speeds as little as a hundredth of an inch per second. Above each pile resided 29 vertical safety rods, normally held in place by electric clutches which would release in the event of a power failure. Given that an earthquake or bombing could damage a pile in such a way as to prevent rods from deploying, a last-ditch safety system was mounted atop each pile: five, 105-gallon tanks filled with a boron solution, an arrangement reminiscent of the CP-1 suicide squad. When released, the fluid would run into the vertical rod holes, but would ruin the pile in the process. In 1953, these were replaced with systems using boron-steel ball-bearings.

The safety systems received a real-world test on March 10, 1945, when an explosives-laden Japanese balloon struck an electrical transmission line in Toppenish, Washington, about 35 miles southwest of Hanford at about 3:30 in the afternoon. The resulting voltage fluctuation caused all three piles to scram automatically. B and D piles were offline for only 10 and 12 min, respectively, but F pile was out for 68 min due to difficulties with raising a vertical control rod. Later the same day, a second balloon drifted into the same area and landed. As it started drifting toward the same power line the earlier balloon had struck, a Hanford security patrolman and an Army MP used their guns to deflate it. Over the course of the war, the Japanese produced some 9,000 such balloons, of which about 300 were eventually found in the United States. These 10-m diameter hydrogen-filled balloons were made of multi-plied paper, with ballast and bombs suspended on 50-foot shroud lines. Remarkably, these balloons were the first weapons in history to possess intercontinental range.

Operators constantly monitored the status of the piles through readouts from over 5,000 instruments. At a glance, they could determine the state of the water pressure at any tube inlet; the water temperature at all inlets and outlets; the water flow rate; the pressure of the helium gas; the temperature of the graphite moderator and the thermal shield; positions of all control rods; and monitor for the presence of any radiation leaks. Safety circuits were programmed to deploy control rods depending on the severity of a problem such as high or low water pressure, high radioactivity in the discharge water, overly high neutron flux, a power failure, or high effluent temperature. The power level was determined by monitoring the temperature difference between inflowing and outflowing water, which in combination with the flow rate gave the heat generated by the pile.

Radiation detectors monitored effluent water, retention basins, ventilating air, discharge areas, and control rooms. Workers who might be exposed to radiation always carried two personal dosimeters via which daily and cumulative doses were tracked; they also wore protective clothing and face masks if needed. Because plutonium tends to collect in bones, urine and blood samples were regularly collected and tested. The dose tolerance for workers was set to a very low level, 0.01 rems per day. Patrol Groups routinely surveyed pile buildings and other areas to check for signs of contamination. As part of monitoring the external environment, Army guards would periodically shoot coyotes, whose thyroids would be examined for iodine, a characteristic fission product. Despite the pressure of wartime work, not a single serious case of radiation exposure occurred at either Oak Ridge or Hanford.

6.9   Startup—And Shutdown

The first fuel was loaded into B-reactor at 5:44 p.m. on September 13, 1944, by Enrico Fermi. During the initial loading phase, all control rods were inserted. The design of the pile was such that only a few hundred fully-loaded tubes would be needed to bring it to criticality, albeit at low power. Initially, only the central-most 1,595 tubes were connected to the cooling system, and 895 of those were filled with dummy slugs.

The first operational benchmark was what reactor engineers term "dry criticality," which is when a pile achieves criticality with no coolant circulating. Given the poisoning effect of water, this is the smallest possible critical size of a pile. If cooling is then activated, criticality will be lost, and more tubes will need to be loaded to restore it. Dry-critical loading of B-pile began with a central area of 22 tubes on a side, and was achieved at 2:30 a.m. on September 15 with 400 tubes loaded. A period of tests and calibrations followed, after which loading was resumed until 748 tubes were charged. At that point, the cooling system was activated, which, as expected, poisoned the reaction. Additional tubes were loaded, and "wet criticality" was achieved at 5:30 p.m. on Monday, September 18, with 838 tubes charged. Loading with control rods inserted continued into the early morning of September 19, by which time 903 tubes were charged, although two had to be shut down due to lack of water pressure. After further tests, control rods were withdrawn until wet-criticality was again achieved with the 901-tube loading. This occurred at 10:48 p.m. on Tuesday, September 26, 1944, and is regarded as the first official operation of the pile. By just after midnight, B-pile was operating at 200 kW, and by 1:40 a.m. on the 27th, 9 MW was achieved.

At first, everything seemed to be operating perfectly. But about an hour after reaching 9 MW, operators noticed that they were having to withdraw control rods to maintain power; the pile appeared to be dying. By 4:00 in the afternoon the power level had fallen to 4.5 MW, at which time it was intentionally reduced to 400 kW in an attempt to halt the decline. This proved unsuccessful, and by 6:30 p.m. the pile had shut itself down completely. There was no obvious problem: the helium atmosphere, water flow, and pressures were all normal, and there was no evidence of any leaks or slug corrosion.

Surprisingly, after a few hours of dormancy the pile spontaneously began coming back to life. The reproduction factor rose back to greater than unity at about 1:00 a.m. the next morning, September 28, and by 4:00 p.m. the power level could again be raised to 9 MW. But as soon as that level had been reached the reproduction factor again began to decline. Sustained operation proved to be impossible, and as Thursday became Friday, B pile was once again effectively dead.

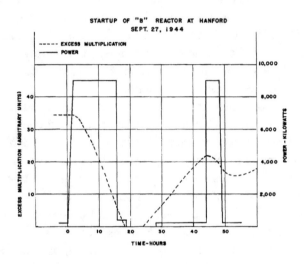

Power output (solid line, right scale) and excess multiplication factor (a measure of the reproduction factor; dashed line, left scale) for B-reactor startup. Time-zero corresponds to about midnight, September 26//27, 1944. Note how the excess multiplication drops as power is increased. From Babcock (1964)

From the pattern of the pile's on-again, off-again reactivity, Enrico Fermi, John Wheeler, and DuPont engineer Dale Babcock determined that the problem was likely a fission product with a poisonously high neutron-capture cross-section. By monitoring the rate at which control rods had to be withdrawn to hold the power steady at 9 MW, they determined that both the parent isotope and its poisoning decay-product isotope both likely had half-lives of the order of several hours. By Friday morning, enough data had been gathered to indicate a half-life for the poison of about 9.7 h. Examination of a table of isotopes showed that the problem was likely an iodine-to-xenon decay chain. The specific culprit was xenon-135, which arises from beta-decay of tellurium-135, which is a direct fission product. Tellurium undergoes a 19-s beta decay to iodine-135, which suffers a 6.6-h beta decay to xenon-135, which has a half-life 9.1 h (modern value) before decaying to cesium and eventually to barium. At over three *million* barns, Xe-135's thermal-neutron capture cross-section is the largest known for any isotope.

The only solution was to increase the amount of fuel in the reactor in order to overcome the poisoning effect. This required plumbing in the initially unused fuel tubes, which necessitated boring holes through the biological shield blocks. Compton presented the bad news to Groves in Chicago on October 3; Groves was highly critical of the scientists for not foreseeing the problem, and was not impressed by Compton's argument that a fundamental new discovery regarding the neutron properties of matter had been made. Compton then left for Hanford to review the situation personally.

Some accounts of the B-reactor startup present the xenon-poisoning episode as a completely unanticipated phenomenon, but this is far from the truth. As described by Babcock in an article published on the twentieth anniversary of the event, the possibility of a severely neutron-capturing fission product had received considerable attention. The reproduction constant achieved in Fermi's CP-1 pile was only slightly greater than unity, and Wigner, Wheeler, and DuPont engineers were well aware that a production reactor would involve many materials not present in CP-1, particularly water and aluminum tubes. Wheeler carried out detailed calculations of how even slight changes in design specifications could affect the reproduction factor, but the numbers were uncertain and not all fission products were known or could be predicted. As early as February, 1942, Wheeler had speculated on the possible effects of fission products, and in April of that year determined that a short-lived fission product could severely affect pile operation if it had a capture cross-section of about 100,000 barns or greater.

As design work progressed, each specification was assigned a plus-or-minus value for how it might affect neutron reproduction: the thickness of a fuel cladding or water jacket, the design of a control rod, the purity of graphite bricks, and so forth. Wheeler's results led him to suggest adding the 504 additional fuel tubes at the periphery of the core, as well as slightly altering the diameter of the tubes; DuPont accepted both suggestions. Wheeler attempted to identify in advance some potential problematic fission products, but the calculations were very sensitive to the assumed distribution of fission products. It does seem to be true, however, that nobody anticipated a poison with a cross-section of millions of barns.

Work on charging an additional 102 tubes began on September 30, and was completed on October 3. With 1,003 tubes loaded, the pile was brought back to criticality and taken to a power of 15 MW, at which it was maintained until October 5. This did not overcome the poisoning, so the pile was shut down to load more tubes. Between October 12 and 15, the number of charged tubes was raised to 1,128 and the pile taken to a power of 60 MW. Poisoning persisted; more fuel would be needed to get to the design power of 250 MW. Another shutdown on October 19 permitted raising the number of charged tubes to 1,300 and the power to 90 MW, and yet another on October 26 brought the number of operating tubes to 1,500. A power of 110 MW was achieved on November 3, but again could not be maintained, so operations were reduced to 90 MW on November 5. The next shutdown came on November 20, following which 1,595 tubes were made active.

B-pile achieved a power of 125 MW (half of its design capacity) on November 30, but it was clear that all tubes would be needed to get to 250 MW. The pile was shut down on December 20 to install a full fuel load; extra reactivity was also obtained by replacing some dummy slugs with active ones. All 2,004 tubes were ready (less the two defective ones) by December 28. A power of 150 MW was achieved on the 29th, and 180 MW the next day. The full design rating of 250 MW was finally achieved on February 4, 1945, with about 1,950 tubes operating. With lessons learned from B pile, the D and F piles started life with full fuel loads. D went critical at 11:11 a.m. on December 17, 1944, with 2,000 tubes loaded, and F on February 25, 1945, with 1,994. Within a day, F-pile was operating at 100 MW, and by March 1 was running at 190 MW. Matthias recorded in his diary that on the morning of March 28, all three piles ran simultaneously for the first time at 250 MW. At this rate, plutonium production would be enough for almost three *Fat Man* bombs per month at about 6 kg per bomb. By May 3, some 1.6 kg of plutonium had been delivered to Los Alamos, and deliveries were taking only two days to get from Hanford to New Mexico. By June 1, Groves was ordering that production be maintained at five kilograms every 10 days. D-pile was also used for polonium production: by May 4, four of its fuel channels had been loaded with 264 bismuth slugs.

For both the *Little Boy* and *Fat Man* bombs, expected availability of fissile material was always the pacing element of when they could be ready. Even while B-pile was undergoing its various reconfigurations in response to the xenon crisis, Groves began pressuring DuPont to find strategies to increase production. In October 1944, the company estimated that production would begin with 200 g of plutonium in February, 1945, and increase to six kilograms per month by August, 1945. At this rate, after allowing time for material cool-down, processing, transport, and fabrication of bomb cores, the first plutonium test bomb would not have been ready until mid-October, 1945, and the first combat bomb not until a month or so later—after the proposed invasion of Japan had begun. Groves wanted five kilograms as soon as possible for a test device, and five kilograms as soon as possible thereafter for a combat weapon.

There were three possible ways to increase production, and all were used in what came to be called the "speed-up program": operate piles at higher power levels; push fuel slugs out of the piles sooner than normal (less plutonium per slug, but more slugs—and more waste); and shorten the post-irradiation thermal and radiological cooling time before slugs were transported to the separation facilities. It had been intended that slugs should remain under water for about 120 days, but the time was first reduced to about 60 days, then to 30, and then, by mid-1945, to as few as 15. By March 1945, Roger Williams advised Groves that DuPont should be able to deliver 5 kg by mid-June, and another five by mid-July. Groves pressed for even more efficiency, and the schedule was tightened to bring the delivery dates to June 1 and July 5. By Independence Day, 13.5 kg had been shipped, with another 1.1 ready to go. During the speed-up, B-pile remained at 250 MW, but by June the D and F-piles would be operated at 280 MW and 265 MW, respectively.

Construction of the Queen Mary separation plants proceeded in tandem with that of the piles. Two Queen Marys, 221-T and 221-U, were completed by December 1944 in the 200-West area; a third reserve unit, 221-B, was constructed in the 200-East area and was completed in the spring of 1945. Ground was broken for 221-T on June 22, 1943, but manpower shortages hobbled full-scale construction until the spring of 1944. Essentially large concrete boxes, these huge buildings were divided internally into cells containing equipment for various stages of chemical processing. The cells were surrounded by seven-foot-thick concrete walls and covered with 35-ton, six-foot-thick concrete lids which could be removed by an overhead crane which ran the length of the building. Each Queen Mary contained 42 cells, most of which measured about 13 feet by 17 feet-8 in in footprint by 22 feet high. In anticipation of inevitable changes in processing chemistry, DuPont provided each cell with, as much as possible, standardized processing vessels and piping and instrumentation connections that would not require subsequent modification. Once operations started, the cells would become intensely radioactive;

operators worked by remote control as they watched through periscopes and early television monitors.

Construction forces left 221-T on October 8, 1944, and operators began test runs the next day, using water and defective slugs which had not been used in piles. The first test run of irradiated fuel was discharged from B-pile on November 6 (while it was being reconfigured); this was much sooner than the nominal irradiation time of 100 days, but slugs were desperately needed to test handling and separation processes. 221-T was ready for the first production-discharge run from B-pile on Christmas Day, and the first pure plutonium nitrate was produced before the end of January. The first Hanford plutonium to go to Los Alamos, a 100-g solution, began its journey south by rail on February 5, 1945; eventually the precious material would make its way to by in heavily guarded trucks in shipping cans containing about one kilogram each. When operations had become routine by mid-1945, the average time for slugs to go from discharge to isolated plutonium was about 50 days, and the processing yield was up to 90%. To receive the 10,000 gallons per day per separation plant of radioactive waste generated by the separation process, 64 underground storage tanks were constructed, many as large as 500,000 gallons. Many of the tanks would begin leaking in the 1950s, and the area is now engaged in a monumental waste cleanup problem. Hanford produced 120 kg of plutonium by the end of 1945.

6.10   Feeding the Beasts

Without a steady supply of uranium ore, the entire Manhattan Project could never have been undertaken. The cache of Union Minière du Haut-Katanga ore was but the first installment in what came to be an extensive program to acquire thousands of tons of uranium-bearing materials and arranging with contractors to process them into forms suitable for use as input materials. By early 1944, the Manhattan Districts's Material Section office boasted a staff of nearly 400.

Manhattan uranium originated not only from the Belgian Congo (now the Democratic Republic of the Congo), but also from the Great Bear Lake region of northern Canada, from the Colorado Plateau region of the United States, from stocks acquired from commercial firms within the United States, and from ores captured in Europe by advancing allied forces. Uranium ores usually occur in the form of oxides, particularly "black oxide" (U_3O_8 to chemists); before the war, they were refined primarily for their radium content, with the oxides treated as by-products or waste. Prewar requirements for radium amounted to about 35–40 g per year, which resulted in a collateral annual production of about 160 tons of uranium compounds. It has been estimated that the total amount of pure uranium metal produced by 1939 was not more than about 10 lb.

Sources, amounts and costs of MED uranium supplies to January 1, 1947. *Source* Reed "The Feed Materials Program of the Manhattan Project" (2014) Table 2.

Source	Uranium oxide (tons)	Pure uranium content (tons)	Cost ($)
Africa	6,983	5,922	19,381,600
Canada	1,137	964	5,082,300
United States	1,349	1,144	2,072,300
Market	270	229	1,056,130
Captured, Europe	481	408	–
Totals	10,220	8,667	27,592,330

Nearly 70% of Manhattan uranium originated from the Union Minière mine. After relocating to New York in 1939, Edgar Sengier arranged for the firm's stock of some 1,250 tons of ore being held in Africa to be shipped to New York, where it was stored in steel drums in a warehouse on Staten Island. In September, 1942, Union Minière's American subsidiary, African Metals Corporation, applied to the State Department for a license to ship the ore to Canada for refining. This came to the attention of Kenneth Nichols, who met with Sengier to purchase the ore already in the United States, and to arrange for shipping to and for the United States to have a prior right of purchase of some 3,000 more tons stored aboveground in the Congo. Colonel Marshall, then the Manhattan District's commander, noted in his diary that the ore being stored on Staten Island was contained in 2,006 drums of dimensions 34 in high by 25 in in diameter, each plainly marked "Product of Belgian Congo" and "Uranium Ore". The markings were painted over before further processing. In recognition of his foresight in securing the original cache of ore in the United States, Sengier was awarded the Medal of Merit in 1946.

General Groves (left) presents the Medal of Merit to Edgar Sengier (1879–1963) at a private ceremony in 1946 while Brigadier General John Jannarone (1913–1995) looks on. Photo courtesy Robert S. Norris

Shipments of Congolese ore to America continued throughout the war, delivered by fast vessels traveling near convoys; only two shipments totaling about 200 tons were lost, one by enemy action and one by accident. These African ores were extraordinarily rich; some samples contained as much as 65% uranium oxide. In comparison, Canadian ores assayed on average at about 1% uranium oxides, and American ores at about 0.25%.

The Canadian ores originated from Port Radium on the eastern shore of Great Bear Lake of what was then the Northwest Territories of that country. Beginning in 1933, a mine operated by Eldorado Gold Mines produced radium, uranium, and silver at this site. Eldorado ores were shipped to a refinery in Port Hope, Ontario, about 100 km east of Toronto on the north shore of Lake Ontario. In the 1930s, the Eldorado mine was ice-bound between November and June; operations continued year-round, but shipments to Port Hope comprising the entire output of the mine for the previous year could occur only between July and October via the Mackenzie river. During the war, ore was continuously airlifted to Fort McMurray, Alberta, for access to surface transport.

American uranium ores were less rich than their Canadian counterparts, but ultimately provided more material. These originated from the Colorado Plateau, which covers parts of Utah, Colorado, Arizona, and New Mexico. Beginning in the early 1900s, these ores were mined principally for their radium content, and to a lesser degree for their vanadium and uranium. Operations declined rapidly after about 1923 as the mines could not compete against richer Congolese sources, but were revived during World War II when vanadium became important as a strengthening agent in steel. Years of vanadium and radium mining had generated a stockpile of nearly 380,000 tons of tailings and by-product sludges containing small but recoverable amounts of uranium. The Manhattan District entered into contracts with, among some smaller firms, the United States Vanadium Corporation (a subsidiary of Union Carbide and Carbon), the Vanadium Corporation of America, and the Metals Reserve Corporation (a government agency) to process the tailings. Because refineries that processed black oxide into intermediate materials needed for production of pure uranium metal were located in the eastern part of the country and it would have been very expensive to ship the tailings to those locations, both US Vanadium and Vanadium Corporatiom built plants to carry out preliminary processing near the locations of the ores before shipping them to the Linde Air Products Company in Tonawanda, New York, for further processing.

Anxious to round up every kilogram of uranium that he could, Groves also acquired small amounts held by various firms. In early 1943, the War Production Board prohibited the sale or purchase of uranium compounds other than for vital military and industrial applications, an action which made available for purchase by the MED supplies equivalent to some 270 tons of black oxide, mostly from the Harshaw Chemical Company (Cleveland, Ohio), African Metals, the Canadian Radium and Uranium Company, and the Vitro Manufacturing Company of Canonsburg, Pennsylvania. Vitro used uranium as a coloring agent in transparency gels made for use in projectors for the theatrical and stage lighting industries.

The capture of uranium ores in Europe was one of the most dramatic episodes of the Manhattan Project. Groves sought intelligence on what German scientists might be doing in the field of bomb research, and established a mission code-named *Alsos* which followed (and sometimes preceded) allied armies advancing through Italy and later Germany with the purpose of rounding up scientists and investigating research sites. In September, 1944, the mission reached a Union Minière refinery in Oolen, Belgium, only to find that some 1,000 tons of uranium ore had been shipped to Germany. Seventy tons remaining at Oolen were shipped out to Britain, and a further 31 tons were soon located in Toulouse, France. The 1,000-ton cache was located in caves in the area of Stassfurt in eastern Germany in April, 1945 (see Chap. 8).

The processing of uranium-bearing ores to final products (uranium metal, uranium hexafluoride, and smaller amounts of some other materials) involved four steps, the last three of which were carried out in parallel. No less than 10 primary contractors were involved, including DuPont, which would use the products of their work to fuel the Hanford reactors they had designed.

The first step was to refine raw ores to produce black oxide or "soda salt" ($Na_2U_2O_7$). This was then refined to produce what were known as brown and orange oxides (UO_2 and UO_3, respectively). The orange oxide, which was produced by Mallinckrodt Chemical, was shipped directly to Oak Ridge for conversion to uranium tetrachloride as feed material for calutrons; this was a temporary measure until the calutrons could be fed with enriched hex from the S-50 and K-25 plants. The brown oxide was converted to "green salt" (UF_4) by reacting it with hydrofluoric acid at high temperature; this was necessary to render it to a form in which it could be reacted with magnesium in a process which resulted in pure uranium metal.

By the end of September, 1943, the Manhattan District had 2,920 tons of ore available and had produced 1,660 tons of black oxide and soda salt; a year later the figures had risen to 5,460 and 3,500 tons, and by the end of September, 1945, stood at 6,600 and 5,150 tons. When the District ceased to legally exist on January 1, 1947, black oxide acquisition totaled to just over 10,000 tons, an amount equivalent to nearly 8,700 tons of pure uranium, or roughly 60 tons of U-235.

By early 1945, all of Groves' fissile-materials production programs were operating. The next task was for scientists and engineers at Los Alamos to turn fissile material into deliverable weapons.

7

Los Alamos, *Trinity*, and Tinian

The Los Alamos Laboratory was the smallest and most secret of the Manhattan Project's sites, and certainly its intellectual center. Its wartime Director, Dr. J. Robert Oppenheimer, would come to be the public face of atomic energy.

Los Alamos' mission of fashioning uranium and plutonium into deliverable weapons sounds almost trivial in comparison to the gargantuan tasks faced at Oak Ridge and Hanford. Simply arrange to bring enough fissile material together at the desired time inside a bomb casing, provide a source of neutrons to initiate the reaction, and then hand the bomb over to a crew trained to transport and drop it, and the job is done. When Oppenheimer took on the directorship of Los Alamos in early 1943, he imagined that he would require only a few dozen scientists, technicians, and engineers. But almost immediately, complexities with fissile materials and the engineering of bomb mechanics demanded expansions of the Laboratory's staff. By mid-1945, Los Alamos employed over 2,000 people. Experimental physicists had to develop and use instruments to measure nuclear parameters for various materials. Theoreticians developed simulations of nuclear explosions over sub-microsecond time increments with slide rules, mechanical calculators, and early computers to turn experimental results into predictions of critical masses to inform optimal bomb design specifications. Chemists refined uranium and plutonium arriving from Oak Ridge and Hanford to purity levels of a few parts per million, after which metallurgists worked to cast them into desired shapes. Reactor-produced plutonium proved to have such a propensity to detonate too soon that ordnance experts had to develop a wholly-new

© Springer Nature Switzerland AG 2020
B. C. Reed, *Manhattan Project*,
https://doi.org/10.1007/978-3-030-45734-1_7

high-speed triggering mechanism that had to operate within microsecond-level tolerances. Fissile materials had to be incorporated into practical bombs that could be carried by existing aircraft in combat conditions. Drop tests had to be conducted to refine bomb-casing designs to ensure stable flight characteristics and reliable fuzing mechanisms. All these tasks, along with aircrew training, aircraft configuration, and preparations for overseas operations, were carried out against an ever-present deadline: when sufficient fissile material became available, a bomb had to be ready. Anticipated production schedules in Tennessee and Washington drove the pace of work at Los Alamos.

7.1 Origins of the Laboratory

The idea of a centralized laboratory to coordinate fast-neutron research and bomb design was circulating well before the formal establishment of the Manhattan Engineer District. In the spring of 1942, the OSRD had contracts with no less than nine universities that had accelerators which could be used as neutron sources, but the work lacked overall coordination. Gregory Breit raised the issue of a centralized laboratory when he resigned from the project in May, 1942. A month later, Vannevar Bush and James Conant suggested in their report to Vice-President Wallace, Secretary of War Stimson, and General Marshall that a special committee take charge of all research and development on military uses of fissionable material. Immediately after the Bohemian Grove planning session, Oppenheimer, Fermi, Lawrence, Compton, Edwin McMillan, and others met in Chicago over September 19–23, 1942, to consider the notion of a dedicated laboratory.

When Leslie Groves was assigned to the project, his letter of appointment made no mention of a design laboratory. He began his new assignment with a familiarization tour of project sites, and met Oppenheimer for the first time in Berkeley on October 8, at which time they discussed the concept of a central laboratory. Groves approved the idea on October 19. In Manhattan District lingo, Los Alamos was known as Project Y.

Groves wanted a site which would be isolated, relatively inaccessible, have a climate that would permit year-round construction and operations, be large enough to accommodate a testing area, and be sufficiently inland to be secure from enemy attack. None of the Project sites at Oak Ridge, Chicago, or Berkeley was sufficiently isolated, and the latter was considered far too vulnerable to Japanese attack. Groves assigned the problem of locating a site to Major John Dudley of the Corps of Engineers. After speaking with some of the scientists involved, Dudley estimated that a staff of some 265 would need to be accommodated. He investigated various locations in California, Nevada, Utah, Arizona, and New Mexico. One possibility near Los Angeles was rejected by Groves on security grounds, and another near Reno, Nevada, was discounted on the basis that heavy snowfalls would interfere with winter operations. The choices narrowed to two sites north of Albuquerque, New Mexico: one about 50 miles north of the city in the Jemez Springs area, and another about 25 miles northeast of Jemez near Los Alamos. Jemez is the Indian name for "Place of the Boiling Springs," and Los Alamos means "the poplars." The latter site, set on a mesa at an altitude of 7,300 feet, was then serving as the home of the Los Alamos Ranch School, a financially-troubled wilderness school for boys. On November 16, Groves, Oppenheimer, Dudley, and Edwin McMillan set out on horseback to inspect the two sites. The Jemez site proved to be in a valley prone to floods, but the Los Alamos mesa was surrounded by deep canyons which would be perfect for test sites. It also had the advantage of 54 ready-to-occupy buildings owned by the school, including 27 houses and dormitories. Oppenheimer owned a ranch not far from Los Alamos, and had spent part of every summer there throughout the 1930s.

Los Alamos was a bargain: about 75 square miles were acquired at a cost of just under $415,000, a tiny fraction of the Manhattan Project's budget; all but some 8,900 acres were federal lands under the jurisdiction of the Forest Service. Groves acquired right of entry to the lands and property of the school on November 23, obtained authority to acquire the site two days later, and authorized the Albuquerque District Engineer to proceed with construction five days after that, just two days before CP-1 went critical in Chicago. To allow students to complete their studies, the Ranch School was given until February 8 before it had to formally relinquish the site. Christmas vacation was cancelled, and the last four students were awarded their diplomas on January 21. To its residents, Los Alamos became known as "The Hill"; Oppenheimer biographers Kai Bird and Martin Sherwin have described the Laboratory as a combination army camp and mountain resort. The entire community would be fenced and guarded, and the Laboratory itself, the "Technical Area," would be built within an inner fenced area that had been the site of the school; 25 outlying test sites were also eventually constructed. Construction costs at Los Alamos ran to some $26 million during the war.

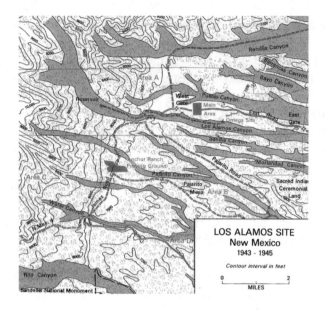

The Los Alamos area. From V. C. Jones, United States Army in World War II: Special Studies—Manhattan: The Army and the Atomic Bomb. Courtesy Center of Military History, United States Army

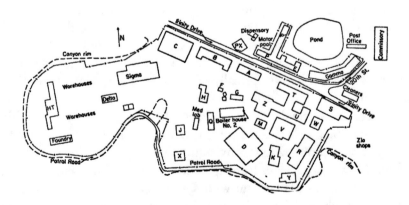

Map of the main Los Alamos "Tech Area". The town proper and residential area were on the north side of Trinity Drive. Source: Edith C. Truslow, Manhattan District History: Nonscientific Aspects of Los Alamos Project Y 1942 through 1946. Los Alamos report LA-5200; http://www.fas. org/sgp/othergov/doe/lanl/docs1/00321210.pdf

Groves wrote in his memoirs that neither he, Bush, or Conant felt committed to Oppenheimer as Director; he was seen to have a number of serious drawbacks. While regarded as brilliant and broadly-educated, Oppenheimer was not an experimental physicist. A quintessential academic, he had no administrative experience such as being a department chair or Dean. His left-wing politics were considered highly suspect, and, unlike Lawrence and Compton, he had not been awarded a Nobel Prize. But neither Lawrence or Compton could be spared from his own work. Lawrence preferred the idea of McMillan as Director, and was outraged when Groves chose Oppenheimer. In his autobiography, Luis Alvarez quotes an unnamed acquaintance of Oppenheimer as considering him incapable of running a hamburger stand. Security officers were so reluctant to clear Oppenheimer that Groves was forced to issue a direct order to them to do so. His July 20, 1943, directive to the District Engineer read

> In accordance with my verbal directions of July 15, it is desired that clearance be issued for the employment of Julius Robert Oppenheimer without delay, irrespective of the information which you have concerning Mr. Oppenheimer. He is absolutely essential to the project.

Oppenheimer's success defied all expectations. Theoretical physicist Victor Weisskopf described his managerial style: "He did not direct from the head office. He was intellectually and even physically present at each decisive step. He was present in the laboratory or in the seminar rooms, when a new effect was measured, when a new idea was conceived. It was not that he contributed so many ideas or suggestions; he did sometimes, but his main influence came from something else. It was his continuous and intense presence, which produced a sense of direct participation in all of us; it created that unique atmosphere of enthusiasm and challenge that pervaded the place throughout its time … The location … gave it a special character by its romantic isolation, in the midst of Indian culture. Living in this unusual landscape, separated from the rest of the world, in walking distance of the laboratories—all this created a community type of living, where work and leisure were not separated. But the special flavor came from the kind of people that were there. It was a large community of active scientists, many of them in their most vigorous and productive years." Another Oppenheimer biographer, Abraham Pais, remarked that "In all my life I have never known a personality more complex than Robert Oppenheimer."

Left: Robert Oppenheimer (1904–1967), ca. 1944; Right: John Manley (1907–1990) in 1957. Sources: http://commons.wikimedia.org/wiki/File: JROppenheimer-LosAlamos.jpg; *Los Alamos National Laboratory, courtesy AIP Emilio Segre Visual Archives, Physics Today Collection*

7.2 Organization

Even before being formally appointed as Director, Oppenheimer was delegated to recruit scientists to staff the new laboratory, and spent the latter part of 1942 and early 1943 traveling around the country doing so. Essentially, he had to recruit a staff for a purpose he could not disclose, all of whom would have to relocate to a place he could not specify, for a duration he could not predict. He could reveal very little of the Laboratory's ultimate purpose. Many leading scientists were already deeply involved in radar and other war work, and considered the bomb an improbable venture.

Oppenheimer particularly wished to recruit two outstanding physicists who were then working on radar at MIT, Robert Bacher and Isidor Rabi. Both were crucial to the radar program, and initially refused to have any connection with a military-directed project. Rabi in particular was concerned that Los Alamos was planned to be a military installation, an arrangement squarely at odds with the scientific tradition of decentralized authority. In a letter to Conant on February 1, 1943, Oppenheimer related that following lengthy discussions with Rabi, McMillan, Bacher, and Alvarez, Rabi felt (and the others concurred) that an indispensable condition was that the Laboratory be demilitarized to avoid the possibility that scientific autonomy would lose out against having to follow military orders. In a letter to Rabi on February

28, Oppenheimer remarked that "I know that you have good personal reasons for not wanting to join the project, and I am not asking you to do so. Like Toscanini's violin, you do not like music." But he went on to ask two things of Rabi: that he participate in an opening conference at the Laboratory to be held in April, and that he use his personal influence to persuade Bacher and Hans Bethe of Cornell University—then also working on radar—to join the project, which they did. Rabi did not formally join Los Alamos, but did visit frequently as a consultant.

Left to right: Robert Bacher (1905–2004); I. I. Rabi (1898–1988) in 1983; Kenneth Bainbridge (1904–1996) holding a photograph of the Trinity explosion, 1945. Sources: http://commons.wikimedia.org/wiki/File:Robert_F._Bacher.jpg; *Photo by Sam Treiman, courtesy AIP Emilio Segre Visual Archives, Physics Today Collection;* http://commons.wikimedia.org/wiki/File:BainbridgeLarge.jpg

Oppenheimer was formally appointed Director on February 25, 1943. As recorded in the appointment letter from Conant and Groves, a compromise had been found on the militarization issue. The Laboratory's work was to be divided into two periods. The first would involve "experimental studies in science, engineering, and ordnance," while the second would see "large-scale experiments involving difficult ordnance procedures and the handling of highly dangerous material." Los Alamos would operate on a civilian basis during the first period, with personnel, purchasing, and business operations to be carried out under an operating contract with the University of California. But when the second part of the work was to begin, which was anticipated as being not earlier than January 1, 1944, the scientific and engineering staff would become commissioned officers. Oppenheimer was authorized to show the letter to individuals whom he was trying to recruit. Ultimately, Los Alamos functioned as a hybrid military-civilian-contractor organization with two heads. Formally, it was a military post with a Commanding Officer who reported to Groves and who was responsible for living conditions and the conduct of military personnel. All residents, civilian and military alike, were subject to military security and censorship regulations. Oppenheimer was responsible for the technical, scientific, and security aspects of the program. Civilian employees never were commissioned, and remained employees of the University of California or other contractors. Los Alamos was formally activated as a military post on April 1, 1943, and the University of California contract became effective on April 20, retroactive to January 1.

Responsibility for overall direction of the Laboratory's scientific work lay in Oppenheimer's hands, but he was always assisted by a number of boards and committees. The first informal group, comprising Oppenheimer, Robert Wilson, Edwin McMillan, John Manley, Robert Serber (a former postdoctoral student of Oppenheimer, then at the University of Illinois) and Associate Director Edward Condon (Westinghouse Electric), met on March 6, 1943, to begin considering when people and equipment would arrive and how the work would be organized. This initial group was superseded a few weeks later by a Planning Board, which met through early April to organize the laboratory's technical operations. The Planning Board was subsequently replaced by a more permanent Governing Board, which comprised Division leaders, administrative officers, and individuals serving in technical liaison capacities. The role of the Governing Board was to consider the work of the laboratory as a whole and to relate it to progress in other parts of the Project. The initial organizational structure of the laboratory comprised an Administrative Division and four Technical Divisions: Chemistry (later Chemistry and Metallurgy) under Glenn Seaborg's Berkeley colleague Joseph Kennedy; Ordnance and Engineering under Navy Commander William S. "Deak" Parsons; Experimental Physics under Robert Bacher; and Theoretical Physics under Hans Bethe. Within each Division were housed a number of individual research groups. Divisions, groups, and committees would come into and go out of existence as the work of the Laboratory evolved, but the basic structure of groups operating within larger divisions remained and is still in place today. Oppenheimer apparently considered that he would lead the Theoretical Division as well as serving as Director, but was dissuaded from that notion by Rabi.

Aside from technical issues, the Governing Board also had its hands full with issues such as housing, construction priorities, water supply, recruitment, security restrictions, procurement bottlenecks, morale, and salary scales. Two later appointments to the Board were George Kistiakowsky and Kenneth Bainbridge, both of whom were recruited from Harvard University. Kistiakowsky was an expert on explosives, and would become intimately involved with the plutonium implosion bomb; Bainbridge, a physicist, would direct the *Trinity* test. The Board remained in place until mid-1944, when it was replaced by separate Administrative and Technical Boards during a reorganization prompted by the plutonium spontaneous fission crisis.

Just as in a university or industrial laboratory setting, the research groups required various support services such as a library, machine shops, photographic and drafting shops, optical shops, business offices, and safety and medical services. The ordnance program alone grew so extensive as to require its own machine shop, capable of handling some 2,000 man-hours of work per week; at one point, more than 500 machinists and toolmakers would be employed at Los Alamos. By July 1945, the library, which was organized by Robert Serber's wife, Charlotte, held some 3,000 books, copies of some 1,500 microfilmed reproductions of articles and parts of books, was receiving 160 journals per month, and served as a repository for some 6,000 internally-generated technical reports. A Patent Office dealt with protection of government interests in any technology that might be developed; about 500 cases were reported to OSRD headquarters in Washington. Directly reporting to Oppenheimer was the Health Group, which bore responsibility for setting health and safety standards and procedures for working with radioactive, explosive, and toxic materials. The work of the Health Group grew substantially in the spring of 1944 when the first shipments of plutonium began to arrive from Oak Ridge. Plutonium is not an external body hazard, but because it tends to collect in bones and kidneys and is only slowly eliminated from the body, the potential harmful dose was set at the very low level of 1 microgram. The scale of safety operations can be gleaned from statistics for the month of July, 1945: 630 respirators decontaminated; 17,000 articles of clothing laundered; and 3,550 rooms being monitored. The Health Group was directed by Dr. Louis Hempelmann, a radiologist recruited from Washington University. No accidental occupational deaths occurred at Los Alamos during the war, but radiation overdoses did lead to two postwar deaths.

7.3   The Primer

One of the first decisions made by the Planning Board was to sanction a series of orientation lectures for arriving personnel. The lectures were delivered by Robert Serber on April 5, 7, 9, 12, and 14, and were recorded by the laboratory's Deputy Director, Edward Condon, whose notes were printed up as a 24-page booklet titled *The Los Alamos Primer*. Designated as Los Alamos' first official technical report, only 36 copies were printed at the time. Declassified in 1965 and published in book form in 1992 with annotations by Serber, the *Primer* is now considered a foundational document in the history of nuclear weapons. The lectures, which were attended by about 30 people, were held in a library reading room, accompanied by background hammering as carpenters and electricians went about their work. In one annotation, Serber recalls that as he began lecturing and used the term "bomb," Oppenheimer, concerned that workmen would overhear, sent John Manley forward to tell Serber to use the term "gadget" instead. Edward Condon, upset with Groves' policy of compartmentalizing information, resigned from Los Alamos before April was out.

The *Primer* still makes for compelling reading. The first section, "Object," gets directly to the point: "The object of the project is to produce a *practical military weapon* in the form of a bomb in which the energy is released by a fast neutron chain reaction in one or more of the materials known to show nuclear fission." Subsequent sections touched on all major aspects of bomb design and operation: reaction cross-sections; the energy released in fission; how a chain reaction operates; the energy spectrum of fission neutrons; why natural uranium is safe against a fast-neutron chain reaction; the use of diffusion theory to estimate the critical mass; how a tamper can be used to lower the critical mass; the expected efficiency of a weapon; the extent of damage expected from blast, thermal, and radiation effects; how a bomb could be triggered; and the probability of low-efficiency "fizzle" explosions arising from effects that could cause the weapon to detonate before the intended moment. Many experimental and theoretical details remained to be filled in, but the basic outline of an overall strategy for the development of fission bombs was fairly clear by the spring of 1943. Immediately after Serber delivered his lectures, a series of conferences were organized to plan the Laboratory's research program. These were held from April 15 to May 6, during which time the laboratory was visited by a special committee that had been appointed by Groves to review plans. The group was chaired by Warren Lewis, who had been involved with the DuPont-initiated review of the entire program in late 1942. The other members were Edwin L. Rose, an ordnance specialist and Director of Research for the Jones and Lamson Machine Company (a precision machine-tool company with ordnance contracts); theoretical physicist John van Vleck (also of the Compton committee); Harvard University physical chemist and explosives expert E. Bright Wilson; and Richard Tolman.

The committee submitted its report on May 10. They approved the Laboratory's proposed program of nuclear research, but recommended major changes in two areas. The first was that final purification of plutonium should be carried out at Los Alamos rather than in Chicago; the rationale for this was that since further purification would likely be required after experimental use of the material at Los Alamos, it might as well be done there. The other was that ordnance development and engineering should be undertaken as soon as possible, and that such work should cover safety, arming, firing and detonating devices, transport of the bomb by aircraft, and studies of bomb trajectories; it was suggested that a Director of Ordnance and Engineering be appointed to coordinate these efforts. These proposals were estimated to require an increase in the number of chemists at the Laboratory by thirty, as well as a two-fold increase in the number of people working on ordnance issues.

The appointment of an ordnance director resulted in a violation of President Roosevelt's admonition to Vannevar Bush to keep the Navy out of the Manhattan Project. At a Military Policy Committee meeting in May, 1943, Groves asked for advice in filling the position. His desire was to find an individual who possessed sound understanding of both the theory and practice of ordnance (high explosives, guns, and fusing mechanisms), but who also had a sufficiently strong scientific background to hold the respect of Los Alamos' professional scientists. Since the appointee might well accompany the eventual bombs into combat, it was also desirable that he be a military officer. As Groves related the story, it was Bush himself who suggested Parsons. Parsons had just completed several years of work on development and testing of proximity fuses, and had met Groves in the 1930s when he was working on radar development for the Navy and Groves was working on infrared technology for the Army. Chemist Joseph Hirschfelder, who worked closely with Parsons, considered him the "unsung hero" of Los Alamos.

To make estimates of critical masses, Los Alamos theoreticians needed accurate measurements of parameters such as cross-sections, secondary neutron numbers, and the spectrum of kinetic energies of fission-generated neutrons. Setting up experiments to obtain such measurements became the first order of business for the Experimental Physics Division. Such a program required large-scale equipment such as particle accelerators, but there was no time to undertake the design and construction of such devices from scratch. As John Manley described it, "What we were trying to do was build a new laboratory in the wilds of New Mexico with no initial equipment except the library of Horatio Alger books or whatever it was that those boys in the Ranch School read, and the pack equipment that they used going horseback riding, none of which helped us very much in getting neutron-producing accelerators." To get work underway, universities sold or loaned the necessary equipment. A cyclotron from Harvard, two Van de Graaff generators from the University of Wisconsin, and a deuteron accelerator from the University of Illinois made their way to Los Alamos. All were used to produce neutrons to bombard various materials; the various machines permitted experimenters to generate neutrons of energies from thermal to millions of volts. No one method was ever relied upon for any particular measurement; overlapping values were always acquired. The bottom pole-piece of the magnet for the Harvard cyclotron was laid on April 14 (the day of Serber's last lecture), and experiments with it began in July. Initially, the Laboratory possessed only about 1 g of U-235 and only micrograms of plutonium; scheduling of experiments and handoff of material between experimental groups had to be carefully monitored. Los Alamos' first experimental result emerged in mid-July, a measurement of the number of neutrons emitted in the slow-neutron fission of a 165-microgram sample of plutonium. At about 2.6, this was about 20% greater than the corresponding number for uranium. Measurements of fission cross-sections for both elements began soon thereafter.

At the time the Lewis Committee was preparing its report, the Governing Board acted on a proposal that would have serious international repercussions years later. On May 6, Hans Bethe suggested holding a regular technical colloquium every week or two. Groves saw the idea as an enormous risk to his policy of compartmentalization, wherein individuals were to have access only to what they strictly needed to know to do their job, but Oppenheimer felt that a colloquium would be the most efficient way to share information among individuals with legitimate need-to-know; Groves relented, although he had Oppenheimer agree to restrict the number of participants and establish a vouching system. Groves raised the issue at a meeting of the Military Policy Committee on June 24, which resulted, via Vannevar Bush, in a letter to Oppenheimer from President Roosevelt on June 29. The President expressed his appreciation for the scientists' work on behalf of the war effort, but made clear the need for very strict secrecy. Groves' concern would prove justified. One of the regular colloquium participants was theoretical physicist Klaus Fuchs, a German-born member of the British Mission to Los Alamos, who later passed detailed design information on the *Fat Man* implosion bomb to the Soviets. Remarkably, security personnel did not follow Fuchs on occasions when he left Los Alamos. His treachery was not discovered until after the war, at which time he was working for the British atomic energy program. In 1950, he was convicted of espionage and jailed; after his release in 1959, he emigrated to East Germany, and lived in Dresden until his death in 1988. We cannot know what Fuchs might have passed to the Soviets had he *not* been privy to the colloquia, but attending them certainly gave him a synoptic view of the laboratory's activities. As John Manley wrote, Fuchs didn't have to *penetrate* Los Alamos, he was an official member of the staff, and a very respected member of the Theoretical Division; by having himself appointed as liaison between the Theoretical Division and the later

Explosives Division that did much of the work on the plutonium bomb, he gained an intimate working knowledge of that device. Two other Los Alamos employees, Theodore Hall and David Greenglass, also passed information to Soviet operatives. Groves' compartments were by no means sealed.

Klaus Fuchs (1911–1988), ca. 1940. Source: http:// commons.wikimedia.org/ wiki/File:Klaus_Fuchs_-_ police_photograph.jpg

7.4   Life on the Hill

Oppenheimer's notion of running the laboratory with a staff of a couple hundred soon collided with the enormity of its task. On average, the working population of Los Alamos doubled about every nine months. By June, 1943, The Hill was home to over 300 officers and enlisted personnel in addition to some 460 civilians. By the end of the year, the total was approaching 1,100. A census in May, 1945, counted 1,055 members of the military Special Engineer Detachment; 1,109 civilians, and 67 Women's Army Corps members, for a total of over 2,200. Like Oak Ridge, one product for which Los Alamos became known was babies. The most probable age of staff members was only 27. Many were recent college graduates starting families, and they wasted no time in doing so. During the war, 208 babies were born at Los Alamos, including Oppenheimer's daughter, Katherine, in December, 1944; nearly 1,000 would arrive between 1943 and 1949. All birth certificates listed addresses as Box 1663, Santa Fe, New Mexico, the Laboratory's official location. By June, 1944, one-fifth of all the married women at Los Alamos were in some stage of pregnancy, and approximately one-sixth of the population were children. The spate of fecundity prompted a poem:

The General's in a stew
He trusted you and you
He thought you'd be scientific
Instead you're just prolific
And what is he to do?

By the time of the *Trinity* test in July 1945, Los Alamos would boast a total population of just over 8,000. By the end of 1946, the number of housing and apartment units for families alone numbered 617, not including 16 ranch houses obtained from the original school, dozens of trailers, and 51 less ostentatious "winterized hutments". Thirty-six dormitories and 55 barracks provided living quarters for 2,700 single personnel. Fuller Lodge, one of the main Ranch School buildings, served as a dining area; eventually it would serve some 13,000 meals per month. No official census was attempted until April, 1946, by which time the mesa was a community of about 10,000.

Even more than Oak Ridge and Richland, Los Alamos was a frontier town. Despite the spectacular surroundings and sense of companionship, life could be arduous. Living conditions for early arrivals often involved several families crowded together or housed in nearby guest ranches. Wartime construction restrictions dictated that new houses were to be equipped only with showers, with the result that the only bathtub-equipped ones were a few which had served as residences for Ranch School teachers; this area became known as "Bathtub Row." Roads were primitive at best. Housing, water, milk, meat, and fresh vegetables were always in short supply. No sidewalks, garages, or paved roads were put in. All houses were painted Army green and known as greenhouses; Los Alamos was regarded as having the worst housing of the entire Manhattan Project. To conserve water, bathers were encouraged to limit showers to a minute or two. At the high altitude, meals could take hours to cook. Turning on a faucet might yield algae, sediments, or worms. James Conant's granddaughter, Jennet Conant, has written that one GI named the place "Lost Almost." Ruth Marshak, wife of physicist Robert Marshak, described her feeling about the place as "akin to the pioneer women accompanying their husbands across uncharted plains westward, alert to dangers, resigned to the fact that they journeyed, for weal or woe, into the Unknown." For new arrivals, the first stop after a long, dusty journey was an unassuming office at 109 East Palace Avenue in Santa Fe. There they would be met by Mrs. Dorothy McKibbin, who arranged for an even dustier ride north to the mesa. McKibbin began work in March, 1943, and managed the office until retiring in 1963.

A serious problem for all Project sites, particularly Los Alamos, was that of securing enough technically-trained personnel. Many scientists' wives were pressed into service in technical, hospital, administrative, and school-system positions; by October,1944, some 30% of the Laboratory's 670 civilian employees were women. To prevent scientifically-educated individuals such as graduate students from being drafted and sent overseas or otherwise lost, the Manhattan District recruited them into a so-called Special Engineer Detachment (SED), which was created on May 22, 1943, as the 9812th Technical Service Unit. By the end of 1943, nearly 475 SED's were assigned to Los Alamos, and by August 1944 they comprised almost one-third of the Laboratory's scientific staff; their number reached some 1,800 by the end of the war. The transition from civilian to military life for SEDs was more than symbolic, however. Housing could not be provided for married enlisted men, and security regulations prohibited them from bringing their wives to Santa Fe or other nearby communities. Each man was allocated only 40 square feet in a barracks, and not until the summer of 1944 were furlough regulations relaxed. Groves does not mention the SEDs at all in his memoirs.

As at Oak Ridge and Richland, all of the services expected by a highly-educated population had to be provided. A Town Council was established, with members elected by popular vote. Nursery and elementary schools had to be set up; by the end of 1946 the elementary school alone enrolled over 350 students. A high school, traffic laws, courts, cafeterias, sewage systems, a fire department (chimney and brush fires were common; Los Alamos sported over 6,800 fire extinguishers), laundry services, a general store, a motor pool, an automobile repair garage, a cleaning and pressing shop, a post office, garbage collection, veterinary services (over 100 horses for the Military Police alone), dental services, and a hospital had to be organized. A policy for housing assignments and rental rates was established which took into account an employee's occupation, family status, and salary. Recreational activities included hiking, horseback riding, skiing, skating, numerous parties, visits to Indian pueblos, and a rough nine-hole golf course. For the first 18 months of the project, security regulations severely restricted personal off-site travel; Groves did not want anyone to use outside facilities. Mail was subject to censorship; all incoming letters had to be addressed to Box 1663, and outgoing ones could not contain last names or information which might provide a clue as to the laboratory's location; the word "physicist" was strictly forbidden. Eventually, every resident over the age of six was issued a security pass. Even at the top administrative levels, Groves kept Los Alamos largely isolated from other branches of the Project; any liaisons with other sites or individuals had to be personally sanctioned by him, with discussions to be limited to a list of approved topics. To the outside world, Los Alamos did not exist.

7.5   Roosevelt, Churchill, and the British Mission

A group that made contributions to the Manhattan Project out of all proportion to its number was a contingent of British and European-born scientists known as the British Mission. The story of how these scientists came to America had as much to do with politics and the personal chemistry between Franklin Roosevelt and Winston Churchill as it did with physics and engineering.

Although Churchill maintained a keen interest in scientific developments that might impact military technologies, the British atomic program suffered chronic mis-handling at the highest political levels despite its early start with the Frisch-Peierls memorandum and the MAUD committee. Churchill's personal scientific advisor was Sir Frederick Lindemann, an Oxford physics professor. Lindemann had taken up his position at Oxford in 1919, and, while he had been very active in aeronautical research during World War I, he soon gave up research to be more of a popularizer of scientific developments. Moving comfortably in high British social circles, Lindemann and Churchill first met in the early 1920s, and developed a strong friendship. When Churchill became First Lord of the Admiralty at the outbreak of World War II, he appointed Lindemann as his private scientific advisor. One of their innovations was to set up the First Lord's Statistical Branch for keeping track of any facts, figures, and economic data relevant to the prosecution of the war. When Churchill became Prime Minister in May, 1940, Lindemann became one of the most influential scientists ever to serve in government; over the course of the war he forwarded some 2,000 briefing papers to Churchill, almost one per day. But his grasp of developments in modern physics was weak; his advice tended to be narrow, and he was regarded as arrogant, amateurish, and disconnected by most scientists. Unfortunately, Lindemann's position enabled him to sideline advice form more informed sources such as Henry Tizard, who resigned in frustration from the Air Ministry the summer of 1940. In October, 1940, Churchill moved to blunt some of the criticism of Lindemann by appointing a Scientific Advisory Committee under the chairmanship of senior civil servant Lord Maurice Hankey, but Lindemann's influence was so strong that it would take some time for Churchill to appreciate the revolution in strategic thinking that nuclear weapons would portend. For his part, Lindemann thrived in his position: In June, 1941, he became Lord Cherwell, named after a stream in Oxford. Hankey would be dismissed from his position by Churchill in March, 1942, when he questioned Lindemann's influence one too many times.

Left: Winston Churchill with his scientific advisor Lord Cherwell (extreme left), Air Chief Marshal Sir Charles Portal, and Admiral of the Fleet Sir Dudley Pound, watching a display of anti-aircraft gunnery, June 1941. Source: https://commons.wikimedia.org/wiki/File:Winston_Churchill_ with_his_scientific_advisor_Lord_Cherwell_(extreme_left),_Air_Chief_ Marshal_Sir_Charles_Portal_and_Admiral_of_the_Fleet_Sir_Dudley_ Pound,_watching_a_display_of_anti-aircraft_gunnery,_June_1941_ H10306.jpg. *Right: Sir John Anderson (1882–1958). Source:* https:// upload.wikimedia.org/wikipedia/commons/c/cd/British_Political_ Personalities_1936-1945_HU59483.jpg

The formal route of the MAUD report was through Hankey's committee, but Lindemann was not about to wait for that. In a six-page memo dated August 27, 1941, he apprised Churchill of its contents. Lindemann stated that while it seemed almost certain that a bomb could be made, he was skeptical of the two-year timeline, giving it odds of no better than even, but advised that the project should go ahead on the grounds that if the German were to acquire such a weapon, they could defeat England or reverse the verdict of the war after England had defeated Germany; he also felt that Britain should undertake the work on its own. Less than two years later, the British would be swimming against a similar exclusionary perspective from the other side of the Atlantic. In a memo to his Chiefs of Staff on August 30, Churchill advocated that no expense be spared to push the project, thus becoming the first national leader in the world to support a nuclear weapons development program. The Chiefs endorsed the project on September 3, assigning Sir John Anderson, a member of the War Cabinet, responsibility for the effort. Anderson's involvement in wartime British administration ran deep: he would also serve as Chancellor of the Exchequer from September, 1943, to July 1945. As a student, he had studied geology, chemistry, and mathematics, and even written a thesis on uranium, but then turned to a career in the civil service where he built a reputation as a master administrator. In one important way, however, his opinion on the project differed from that of Churchill and Lindemann: he was firmly of the opinion that the bomb should be built in America. In the British program, likely only Churchill, Lindemann, and Anderson knew the full story; Churchill otherwise kept his Cabinet in the dark. Lindemann briefed Hankey's committee on September 17, but in their own report a week later, that group advocated that a gaseous diffusion plant be built in Canada to remove it from the threat of German bombs.

Less than a month later, on October 12, a golden opportunity landed on Churchill's lap in the form of a private letter from Franklin Roosevelt, who suggested that the two leaders "correspond or converse concerning the subject which is under study by your MAUD Committee and by Dr. Bush's organization in this country, in order that any extended efforts may be coordinated or even jointly conducted...." Roosevelt sent his letter just two days after receiving a critical briefing from Bush. Churchill, however, was wary of sharing technical secrets with America, at least so long as it remained stingy in its support of Britain's war effort. Historian Barton Bernstein has suggested that British scientific advisors, realizing that the bomb could be a revolutionary weapon, may have been reluctant to tie their efforts to an outside party.

While it was short on details such as which countries would possess bombs and how they might be used, Roosevelt's offer represented an opportunity for the British to enter the Manhattan Project on an almost equal footing with America. But the British response was slow and noncommittal. Churchill let several weeks elapse before offering a perfunctory response that Anderson and Lindemann had been delegated to speak with an OSRD representative in London. That meeting took place on November 21, with the British representatives giving the distinct impression that they believed themselves to be in the dominant position. They were also critical of American security, an ironic position given that they would later clear Klaus Fuchs to work at Los Alamos and that Hankey's private secretary had passed on the MAUD report to Moscow in October. Anderson advised Churchill to give Roosevelt a general assurance of the British desire to collaborate. Just over two weeks later, the critical S-1 meeting of December 6 would occur, to be followed the next day by the Japanese attack at Pearl Harbor, events which propelled the American program into high gear. Churchill and Roosevelt soon met for the First Washington Conference to discuss war strategy (22 Dec 1941–14

Jan 1942), but they do not appear to have discussed the bomb at that time; in the immediate pressure of the war, such a thing must have seemed a distant possibility at best. Soon after Churchill's departure from Washington, Roosevelt approved Vannevar Bush's proposal for a much-expanded and reorganized nuclear project. Wallace Akers toured American project sites in early 1942, and realized how far ahead American scientists were on the experimental side of things. On his return to Britain, he proposed to Anderson and Lindemann that the British project should be merged with the American one. Anderson wrote tepidly to Bush that he felt it desirable to continue complete collaboration; Bush responded with a description of his rearrangement of the S-1 administrative structure, but made no commitments.

While the American assessment of cooperation with Britain on atomic matters cooled considerably between the fall of 1941 and late 1942 when Bush informed Roosevelt that there would be no unfairness to the British if all interchange were to cease, the British made valiant if ultimately futile efforts to restore Roosevelt's equal-partnership offer after their initial fumbling of it. In June, Churchill traveled to the United States for the Second Washington Conference, and discussed the issue with Roosevelt in a private meeting at the Presient's family estate in Hyde Park, New York, on the afternoon of June 20. Churchill urged that Britain and America should pool their information, work as equal partners, and share whatever results might emerge, despite the fact that production plants would be located in the United States. Three weeks later, Roosevelt informed Bush that he and the Prime Minister were "in complete accord," but no written agreement had been drafted nor any details specified. Roosevelt, a master political tactician, was famous for telling listeners what they wanted to hear; his words must have sounded reassuring to Churchill, but had no force of law. By this time, Wallace Akers had managed to convince Anderson and Lindemann of his point of view, and on July 30 Anderson wrote a pleading memo to Churchill in which he advocated merging the projects to capitalize on what assets the British could still contribute. Churchill agreed, and on August 5, Anderson attempted to formalize the discussions in letters to Bush, suggesting that a British-designed diffusion plant be built in America, that a heavy-water pile program be transferred to Canada, that a common patent policy be developed, and that a joint nuclear energy commission be established. But by this juncture the initiative had been lost: the American program was in the middle of its transfer to military authority and its attendant secrecy; for Groves and Bush, international negotiations could only be a nuisance and a security risk. While the British had made some progress with diffusion, research on all of the other production methods—electromagnetic, piles, and centrifuges—were strictly

American affairs. On October 1, Bush informed Anderson of the evolving arrangements in America, evasively refering to keeping up contact on how best to put the resources of both countries to work. Henry Stimson discussed the issue with President Roosevelt on October 29, and suggested that matters be allowed to go along for the time being without sharing any more information than was necessary. The British were effectively in limbo.

The rapidly-diverging viewpoints of British and American atomic-project leaders became clear during late 1942 and early 1943, when Akers traveled to America to confer with James Conant. During a meeting on December 11, Conant presented the American perspective, which was that interchange should be restricted only to information that Britain could use during the war. Akers argued that Roosevelt and Churchill intended collaboration in both research and production, and felt that British scientists should have access to all large-scale American developments. Conant reported back to Bush the next day; four days thereafter, Bush carried to Roosevelt the 29-page December 15 Military Policy Committee report which recommended no or only limited interchange. But other factors were in play. On September 29, Britain and Russia had concluded an agreement on exchange of new weapons which covered both those in use and any that might be developed. Roosevelt and Stimson had apparently known nothing of this until around Christmas, when a copy of the agreement, which would cast into doubt the security of any atomic information passed on to the British, reached Stimson. Roosevelt initialed the MPC report on December 28, setting the policy at limited interchange: cooperation in the design and construction of the diffusion plant, research-level information interchange on plutonium and heavy water, and no sharing of information on the electromagnetic method or Los Alamos. Akers was bluntly informed of the new policy in a meeting on January 13, 1943. Churchill brought up the issue with Roosevelt again when the two met at the Casablanca Conference (January 14–24), and further protested to Roosevelt aide Harry Hopkins in late February that interchange restrictions were contrary to the idea of a jointly-conducted effort. Hopkins took no action until prodded again by Churchill by cable on April 1; Roosevelt left it to Bush to develop a reply, which was that there was no reason to change the American position.

Churchill raised the issue yet again during a visit to Washington in late May, 1943, during which Bush was brought into discussions with Hopkins and British advisors. On the rationale that since a weapon might be developed in time for use in the war (in which case the "direct use" scenario would hold), Churchill departed with the understanding that he had secured a promise from Roosevelt that the work was to be joint and that interchange would be resumed. Bush met with Roosevelt on June 24 to review the situation. Roosevelt had apparently not been apprised of Bush's discussion with Hopkins and the British advisors, and did not speak of his promise to Churchill; Bush left the meeting with the impression that Roosevelt had no intention of going beyond the standing limited-interchange policy. Churchill raised the issue with Roosevelt again on July 9, at which point the President acquiesced: on the 20th he wrote to Bush to instruct him to renew full interchange with the British.

At the time of Roosevelt's July 20 directive, Bush was in London conferring with counterparts there on scientific aspects of the war. Roosevelt's note had not arrived when Bush met with Churchill on the 15th, and the Prime Minister was furious that interchange seemed to have stalled. Unaware of the Roosevelt's directive, Churchill, Hopkins, Bush, and Anderson met again on July 22, at which time Churchill offered a five-point proposition that would form the basis of the Quebec Agreement that would be signed a month later. The essential points were that the enterprise would be joint with free interchange; neither government would employ nuclear weapons against the other; neither would pass information to other countries without the consent of the other; use of the bomb in war would require common consent; and the President might limit commercial or industrial uses by Britain in such a manner as he considered fair in view of the expense being borne by the United States. Back in Washington in early August, Bush and Conant met with Anderson at the British Embassy. Working from Churchill's draft proposal, the latter points would go into the Quebec Agreement essentially unchanged, but the interchange issue was still sticky. As a compromise, Anderson suggested the establishment of a "Combined Policy Committee" to coordinate what work would be done in each country and to serve as a focal point for exchanging information. Interchange on scientific research and development was to be "full and effective," but interchange in the area of design, construction and operation of plants was left on an ad hoc basis to be decided by the Committee. Stimson, Bush, and Conant were specified as the American members of the Committee; the others were two British military officers and the Canadian Minister of Munitions and Supply. The formal agreement was signed by Roosevelt and Churchill on August 19 during a meeting in Quebec City, and the Committee met for the first time in Washington on September 8. In effect, the Committee was a bureaucratic dodge that let the Americans decide what information they would release; functionally, it did not set policy, and would meet only eight times in two years. As might be imagined, Lindemann, although skeptical that the bomb would work, was not happy with the agreement. The Anglo-American partnership on atomic energy was an uneasy one, with the leaders of both countries likely thinking beyond immediate wartime objectives to the power balance of the post-war world.

As part of the interchange program, groups of British scientists, both native and newly-naturalized, went to America and Canada. In particular, they became involved with a pile-research program in Montreal, the diffusion and electromagnetic projects, and Los Alamos. James Chadwick headed the "British Scientific Mission in USA," and spent most of his time in Washington. That role was to go to Wallace Akers, but Bush and Groves objected on account of his ties to Imperial Chemical. Nineteen individuals would ultimately be appointed to work at Los Alamos, including Rudolf Peierls, Klaus Fuchs, and Otto Frisch; apparently Groves was allowed no security vetting of appointees. The first two members of the contingent, Frisch and Ernest Titterton, arrived on December 13, 1943.

Hans Bethe was of the opinion that

> For the work of the Theoretical Division of the Los Alamos Project during the war the collaboration of the British Mission was absolutely essential. . . It is very difficult to say what would have happened under different conditions. However, at least, the work of the Theoretical Division would have been very much more difficult and very much less effective without the members of the British Mission, and it is not unlikely that our final weapon would have been considerably less efficient in this case.

In addition to the Los Alamos contingent, over 60 other British scientists worked in Montreal on reactor theory and development. The Canadian-British collaboration began in September, 1942, when the Canadian Minister of Munitions and Supply, Clarence D. Howe, agreed to receive the scientists, provide laboratory facilities, and administer the project as a division of the National Research Council of Canada. This group, which would grow to encompass a staff of over 300 (about half of whom were Canadians), was initially under the administration of Hans von Halban, but his unpopularity and ineptness as an administrator combined with Groves' suspicion of him lead to his being replaced by John Cockcroft. In April, 1944, the Combined Policy Committee decided to proceed with the construction of a heavy-water-moderated reactor in Canada, which was located along the banks of the Ottawa river in Chalk River, Ontario, about 200 km northwest of Ottawa. The resulting Zero-Energy Experimental Pile, 'ZEEP', went critical on the afternoon of September 5, 1945, becoming the first reactor to operate outside of the United States.

7.6   Criticality and Critical Assemblies

A major part of the work for theoretical physicists and mathematicians at Los Alamos was to refine estimates of critical mass and bomb energy yields as experimental data were accumulated.

The fundamental idea behind a critical mass is to assemble a great enough mass of fissile material so such that once fissions have been initiated, more neutrons will cause subsequent fissions every second than will escape from the mass during the same time. The mass will ultimately disrupt itself after a few microseconds, but the goal is to obtain, at least for this brief time, a growing population of neutrons.

The precise value of the critical mass depends on the density of the material, the number of neutrons liberated per fission, the distribution of speeds of neutrons emitted in fissions, and the cross-sections for fission and scattering. The mathematics involved in the analysis relies on diffusion theory, not unlike that involved with the diffusion facilities at Oak Ridge. In the case of criticality, an expression known as the diffusion equation can be used to model the travel of neutrons from where they are created to when they encounter another nucleus and either cause a fission or are scattered; this provides a way of calculating the critical *radius* for a given set of nuclear parameters, which can be transformed into an equivalent mass via the density of the material. But results can be sensitive to slight changes in values of the input parameters, and the complexity of the equations involved is such that solutions can only be determined by numerical approximations.

For uranium-235, the critical mass of a sphere of untamped ("bare") material is about 50 kg (110 lb). For plutonium-239, which has a greater fission cross-section and emits more neutrons per fission than uranium, it is about 16 kg (35 lb). But the densities of these elements are so great, 15–20 times that of water, that a critical mass of either would fit easily in one hand. A regulation softball has a radius of about 5 cm and a mass of about 180 g; a bare critical mass of plutonium is only slightly larger but some 90 times heavier.

A bomb which contains only a single critical mass of fissile material will give only a very inefficient explosion, however. This is because the mathematics of diffusion shows that the condition of criticality can be expressed in one essential parameter: the product of the radius of the core of material (assumed to be a sphere) and the density of the material. If this product exceeds a critical value for a given material, the explosion will be self-sustaining for at least a brief period, but if it falls below the critical value, the chain reaction will shut itself down very abruptly. For a given mass of material, the density is inversely proportional to the cube of the radius. As soon as the explosion starts, the core will begin expanding and the density will drop, but, because of the cubic dependence, the density will drop faster than the radius increases, eventually leading to shutdown. If a bomb comprises only a single critical mass, this will happen as soon as the explosion begins. To overcome this limitation, it is necessary to start with a *supercritical* core containing more than one critical mass: this creates an exponentially-growing reaction and allows for accommodation of some expansion distance, perhaps only a centimeter or two, before shutdown occurs.

In view of this shutdown effect, it would seem that there would be no hope of making an effective nuclear explosion if less than a critical mass of material is available. But there is a loophole in the analysis: if the material can be crushed from its normal density to a higher density before triggering the chain-reaction, you will be starting with a value of the criticality parameter greater than that needed for "normal" criticality. Engineering such an implosion is incredibly challenging, but opens up the possibility of making a bomb with considerably less fissile material than would be the case for a non-implosion bomb. At Los Alamos, implosion would prove essential to developing a successful plutonium bomb.

Further gains in efficiency can be had by surrounding the bomb core with a metal tamper. This boosts the efficiency in two ways. Most importantly, if the tamper material reflects escaping neutrons back into the core (without capturing them), this effectively gives more neutrons per fission and so lowers the critical mass; a tamper just a few centimeters thick can achieve a factor-of-two reduction in critical mass. Less important but still a contributing factor is that by briefly retarding the core expansion by its sheer inertia, the tamper allows criticality to persist a little longer. A weapon that utilizes both implosion and a tamper will be more efficient yet. A tamper is dead weight as far as transporting a bomb to a target goes, but is vital to its efficient functioning.

The question of how to assemble a supercritical mass was considered very early on in the Los Alamos project. In the *Primer*, the first and most straightforward system described is the "gun" method; Serber referred to it as "shooting." As ultimately realized in the Hiroshima *Little Boy* bomb, the concept is to begin with two sub-critical pieces of fissile material placed within the barrel of an artillery cannon. A cylindrical target piece is held fixed within a tamper casing at the nose end of the barrel, while a mating sleeve, the projectile piece, is fired toward the target piece from the tail end. When fully mated, the two comprise more than a critical mass. Since the average density of the projectile piece is fairly low because it is hollow, it can by itself comprise by itself the equivalent of more than one "solid" critical mass, giving the completed assembly over two critical masses. By surrounding the target piece with a heavy tamper, the completed assembly can comprise several tamped critical masses.

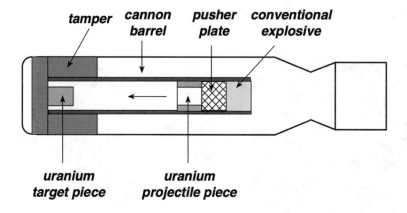

Schematic illustration of a gun-type weapon. The uranium projectile is fired toward a mating target piece in the nose

In World War II, the greatest muzzle velocity achievable with artillery pieces or naval cannons was about 1,000 m/s. Since the target and projectile pieces are on the order of 10 cm in size, this speed implies that about 100 microseconds will elapse between the time that the leading edge of the projectile piece first encounters the target piece and when full assembly is achieved. This 100-microsecond assembly timescale—much shorter than the blink of an eye—will prove to be extremely important.

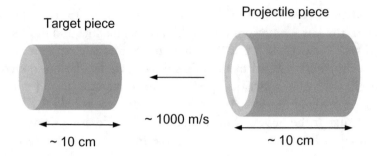

Assembly process for a gun-type fission weapon

Another method described by Serber contained the genesis of implosion. As described in the *Primer*, the idea was to mount pieces of fissile material on the inside of a ring, with explosive material distributed around the outside. When fired, the fissile pieces would be blown inward to form a cylinder or sphere and hence form a critical assembly.

Left: Seth Neddermeyer (1907–1988) in his later years. Source: Photograph by David Azose, courtesy AIP Emilio Segre Visual Archives, Physics Today Collection. Right: Sketch of early implosion concept adapted from Robert Serber's Los Alamos Primer. The four triangular-shaped wedges are pushed together by detonating an enclosing ring of explosives; the fissile material (shaded) consequently forms a cylindrical critical assembly. If the wedges are three-dimensional pyramids, a spherical assembly results. Serber's original sketch did not include the surrounding tamper

Conception of the implosion method is often attributed to Seth Neddermeyer, a Caltech physics Ph.D. whom Oppenheimer recruited from the National Bureau of Standards. However, Serber has dismissed this attribution as "television history," stating that the idea had been raised by Richard Tolman at the 1942 Berkeley summer conference. Serber and Tolman wrote memos on the subject which apparently went up to Bush and Conant, and Tolman wrote Oppenheimer on March 27, 1943 (just before Serber's first orientation lecture) to describe how it might be possible to blow a shell of "active material" inward upon itself with an ordinary explosive. Neddermeyer apparently conceived of modifying the idea to surround a thick but initially centrally-hollow cylindrical or spherical core with a tamper, which would be surrounded by a layer of explosive. When detonated at many points simultaneously, the explosive would push inward at several kilometers per second, crushing the core to critical density in much less time than a gun mechanism could assemble subcritical pieces. However the idea originated, Neddermeyer was struck by the concept, and records indicate that it was discussed at Planning Board meetings held on March 30 and April 2, 1943. Neddermeyer soon developed calculations on the velocities that might be achieved, and was assigned by Oppenheimer to lead a small group devoted to implosion research within the Ordnance Division. While implosion was given low priority at first and considered a backup scheme in case the gun method failed to work, preliminary estimates of the size and weight of a spherical implosion weapon were generated in order to investigate how test-drop mockups would fare in comparison to models of the more conventionally-shaped gun bomb. Neddermeyer carried out his first test-shot on July 4, 1943, using tamped TNT surrounding hollow steel cylinders. The symmetry of the implosion was poor, but the shot did demonstrate the fundamental feasibility of using an explosion to crush something.

7.7   Predetonation Physics

The 100-microsecond assembly timescale for a gun-type bomb is one of three timescales that are involved in the efficient functioning of such a nuclear weapon. The assembly time is purely mechanical in origin; the other two involve the physics of the fission process and the rate at which the core expands, and it is their characteristic values that highlight the importance of the assembly time.

The first fission timescale is how much time is required for the entire core to fission once the chain reaction has been initiated. For a core of mass 50 kg, this is only about one to two microseconds. The incredible brevity of this process is a reflection of the tremendous speeds with which neutrons emerge from fissioning nuclei; they travel for only about ten nanoseconds before causing another fission. The second physics timescale concerns the amount of time required for the core to expand to the point where its decreasing density results in criticality shutdown. The evolution of the core to this condition, which is known as "second criticality" ("first criticality" is when conditions for supporting a chain reaction are first achieved), has to be determined via numerical simulations of exploding cores. Calculations by Los Alamos theoreticians indicated that this expansion phase takes about the same microsecond or two as is required to fission the entire core once the chain reaction has started. The similarity of these timescales means that there is a neck-and-neck competition between the exponential growth of the explosion and the onset of criticality shutdown. As Serber wrote in the *Primer*, "Since only the last few generations will release enough energy to produce much expansion, it is just possible for the reaction to occur to an interesting extent before it is stopped by the spreading of the active material."

The shortness of these times in comparison to the 100-microsecond assembly time brings up a potential problem with the latter. At some point during core assembly, a critical mass will come to be present in the partially-assembled system: "first criticality." If a stray neutron should initiate the first fission soon after first criticality, the exploding core may well reach shutdown before assembly is completed. The result would be an explosion of much lower efficiency than what the weapon was designed to achieve on the presumption of the reaction not being initiated until assembly *was* completed. Because the chain reaction can begin at any time between first criticality and full assembly, there will be a range of possible weapon efficiencies; the worst-case scenario is if the chain gets initiated just at the moment of first criticality. Such an extreme predetonation is known to weapons engineers as a "fizzle."

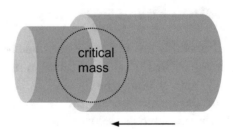

Core achieves first criticality before assembly is completed

There are two possible sources of stray neutrons, and their effects are additive. Both are controllable to some degree, albeit with difficulty.

7.8   The Alpha-n Problem

The first source of possible stray neutrons is that if the fissile material contains any light-element impurities, especially trace amounts of elements such as beryllium or aluminum, neutrons will be generated as a result of alpha-bombardments originating from alpha-decays of the fissile material. Chemical processing will inevitably introduce some level of impurities, and since uranium and plutonium are both natural alpha-emitters, the problem is unavoidable: a stray neutron could initiate a premature reaction as soon as first criticality is achieved. The rates of alpha-emission per gram of material are fixed by nature, so minimizing the probability of a predetonation during core assembly means minimizing the level of impurities and making the assembly time as short as possible. The probability can never be reduced to zero, but can be made acceptably small. Weapons engineers refer to this as the "alpha-n" problem.

Alpha-decay rates or U-235 and Pu-239 are huge: about 80 million per kilogram per second for the former, and about 2.3 *trillion* per kilogram per second for the latter. But so far as the alpha-n problem goes, these are mitigated by two factors: the typical *yield* of light-element reactions, and the distance alpha particles typically travel within the fissile material, their so-called *range*. The yield of a reaction is a reflection of the fact that atoms are mostly empty space: not all alpha-particles will strike a light-element nucleus. Technically, the yield is the fraction of alphas that are likely to strike a light-element nucleus in a well-mixed sample of the alpha emitter and the impurity. For light-element impurities, this is typically about one per ten thousand. As might be surmised from the jargon, the range of a particle is a measure of how far it will travel through some material before coming to rest due to losing energy by causing successive ionizations and suffering collisions in the material.

At Los Alamos, the alpha-decay rate of plutonium was flagged early on as a potential problem, and the most serious potential contaminant was expected to be beryllium. Measurements of yields and ranges indicated that for a 10-kg core of Pu-239, it would be necessary to keep the level of beryllium contamination to no more than about one atom per four to five million of plutonium to keep the probability of a neutron emission over 100 microseconds to less than 1%. Chemists were ultimately able to reduce light-element impurities in plutonium to the level of a few parts per million, which was good enough in view of the second source of neutrons. In the case of U-235, the impurity issue was much more forgiving: a purity of a part per thousand, well within normal chemical processing tolerances, would be entirely adequate. An artillery cannon capable of accelerating a projectile piece of plutonium to the maximum achievable 1,000 m/s was anticipated to be about 17 feet long, a figure which set the initial design length of the gun bomb. On the upside, if the gun design could be made to work for plutonium, it would surely work for uranium as well.

An issue connected to alpha-bombardment yields is the that of how to trigger a nuclear explosion. At Los Alamos, this was accomplished by fabricating devices known as *initiators*; they were also known as "Urchins." Placed within the bomb core, these spheres, approximately the size of a golf ball, contained polonium and beryllium. The two were initially separated, but when the initiator was crushed by the incoming projectile piece (uranium bomb) or implosion (plutonium bomb), they would mix. Polonium is a copious natural alpha-emitter, and these alphas would strike beryllium nuclei and liberate neutrons to trigger the detonation. The idea of such initiators was apparently conceived by Hans Bethe. Manhattan Project initiators used about 50 Curies of polonium-210, which was created by neutron bombardment of bismuth in the X-10 reactor at Oak Ridge and in the production reactors at Hanford. This amount of polonium, equivalent to about 10 milligrams, guaranteed the emission of about 100 neutrons over about a microsecond once mixture was underway.

Initiator manufacture was a difficult business. Polonium is hazardous not only because of its high alpha-activity, but also because it is one of the most motile elements known; it is virtually impossible to work with and avoid its entrance into the human body, although it does get eliminated rapidly and does not collect in bones as do radium and plutonium. Bismuth has a small thermal-neutron capture cross-section, so hundreds of pounds of it had to be irradiated for on the order of a hundred days to make even a small amount of polonium. In a letter to Groves on June 18, 1943, Oppenheimer related that one hundred pounds of bismuth placed near the center of the X-10 pile would create only nine Curies of polonium every four months if the pile were operated at 20 kilowatts per ton of fuel; a full fuel load for X-10 was about 120 tons. Much greater production was anticipated from the Hanford piles when they went into operation; the same 100 lb of bismuth would yield almost five Curies of polonium *per day*.

The polonium was separated from the parent bismuth material at a Monsanto Chemical Company facility located outside Dayton, Ohio, where

Charles Allen Thomas, Director of Research for Monsanto, set up a laboratory in the indoor tennis courts of the estate of his mother-in-law. The first batch of irradiated bismuth reached Dayton in January, 1944, and Los Alamos received its first shipment of polonium, about two Curies worth, by mid-March. The uranium gun bomb used four initiators mounted within the projectile piece; the implosion bomb used one at the very center of the core. The first "production" Urchin unit was completed on June 21, 1945, only about three weeks before the *Trinity* test.

Another aspect of the impurity issue is that plutonium is brittle at room temperature, and consequently difficult to form into desired shapes unless alloyed with another metal. But common light alloying metals such as aluminum cannot be used because of the alpha-n problem; something heavier must be used so that the alphas get repelled by the highly-charged nuclei of the alloying element. Los Alamos metallurgists found that by alloying plutonium with 3% gallium by weight, they could avoid the alpha-n problem while also lowering the melting point of plutonium so that it could be worked at room temperature.

In comparison to the high-profile work in physics and engineering carried out at Los Alamos, the work of the metallurgy group has tended to be overlooked. From a complement of about twenty in June, 1943, the staff of the Chemistry and Metallurgy Division would grow to number some 400, about one-sixth of the Laboratory personnel. Much of the research on the properties of plutonium was carried out by Charles Thomas and Cyril Smith, a metallurgist employed with the American Brass Company who was working with the NDRC in Washington. In a 1981 reminiscence, Smith reflected on the importance of chemistry and metallurgy in the Project:

Of course the nuclear bomb was a physical concept, stemming from physical theory and experiment of the most magnificent kind, but the design would have been nothing without fantastic chemistry, without stupendous achievements in engineering both chemical and mechanical, or if the metallurgists had not been able to fabricate fantastic materials into many tricky shapes. Before any nuclear cross-section could be measured or before any critical assembly could be achieved, something had to be *made*.

Cyril Stanley Smith (1903– 1992) in 1948. Source: Allen M. Clary, Camera Portraits, courtesy AIP Emilio Segre Visual Archives

7.9   Spontaneous Fission

The second source of possible pre-detonation-initiating neutrons arises from the fact that both uranium and plutonium suffer spontaneous fissions. This problem nearly doomed the plutonium-bomb program.

Spontaneous fission operates under the same random-process mathematics as alpha-decay, but differs from the impurity issue in two critical ways: neutrons are emitted directly, and, unlike electrically-charged alpha-particles, they do not suffer any range limitation due to ionization-energy losses. Further, two to three neutrons are typically emitted per fission, as opposed to just one in an alpha-n reaction.

With either *pure* U-235 or *pure* Pu-239, spontaneous fission is not an overly serious concern for bomb engineers. Over 100 microseconds, 10 kg of pure Pu-239 will suffer on average less than 0.01 spontaneous fissions, so the danger of a predetonation is remote. Even if supercriticality for such a core lasts for a full 100 microseconds, the probability that the bomb will achieve its full design yield is greater than 99% *if* the separate issue of light-element impurity levels is addressed successfully. The situation is even more relaxed for uranium.

The problem with plutonium concerned the way it was—and still is— produced. If an already-synthesized Pu-239 nucleus is stuck by a thermalized neutron within a reactor, it has about a one-in-four chance of capturing the neutron and becoming Pu-240 as opposed to undergoing fission. Reactor-produced plutonium *inevitably* contains some percentage of Pu-240. Leaving a uranium fuel slug in a reactor for a longer time will give more Pu-239, but will also lead to more Pu-240 in the bargain. But Pu 240 has a spectacular spontaneous fission rate, almost 500,000 per second per gram: the presence of even a small percentage of it in an otherwise very pure Pu-239 core can be disastrous. The *Trinity* and Nagasaki bombs each used about 6 kg of plutonium, and a 6-kg core contaminated with even only 1% Pu-240 will suffer on average nearly three spontaneous fissions over the course of a 100-microsecond supercriticality time. The probability of *not* suffering a pre-detonation in this case can be calculated to be only about 10%, not terribly compelling odds if hundreds of millions of dollars have been invested in synthesizing plutonium in the first place. The supercriticality period has to be reduced to about 30 microseconds to have a 50% chance of no predetona-tion, and down to about five microseconds for a 90% chance. Aside from the virtually impossible task of removing the offending Pu-240, the only option is to speed up the assembly process to on the order of a few microseconds. Unfortunately, this was and still is impossible with any gun-type mechanism.

The potential for a problem with spontaneous fission was not wholly unap-preciated when Los Alamos was established. Serber discussed the issue in the *Primer*, but the only data then available pertained to natural uranium. Spon-taneous fission in natural uranium had been discovered by Georgi Flerov and Konstantin Petrzhak in the Soviet Union in 1940; they published the discov-ery openly in the *Physical Review*. It was certainly anticipated that plutonium would likely suffer the same effect; in March, 1943, Glenn Seaborg specu-lated in a diary entry that if Pu-240 were created it could complicate the purity issue.

A group at Berkeley led by Emilio Segrè began spontaneous fission research around late 1941. Using plutonium created in Ernest Lawrence's 60-inch cyclotron, they had determined by June, 1943, a spontaneous fission rate for the new element of 18 per gram per hour. Serber relates in the annotated *Primer* (published 1992) that the last time he saw Segrè in Berkeley, he was driving a beat-up old car with a bumper sticker which read "My Owner has A Nobel Prize"; Segrè had shared the 1959 Nobel Prize for Physics for the discovery of the antiproton.

Oppenheimer invited Segrè to move his work to Los Alamos, which he did in June, 1943. Because counting statistics can be confounded by small fluctuations in background radiation, a special remote field station was set up in a Forest Service cabin at Pajarito Canyon, a 14-mile drive from the main area of the Laboratory. Segrè described the cabin as being in one of the most picturesque settings one could dream of; Fermi was fond of visiting the site. Segrè and his group were set up by August, working with five 20-microgram samples of Pu-239 that had been prepared at Berkeley. These proved too small for any reliable determination of the spontaneous fission rate, but they were able to measure the number of neutrons per spontaneous fission as about 2.3. The samples slowly accumulated spontaneous-fission counts, a grand total of six over the course of five months to January 31, 1944. However, small-number statistics are inevitably subject to uncertainty; real confidence would come only with the arrival of larger quantities of pile-produced plutonium from Oak Ridge.

Soon after getting set up, the group noticed a curious effect with spontaneous fission of uranium. This was that the rate for U-238 agreed with what they had measured in Berkeley, but the rate for U-235 was *higher* at Los Alamos than at Berkeley. Since the 235 itself should not have changed, the cause had to be something external. It was soon determined that at the higher altitude of Los Alamos, cosmic rays were inducing more fissions than they could at sea level in Berkeley, where many of them were absorbed by the thicker intervening atmosphere. The cosmic rays were not energetic enough at either location to induce fission in U-238, so their effect on 235 at sea level had gone unappreciated. The rate of natural spontaneous fissions in 238 greatly exceeded the rate of cosmic-ray induced fissions, which led to the realization that the majority of spontaneous fissions in uranium arises from the heavier, non-fissile isotope. The very low rate deduced for 235 in combination with the relatively gentle purity requirements for that isotope led to the relaxation of the assembly velocity requirement for the uranium bomb, with the result that the cannon could be shortened to a length of 10 feet.

The first X-10 plutonium arrived at Los Alamos in the spring of 1944, and was placed in detection chambers on April 5. More arrived over the following week, and it soon became clear that a problem was developing. During the first three days of observations, the pile-produced material exhibited a rate of spontaneous fissions five times that of samples which had been prepared with the cyclotron. In an April 15 report, Segrè reported a tentative rate of 200 spontaneous fissions per gram per hour, or eight times the rate of pure Pu-239. By May 9, the estimate had risen to 261 per gram per hour, which corresponds to a Pu-240 contamination level of only 0.01%. For a 10-kg core that is supercritical for 100 microseconds before assembly becomes complete, a contamination level only ten times greater would reduce the probability of achieving the weapon's full design yield to only about 67%. This portended disaster: Hanford-produced plutonium would contain far more than 0.1% Pu-240 due to the much greater neutron flux of the 250-megawatt production-scale reactors unless the fuel was withdrawn much earlier than planned. Robert Bacher reported the news to Arthur Compton during an early-June visit to Chicago, and related that Compton turned as white as a sheet of paper.

Oppenheimer presented the evidence at a Laboratory colloquium on July 4, which was attended by James Conant. It was clear that the gun-assembly mechanism for the plutonium bomb would have to be abandoned. On the 17th, a conference was held in Chicago with Compton, Oppenheimer, Charles Thomas, and Conant present; Fermi, Groves, and Kenneth Nichols attended another meeting of the same group later that evening. Excerpts from a handwritten summary prepared by Conant give a sense of the severity of the situation (slightly edited):

> The disquieting prospect first discussed with Conant by Oppenheimer on the visit to L. A. on July 4 … was considered. It was concluded that the evidence was now so clear that "49" prepared at Hanford could not be used in the gun method of assembly …. Dr. Oppenheimer was not very optimistic about a speedy solution of the implosion method which is now left as the only hopeful way of using 49.

The next day, Oppenheimer summarized the situation in a letter to Groves: "At the present time the method to which an over-riding priority must be assigned is the method of implosion." John Manley summed up the situation in one sentence: "The choice was to junk the whole discovery of the chain reaction that produced plutonium, and all of the investment in time and effort of the Hanford plant, unless somebody could come up with a way of assembling the plutonium material into a weapon that would explode."

Oppenheimer promptly developed a plan to drastically reorganize the laboratory. The new structure was approved at an Administrative Board meeting on July 20, and was formally instituted on August 14. The shakeup was extensive. The Governing Board was abolished, replaced by separate Administrative and Technical Boards. The plutonium program was removed from the original Ordnance Division and divided between two new Divisions. The first of these, X Division ("Explosives"), headed by George Kistiakowsky, would be concerned with experimentation involving explosives; methods of initiation; development, fabrication and testing of implosion systems; and developing a suitable design for assembly of the explosives and the initiating system. The other new arrival was G Division ("Gadget"; also known as the Weapon Physics Division), which would be under the leadership of Robert Bacher. G Division was to be responsible for developing methods for investigating the hydrodynamics of implosion, with particular emphasis on symmetry, compression, behavior of materials, and developing design specifications for the tamper, core, and neutron-initiating source. G-Division absorbed several groups which had been part of the Experimental Physics Division, which was re-named R (Research) Division; this was led by Robert Wilson of Princeton University and was responsible for criticality experiments, measurements of parameters such as cross-sections and spontaneous fission rates, and for developing instrumentation for the *Trinity* test. The Ordnance Division (now O-Division) retained responsibility for the uranium gun bomb, and remained under William Parsons' leadership; Kistiakowsky and Bacher were to keep each other and Parsons closely informed of their work.

Implosion experimentalists were aided by a Theoretical Division group under the direction of Edward Teller. This group had actually been established in January, 1944, to address analyses of estimating the properties of metals imploded under millions of atmospheres of pressure. This work involved analyzing equations using early computers fed information via punch-cards, making the Manhattan Project the first major scientific endeavor where large-scale simulations complemented experiment and theory. Unfortunately, Teller remained so distracted by the idea of the fusion-based "super" bomb that Oppenheimer had to replace him with Rudolf Peierls in June, 1944. Overall, implosion would come to be the concern of over 14 groups within the T, G, and X-Divisions. Other reorganizations also came into effect at the same time. Enrico Fermi, who had frequently consulted at Los Alamos, arrived on a full-time basis to head the new F-Division after completing his work at Hanford. Named after him, this division took responsibility for problems that did not fit into the work of other divisions, including investigation of the hydrogen bomb; Fermi and Parsons also became Associate Directors of the Laboratory. As the implosion program grew in complexity over subsequent months, other groups arose. The most important of these were the Intermediate Scheduling Conference (under Parsons), the Technical and Scheduling Conference, and the "Cowpuncher" Committee. Both of the latter were under the leadership of Fermi's Chicago colleague Samuel Allison, who arrived in November, 1944. Parsons' committee was responsible for coordinating aspects of the "packaging" of the gun and implosion bombs for testing and eventual delivery to their combat bases, while Technical and Scheduling took on responsibility for scheduling experiments, shop time, and use of fissile material. The Cowpuncher committee was more concerned with the later *Trinity* test.

7.10   The Gun Bomb: Little Boy

Responsibility for the design, engineering, drop tests, and assembly of the uranium gun bomb lay with the Gun Group of the Ordnance Division, which was directed by Albert Francis Birch, a Harvard University geophysicist and Navy Commander who had an extensive background in physics, electronics, and mechanical design. While the gun method was straightforward in principle, it faced a number of unique engineering issues which were unlike typical military ordnance problems. The neutron-reflecting properties of the steel used in the guns had to be determined; if it should prove reflective, it could contribute to lowering the critical mass. The most unusual aspect of the design was that rather than exiting the gun as would a normal shell, the projectile piece was to be *stopped* by the tamper after mating with the target piece so that the chain reaction could proceed for enough time to give reasonable efficiency. Fortunately, U-235 proved to be strong enough to be able to withstand such deceleration without disintegrating.

When the 17-foot-long *Thin Man* configuration was under consideration for both the uranium and plutonium bombs, established ordnance practice indicated that a gun designed to achieve a 1,000-m/s muzzle velocity would weigh five tons and have to be able to withstand a breech pressure of 75,000 lb per square inch from the chemical explosive used to propel the projectile piece. These requirements created a potentially serious problem for delivering such a weapon. The payload of a B-29 bomber depended on the duration of the mission: up to 10 tons could be carried to combat radii of about 1,600 miles, which was about the distance from Tinian Island in the Pacific (from where the bombs would be delivered) to Tokyo. But massive artillery cannons surrounded by heavy tampers and carrying sophisticated triggering electronics were not normally configured as bombs. Could *Thin Man* be configured into a safely deliverable weight? The answer was yes: Edwin Rose realized that regular artillery pieces are designed to withstand thousands of firings, but that such a requirement was entirely unnecessary for a weapon which would be fired only a few times in tests and only once in combat. By sacrificing durability, the otherwise prohibitive weight of a gun bomb could be reduced to a point where it could be made into a practical weapon.

Los Alamos established a gun testing area, the Anchor Ranch Proving Ground, about three miles from the main laboratory area. The first true experimental gun units did not arrive until March, 1944, although the first test shots were fired on September 17, 1943. All Los Alamos guns were fabricated by the Naval Gun Factory at the Washington Navy Yard. The first two delivered were 1,000-m/s prototypes, but they arrived just as the pluto-nium spontaneous fission crisis was emerging, and were abandoned. Three new *Little Boy* guns designed for 300-m/s operation were ordered. Since the Gun Group could not do any test-firings using "active" U-235 components, they had to find a proxy material whose mechanical properties mimicked that of U-235; natural uranium proved adequate for this purpose. Since guns designed for lightness could not be repetitively test-fired, proof-testing con-sisted of a few instrumented firings at 300 m/s with a 200-pound projectile, after which the guns were greased and stored for future use. A number of dummy models made from discarded naval guns were used in drop tests of simulated assembled bombs; these units were not intended for test firing.

An important element in both the gun and implosion designs was the choice of tamper material. The best option would be a heavy metal which elastically scattered neutrons. Responsibility for investigating tampers was assigned to the Radioactivity Group of the Experimental Physics Division. By October, 1943, the list of possible tamper materials had been narrowed to tungsten carbide (steel), natural uranium, beryllium oxide, iron, and lead, although measurements would be made on over two dozen elements, includ-ing gold and platinum. As Robert Serber wrote, "The active material seemed so precious that everything else in contrast seemed cheap. The notion of vaporizing a few hundred pounds of gold in the explosion did not strike us as odd." Ironically, beryllium generates neutrons when struck by alpha parti-cles, but is an excellent reflector of neutrons and would have made an ideal tamper material but for the fact that such use would have virtually exhausted the country's supply at that time.

Tungsten carbide was chosen as the tamper for *Little Boy* on account of its high elastic-scattering cross-section, while natural uranium was used in the *Fat Man* design in view of its inertial and nuclear properties. The first gun-bomb target case to be test-fired proved to be one of the best made. Known as "old faithful," it was tested four times at Anchor Ranch, and was incor-porated into the bomb dropped at Hiroshima. *Little Boy's* 28-inch diameter target case was three feet long and weighed over 5,000 lb. Within the target case resided a 13-inch diameter tungsten-carbide liner (the tamper material proper), which surrounded the 6.5-inch diameter gun tube. The chemical symbol for tungsten carbide, WC, led to its becoming known as "Water-cress."

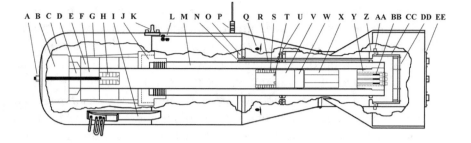

Cross-section drawing of Y-1852 Little Boy showing major components. Not shown are radar units, clock box with pullout wires, barometric switches and tubing, batteries, and electrical wiring. Numbers in parentheses indicate quantity of identical components. Drawing is to scale. Copyright by and used with kind permission of John Coster-Mullen. (A) Front nose elastic locknut attached to 1-inch diameter Cd-plated draw bolt. (B) 15-inch diameter forged steel nose nut with 14-inch diameter back end. (C) 28-inch diameter forged steel target case. (D) Impact-absorbing anvil surrounded by cavity ring. (E) 13-inch diameter 3-piece WC tamper liner assembly with 6.5-inch bore. (F) 6.5-inch diameter WC tamper insert base. G) 18-inch long K-46 steel WC tamper liner sleeve. (H) 4-inch diameter U-235 target insert discs (6). (I) Yagi antenna assemblies (4). (J) Target-case to gun-tube adapter with four vent slots and 6.5-inch hole. (K) Lift lug. (L) Safing/arming plugs (3). (M) 6.5-inch bore gun. (N) 0.75-inch diameter armored tubes containing priming wiring (3). (O) 27.25-inch diameter bulkhead plate. (P) Electrical plugs (3). (Q) Barometric ports (8). (R) 1-inch diameter rear alignment rods (3). (S) 6.25-inch diameter U-235 projectile rings (9). (T) Polonium-beryllium initiators (4). (U) Tail tube forward plate. (V) Projectile WC filler plug. (W) Projectile steel back. (X) 2-pound Cordite powder bags (4). (Y) Gun breech with removable inner breech plug and stationary outer bushing. (Z) Tail tube aft plate. (AA) 2.25-inch long 5/8-18 socket-head tail tube bolts (4). (BB) Mark-15 Mod 1 electric gun primers with AN-3102-20AN receptacles (3). (CC) 15-inch diameter armored inner tail tube. (DD) Inner armor plate bolted to 15-inch diameter armored tube. (EE) Rear plate with smoke puff tubes bolted to 17-inch diameter tail tube

The altitude at which combat bombs would be detonated had to be given careful consideration. In addition to liberating great quantities of electromagnetic radiation and billions of Curies of radioactivity, a nuclear explosion differs from a conventional one of the same energy in that the pressures generated are higher at closer distances. Based on the results of the *Trinity* test, the detonation heights for the Hiroshima and Nagasaki bombs were set at 1,850 feet. This was chosen to maximize destruction by the shock wave created by the bombs, while minimizing the amount of localized fallout that would be created if they were detonated near ground level and irradiated tons of dirt and debris. The Ordnance Division's concern with the altitude issue was that most combat bombs detonate near ground level; little thought had been given to mechanisms designed for high-altitude operation. Extreme reliability was the paramount consideration. In a conventional mission where thousands of bombs might be dropped, the failure of a few percent will likely not affect the outcome of the operation. But any type of fuse that failed even one percent of the time would be unacceptable for a single bomb whose development had consumed hundreds of millions of dollars. Fuse specifications called for a less than one in ten-thousand chance of the bomb failing to fire within about 100 feet of the desired altitude.

Two major lines of fuse development were investigated. One was to use barometric switches which would be sensitive to changing air pressure as a function of altitude. The other was to adapt electronic techniques such as proximity fuses or fighter-plane tail-warning radar sets for use with the weapons, presuming that a reliable signal could be obtained with a falling bomb. For both *Little Boy* and *Fat Man*, a redundant series-parallel system of clocks, barometers, and four modified tail-warning radars known as "Archies" was adopted. The first stage in the firing process was that when the bombs were released, pullout switches activated timers that counted off a 15-second delay before the arming system became activated; this was to ensure safe separation from the aircraft. Following this, barometric switches activated the radar units at an altitude of 17,000 feet. These were designed to close a relay at a predetermined altitude when any two of them detected the desired firing altitude. To lessen the possibility of failure due to Japanese jamming, each radar operated on a slightly different frequency.

The final *Little Boy* bomb was ten feet long, 28 inches in diameter, and weighed about 9,700 lb. The gun barrel itself was six feet long and weighed 1,000 lb. The target and projectile pieces were not cast as solid wholes; rather, they each comprised a number of washer-like rings which were cast as uranium became available from Oak Ridge. The projectile was made up of nine rings totaling 7 inches in length, with inside and outside diameters of 4 inches and 6.25 inches. Because the amount of uranium received from Oak Ridge varied from shipment to shipment, none of the individual rings were of the same thickness (nor, likely, of the same enrichment). The assembled projectile rings totaled 39 kg, or about 85 lb. The target consisted of six rings, also of 7 inches total length, but with inside and outside diameters of one and four inches, for a mass of 25 kg, or about 56 lb. The projectile piece traveled just over four feet before meeting the target piece, which resided about 20 inches to the rear of the nose of the target case. The target assembly and tamper liner were secured to the front of the bomb with a nut which itself weighed several hundred pounds.

By December, 1944, Groves was sufficiently confident of anticipated uranium production schedules that he ordered all research and development on the gun bomb to be complete by July 1, 1945. The design was frozen in February, and *Little Boy* was ready for combat by May. Deployment awaited only enough U-235, which was expected to be ready about August 1. Robert Serber's "shooting" concept would underlie the first nuclear weapon used in combat.

7.11 The Implosion Bomb: Fat Man

Despite starting out with priority lower than the gun-bomb project, the implosion program under Seth Neddermeyer did enjoy an increasing measure of attention and resources from the fall of 1943 onwards. Neddermeyer made some progress, but achieved only very rough implosion symmetry due to a problem that cropped up well before the spontaneous fission crisis: the presence of "jets" of material which traveled ahead of the main mass of compressed material, asymmetries which would render the method too inefficient for a practical weapon.

The jet problem appeared insuperable until mathematician John von Neumann of the Institute for Advanced Study in Princeton, New Jersey, visited Los Alamos in late September, 1943. von Neumann had been studying shock waves for the NDRC, and had considerable experience with analyzing shaped-charge explosives used in armor-piercing projectiles. His work had convinced him that more symmetric implosions could be obtained if higher material velocities than what Neddermeyer had been working with could be achieved. Neddermeyer's superior, William Parsons, saw the advantage of von Neumann's approach, and the decision was made at a Governing Board meeting on October 28, 1943, to strengthen the implosion program.

Unfortunately, Neddermeyer and Parsons were of almost completely opposite personalities. Neddermeyer preferred the academic tradition of working alone or in a small group, and chafed under Parsons' more regimented military approach. Oppenheimer's solution was to bring in Harvard University explosives expert George Kistiakowsky to oversee the work. Kistiakowsky had visited Los Alamos as a consultant while serving as chief of the NDRC Explosives Division, and joined the Laboratory full time in February, 1944, to serve as Parsons' deputy. This made him Neddermeyer's superior, but since he was a scientist like Neddermeyer, he served as an effective buffer between Neddermeyer and Parsons. Oppenheimer formally relieved Neddermeyer of his leadership of the Implosion Experimentation group on June 15, 1944, although he remained on as a technical advisor and as a member of an implosion steering committee. Another valuable recruit at the time of the reorganization was Lieutenant Commander Norris Bradbury, a Stanford physicist and naval reserve officer. Bradbury had been carrying out research in projectile ballistics at the Dahlgren Proving Ground in Virginia, and was brought to Los Alamos to assist with implosion lens research; he also headed the implosion field-test program.

The spring and summer of 1944 marked the low point of implosion research. In an official history, David Hawkins summarized the situation as: "at that time there was not a single experimental result that gave good reason to believe that a plutonium bomb could be made at all." In a report prepared in the spring of that year, Kistiakowsky outlined work yet to be carried out, and summarized his pessimism with a prediction that "the test of the gadget failed … Kistiakowsky goes nuts and is locked up."

Implosion development took a significant step forward with a suggestion by British Mission member James Tuck, who conceived of modifying von Neumann's shaped-charge concept into a system of three-dimensional implosion "lenses." In combination with electric detonators, this idea was key to the eventual success of the implosion bomb. Tuck, Neddermeyer, and von Neumann filed for a patent on the concept, which has never been made public.

The fundamental idea of an implosion lens is sketched in the accompanying figure, which shows a single lens in side-view cross-section. In three dimensions, imagine a somewhat pyramidal-shaped five or six-sided block about a foot across and a foot and a half from end-to-end (left to right in the sketch, which is not to scale). Each block comprises two castings of different explosives that fit together very precisely, and which interlocks with neighboring blocks to form a complete sphere. The outer casting of each block is a fast-burning explosive known as Composition B (Comp B), which had been developed by Kistiakowsky. The inner lens-shaped casting is a slower-burning material known as Baratol, a mixture of barium nitrate and TNT. A detonator at the outer edge of the block of Comp B triggers an outward-expanding detonation wave, which progresses to the left in the sketch. When the detonation wave hits the Baratol, it too begins exploding. If the interface between the two materials is of just the right shape, the two waves can be arranged to combine as they progress along the interface to create an inwardly-directed *converging* burn wave in the Baratol.

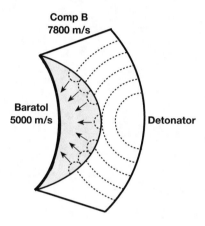

Schematic illustration of a binary-explosive implosion lens segment. Not to scale. Sketch by author

In the *Trinity* and *Fat Man* devices, 32 such "binary explosive" assemblies interlocked to create a complete spherical shell. This surrounded an inner spherical assembly of 32 blocks of Comp B, which surrounded the tamper/core assembly. The choice of 32 assemblies was dictated by the fact that this is the number of pentagonal and hexagonal-shaped blocks that can be fitted together to give nearly regular outer faces; think of the patches on a soccer ball. The *Trinity* and Nagasaki weapons used 12 pentagonal and 20 hexagonal sections, which respectively weighed about 47 and 31 lb each.

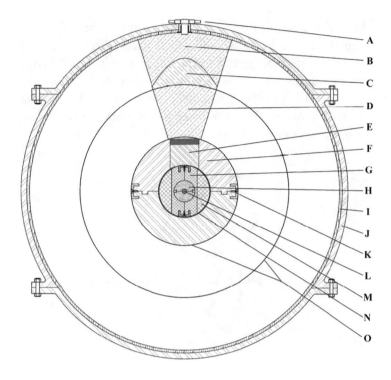

Cross-section drawing of the Y-1561 Fat Man implosion sphere showing major components. Only one set of 32 lenses, inner charges, and detonators is depicted. Numbers in parentheses indicate quantity of identical components. Drawing is to scale. Copyright by and used with kind permission of John Coster-Mullen. (A) 1773 Electronic Bridge Wire detonator inserted into brass chimney sleeve (32). (B) Comp B component of outer polygonal lens (32). (C) Cone-shaped Baratol component of outer polygonal lens (32). (D) Comp B inner polygonal charge (32). (E) Removable aluminum pusher trap-door plug screwed into upper pusher hemisphere. (F) 18.5-inch diameter aluminum pusher hemispheres (2). (G) 5-inch diameter U-238 ("Tubealloy") two-piece tamper plug. (H) 3.62-inch diameter Pu-239 hemisphere with 2.75-inch diameter jet ring. (I) 0.5-inch thick cork lining. (J) 7-piece Y-1561 Duralumin sphere. (K) Aluminum cup holding pusher hemispheres together (4). (L) 0.8-inch diameter Polonium-beryllium "Urchin" initiator. (M) 8.75-inch diameter U-238 tamper sphere. (N) 9-inch diameter boron plastic shell. (O) Felt padding layer under lenses and inner charges

The purpose of the inner layer of Comp B blocks is to achieve a high-speed symmetric crushing of the tamper and core in order to beat the spontaneous-fission predetonation problem. A trap-door arrangement with a plug of tamper material allowed for insertion of the core while the bomb was being assembled. The total weight of the high-explosive assembly alone was about 5,300 lb, just over half the bomb's total weight of about 10,200 lb; the inch-thick outer casing contributed 1,100 lb alone. The implosion bomb was a hand-assembled three-dimensional jigsaw puzzle with explosive pieces that had to fit together very precisely.

Estimating the expected yield of the *Fat Man* design was difficult on account of the nested aluminum/uranium tamper-spheres configuration. Geoffrey Taylor of the British Mission had determined that when a heavy metal is accelerated against a light metal, the interaction is stable. But if the acceleration is done with the light metal moving into a heavy metal, the inter-face becomes unstable, and gives rise to jets of light material spurting ahead of the main mass of that material as Neddermeyer had discovered. This jetting effect is now known as a Rayleigh-Taylor instability; to avoid it, the implosion must be extremely symmetric. Despite this, the lighter aluminum shell was imploded into the heavier uranium shell to enhance the efficiency; it has been estimated that some 30% of *Fat Man's* yield was due to fast neutrons fissioning U-238 in the tamper sphere.

The pressure created in the bomb core was estimated to be similar to that at the center of the Earth, and the properties of materials in such circum-stances were not well known. Detonation waves can interfere with each other unless they are arranged to be perfectly converging, which required simulta-neous multi-point triggering; variations in the velocity of the implosion had to be held to less than about 5%. In the original conception of the implo-sion scheme, the jetting problem was aggravated by the intent of trying to compress a thin shell of fissile material to many times its normal density. But there was little confidence that the necessary symmetry could be maintained. In September, 1944, Robert Christy, a former student of Oppenheimer's and one of the first staff members at Los Alamos, proposed a configuration with a core which was solid except for a small central void to hold the initiator. This came to be known as the "Christy core," and was adopted for the *Trinity* and Nagasaki bombs. As Christy described it:

> Earlier designs of the implosion bomb had been a relatively thin shell of plu-tonium, which would then be blown in by the implosion. It was assembled in the center with ideally very high density and spherical shape. But, there were constant worries at the time that, because of irregularities in the explosive, it would end up in a totally unacceptable form. They were worried it wouldn't be spherical and that it might end up with jets coming in and it wouldn't even go off. These worries were very real. They wanted to be sure it would not fail. It would be a very bad thing if they had a failure. So I suggested if they took the

hole out of the middle, and just made it solid, it couldn't very well be made non-spherical. There was a very small hole for the initiator that was required.

Robert F. Christy (1916– 2012) ca. 1959. Source: AIP Emilio Segre Visual Archives

Responsibility for developing the explosive components of the implosion bomb lay with George Kistiakowsky's X-Division, which eventually came to have a staff of some 600. This meant investigating methods of detonating high explosive components, improving the quality of castings, developing and testing the lens system, and fabricating explosive charges. Kistiakowsky organized an extensive series of test shots to investigate the best number of detonation points, the types and arrangement of explosives, and the material to be collapsed. The testing program generated some 20,000 castings of acceptable quality over a period of 18 months, with many more rejected. Some 100,000 lb of high explosive were used per month. Casting operations became so extensive that a separate site (Sawmill, or "S" site), staffed largely by SEDs, was set up in May, 1944, for that purpose. McAllister Hull, a 21-year old SED, arrived at Los Alamos in the fall of 1944 to, as he put it, "figure out how to cast the lenses to the specifications required." Hull had worked at an ordnance plant where TNT was cast into shells, and had much practical experience with such operations. The lens castings were done as slurries in modified commercial candy-making machines.

An original implosion-lens casting machine on display at National Museum of Nuclear Science and History in Albuquerque, NM. Photo by author

Three main problems cropped up in casting operations. As cast explosives cooled, internal voids would form, surfaces tended to have bubbles, and the chemicals from which the explosives were made could separate during cooling. Hull and his group managed to lick each problem in turn. The molds for the lenses had double walls, which housed cooling-water coils. By pumping water through the coils and gradually lowering the temperature as molten explosive was poured into the mold, formation of surface bubbles could be eliminated. The void and separation problems took longer to solve. As a student, Hull had worked for a time as a waiter, where he had gained experience in producing smooth, well-blended milkshakes. Working from that experience, he developed a protocol for placing a stirrer in the setting explosive and withdrawing it vertically just ahead of the solidification line. This prevented void formation while keeping the chemicals well-mixed. By late 1944, a reliable if labor-intensive system was in place, with men usually working three shifts per day, even, against regulations, during thunderstorms. As Hull related:

> I used a stirrer, gradually pulled up as the casting cooled from the outside, to keep the Baratol mixture uniform and the interior cavity free. I determined the rate at which to pull up the stirrer by casting 10 inner lenses simultaneously, then sawing them in half at five-minute intervals to see where the solid line had reached. The stirrer was pulled up at a rate to keep the blades ahead of the solidification curve inside the lens.

In his memoirs, Hull relates a humorous story involving Groves. Groves and Oppenheimer came by one day to witness the casting operation, and Groves accidentally stepped on a water line. The line popped away from its wall connection, and a jet of near-boiling water struck Groves on his backside. Hull, in uniform, suppressed his laughter while turning off the water supply, but broke up when Oppenheimer said "It just goes to show the incompressibility of water."

After being removed from their molds, all castings were checked for uniformity by X-raying them. Those containing voids were repaired by drilling into them with non-conducting tools to get to the void and then pouring in molten explosive, akin to a dentist filling a cavity; Kistiakowsky put things in perspective with the remark that one *gram* of such explosives could "finish off a hand". After repairs, castings were machined to remove any flashings or roughness; thousands of machining operations were conducted without a single accidental detonation. Design and manufacture of molds and producing enough castings of acceptable quality were always pacing elements in implosion testing.

An enormously challenging but critically important area of work was the development of diagnostic routines to test implosion symmetry. Essentially, the problem was to obtain information on events *inside* an explosion to time resolutions on the order of a microsecond. Multiple complementary methods were developed: "terminal observations" involved examining the imprint left on a metal plate by a two-dimensional version of an implosion lens; cameras were developed where a shutter remained open while film was advanced at high speed on a rotating drum or where images were recorded along a fixed film by a rotating mirror; blocks of imploding explosive were imaged with brief but intense bursts of X-rays detected by banks of Geiger counters; currents induced in wires caused by a changing magnetic field induced by the implosion gave information on the velocity of the external surface of an imploding sphere; electrical contacts formed between an imploding sphere and a network of prearranged pins gave information on implosion velocities;

and the intensity of gamma-rays emitted by a radioactive source placed within the sphere could be used to track density as the sphere imploded.

In Robert Bacher's G-Division, implosion work centered on the problem of developing detonators that would fire with sufficient simultaneity to initiate a symmetric implosion. The simultaneity required was far beyond that available in any commercial detonators, which were normally required to trigger only one explosion at a time. Much of the work on detonator design fell to Luis Alvarez and Donald Hornig, who took a trial-and-error approach. At Harvard University, Hornig had written a thesis on shock waves produced by explosions, and had come to Los Alamos from the Underwater Explosives Laboratory of the Woods Hole Oceanographic Institute in Massachusetts. Hornig worked with "spark-gap" switches, wherein a high voltage causes a spark to jump between two pieces of metal placed a small distance apart within an explosive. The explosive triggered by the detonator did not directly initiate explosion of an implosion lens, but rather drove a copper "slapper" plate into the lens which triggered the implosion via a high-pressure pulse. The first test of multiple electric detonators was carried out in May, 1944, and by late that year Horning had managed to reduce the timing spread for firing down to several hundredths of a microsecond. Refinements to detonator design continued right up to the time of the *Trinity* test.

Left: Luis Alvarez's (1911–1988) Los Alamos ID-badge photo. Right: Donald Hornig (1920–2013) in 1964. Sources: http://commons.wikimedia.org/wiki/File:Luis_Alvarez_ID_ badge.png; *AIP Emilio Segre Visual Archives, Physics Today Collection*

The most dramatic work of the R and G-Divisions involved criticality experiments, assemblies of varying amounts of U-235 or Pu-239 arranged to approximate critical masses. Short of a real explosion, there was no way to determine the extent of supercriticality that would be achieved with a full-scale gun or implosion assembly, but data from subcritical and barely-critical experiments could be extrapolated to give checks on theoretical estimates. Initially, these experiments involved assembling blocks of uranium hydride, on the premise that the hydrogen would slow down neutrons and hence give researchers experience with slower reactions before moving to fast-neutron configurations. By surrounding a subcritical assembly of hydride blocks with neutron-reflective beryllium tamper blocks, the number of fissions could be enhanced; these were known as "Godiva" assemblies, where an otherwise bare core would be clothed by the tamper blocks. Some hydride assemblies were so near-critical that the neutron-reflecting effect of the body of a person hovering over the assembly could make it supercritical; the experimenter would hop away just as criticality was approached.

By September, 1944, enough pure uranium metal was becoming available to begin criticality experiments without hydration. The first such experiments used a 1.5-inch diameter sphere (two hemispheres) of uranium enriched to 70% U-235; later experiments involved spheres up to 4.5 inches diameter of 73%-enriched material. (The bare critical diameter for pure U-235 is about 6.6 inches). When a neutron source was placed within the sphere, the number of neutrons emerging from the sphere would be greater than from the neutron source alone due to the effect of induced fissions; by extrapolating to infinite neutron multiplication, the critical mass could be determined. By March, 1945, enough uranium had been accumulated to make a tamped critical mass, and, on April 4, a combination of 4.5-inch hemispheres and tamper cubes was brought to within one percent of criticality. The first critical plutonium assembly was achieved in April, 1945, using a plutonium-water solution with a beryllium tamper.

Criticality experiments resulted two postwar fatalities at Los Alamos. On the night of August 21, 1945, Harry Daghlian was working alone (against regulations) with a plutonium sphere and tamper blocks when a block slipped out of his hand and caused a brief chain reaction. Daghlian had to partially disassemble the pile to halt the reaction, but received a radiation dose estimated at 500 rems, which, as explained in Chap. 5, is considered to be the single-shot dose that will cause 50% of exposed individuals to die within 30 days. Daghlian died on September 15; his hands, which had been the closest parts of his body to the assembly, became gangrenous, and his kidneys were unable to remove decomposition products from his blood. A similar accident took the life of Louis Slotin on May 21, 1946. Slotin was demonstrating how to make criticality measurements using the same hemispheres Daghlian had used; they would become known as the "demon core."

Slotin was gradually decreasing the separation between the hemispheres with a screwdriver, but the screwdriver slipped and they came together. Thermal expansion quickly halted the reaction, but Slotin received a radiation dose estimated at over 2,000 rems, and died nine days later. Seven other people were in the room at the time; two suffered acute radiation symptoms, but recovered. The Slotin accident permanently ended all hands-on criticality work at Los Alamos.

Less hands-on but also potentially dangerous were "Dragon drop" experiments. In October, 1944, Otto Frisch proposed constructing a device where a slug of uranium hydride would be dropped through the center of an almost-critical assembly of the same material. When the slug passed through, the assembly would become supercritical for a brief time. Richard Feynman, a Los Alamos theoretician and future Nobel Laureate, described this as "tickling the dragon's tail," and Frisch's machine became known as the Dragon machine. As realized, the Dragon machine stood about 6 m high. Designed to be operated largely by remote control, the operator could not activate the "Here We Go" button until various safety interlocks had been activated. A steel box which contained uranium hydride rode on guide wires and was dropped from the top of the device, which looked like an oil-well derrick. The box would pass through a lower table on which had been mounted more hydride, producing a slightly supercritical assembly for about a hundredth of a second. Frisch estimated that even if the box became stuck, the resulting explosion would be equivalent to only a few ounces of high explosive.

The Dragon machine. Note chair for scale. From Malenfant (2005)

Frisch was ready by mid-December, and began with tests using dummy materials before moving to active material. On January 20, 1945, the Dragon machine produced the world's first fast-neutron chain reaction. The reactions were brief, but power releases of up to 20 million Watts over about three milliseconds were recorded; there was not a single accident or instance of material hanging up in the drop mechanism. Because other experimental groups needed the hydride, experiments ceased in February, and the machine was dismantled. Dragon experiments contributed information on the generation time between fissions and the exponential growth rate of the chain reaction.

Prospects for implosion slowly improved through the latter half of 1944 and the first half of 1945. For James Conant, pessimism began to give way to guarded optimism. By October, he was giving a lensed device a 50:50 chance of working for a test on May 1, 1945, and three-to-one odds for a test on July 1. Conant was visiting the Laboratory at the time of an implosion test shot on December 14, and concluded that while that method had the possibility of giving relatively high efficiency (a few percent), it still faced enormous difficulties: "Further experiments which may be completed by March 1 will show the chances of doing this in 1945. My own bets are very much against it." He judged that an implosion bomb would likely yield less than 850 tons TNT equivalent, and perhaps only 500 tons.

On February 17, 1945, a schedule for working toward to a full-scale test was developed at a Technical and Scheduling Conference meeting. Full-scale lens molds were to be available for casting by April 2, and full-scale lens shots to test the timing of multi-point detonations were to be ready by April 15. Shots with hemispheres of explosives were to be ready by April 25. Detonators should come into routine production between March 15 and April 15. A full-scale test of implosion without fissile material but using the magnetic diagnostic method should be made between April 15 and May 1. Between May 15 and June 15, plutonium spheres were to be fabricated and tested for criticality. Fabrication of implosion lenses for the full-scale test were to be underway by June 4, and fabrication and assembly of the implosive sphere should begin by July 4. The target date for the test itself was set as July 20. On February 28, just eleven days later, Oppenheimer and Groves decided on the

Christy-core design. Characteristic of so many decisions in the Manhattan Project, the choice was a gamble: few implosion lenses had by then been tested in the diagnostic program.

7.12 The Delivery Program

Had the scientists and engineers of Los Alamos merely constructed and tested a nuclear weapon, they would have left their job undone. A laboratory experiment is one thing; a deliverable military weapon is quite another. To make a bomb ready for combat meant modifying aircraft to carry it, training crew members, designing a bomb casing that gave stable flight, and ensuring that electronic systems would function reliably in combat conditions and be immune from enemy interference.

William Parsons recognized these needs early on, and saw that a "delivery" program was organized as soon as possible. In October, 1943, a group within his Ordnance Division was charged with responsibility for integrating the design and delivery of weapons. This group was headed by physicist Norman Ramsey, who possessed the ideal combination of familiarity with military operations and understanding of the science of the project. The son of an Army general, Ramsey had earned a doctorate from Columbia for work in the area of molecular-beam physics, and had been serving as a consultant to the Secretary of War on microwave radar when he was recruited to Los Alamos. In his postwar career at Harvard, Ramsey pioneered methods for precisely measuring the electron-transition frequencies of atoms and molecules, work which would lead to the development of atomic clocks and magnetic-resonance imaging medical scanners, and which earned him a share of the 1989 Nobel Prize for Physics. The formal name of the delivery program was Project Alberta, or simply Project A. Ramsey's first task was to undertake a survey of the sizes, shapes, and weights of bombs that could be carried by Army Air Forces aircraft. (The Air Force did not become a separate branch of the armed services until 1947.)

Left: Thin Man (front) and Fat Man test bombs (rear) at Wendover Army Air Base (Utah); Right: Norman Ramsey (1915–2011). Sources: http://commons. wikimedia.org/wiki/File:Thin_Man_plutonium_gun_bomb_casings.jpg   *AIP Emilio Segre Visual Archives, W. F. Meggers Gallery of Nobel Laureates*

Initially, the gun bomb was anticipated to be 17 feet long by 23 inches in diameter. These dimensions dictated the necessary size of the bomb bay, and Ramsey zeroed in on the long-range B-29 "Superfortress" bomber, which at the time was still undergoing flight tests. Powered by four 18-cylinder engines that each developed 2,200 horsepower, the B-29 would be the largest combat aircraft of World War II. The first production model came off the assembly line in July, 1943, and their first mission against the Japanese home islands occurred in June, 1944.

B-29's were equipped with two 150-inch-long, 64-inch-wide bomb bays, one forward and one aft of the wings. If the two bays could be joined together, they could accommodate the gun bomb under the main wing spar. On December 1, 1943, Army Air Forces Headquarters directed the Materiel Command at Wright Field in Dayton, Ohio, to undertake a high-priority modification of a B-29. This directive was named "Silver Plated Project," which eventually evolved to "Silver Plate," and finally to "Silverplate." The first modified bomber was a prototype with a single 33-foot long bomb bay and release mechanisms for both *Thin Man* and *Fat Man* designs. A total of 46 Silverplate B-29's were produced by the end of 1945; the total would come to 65 by the time the project was terminated in December, 1947. The cost of each Silverplate aircraft has been estimated at about $815,000 in 1945 dollars.

Parsons arranged for Ramsey to supervise a drop-test program at the Dahlgren Naval Proving Ground in Virginia. To prepare a mockup 14/23-scale model of the gun bomb, Ramsey had a standard 23-inch diameter, 500-pound bomb split in half and the halves joined by a length of 14-inch diameter pipe. Known as the "sewer-pipe" bomb, the first drop test was conducted on August 14, 1943, and proved, in Ramsey's words, "… an ominous and spectacular failure. The bomb fell in a flat spin the like of which had rarely been seen before." Adjustments to the tail-fin design and moving the bomb's center of gravity forward soon resulted in more stable flight. The eventual boxlike tail of the gun-type bomb was square-shaped and thirty inches on a side.

By the fall of 1943, Ramsey was ready to begin tests with full-scale models. He and Parsons selected two external shapes and weights as representative of the bombs then under development: the 17-foot/23-inch gun model, and an ellipsoidally-shaped implosion model just over 9 feet long and 59 inches in diameter. Fifty-nine inches was the largest diameter that could be squeezed into a B-29 bomb bay, a constraint which set an absolute limit to *Fat Man's* girth. The nested structure of the implosion design meant that any change in the dimensions of any component propagated throughout the design: The diameter of the neutron-generating initiator at the center of the weapon dictated the dimensions of the fissile core, which dictated the dimensions of the surrounding tamper sphere and high-explosive assembly, all of which was housed within a spherical metal case held within an outer ballistic ellipsoid, with enough space between the casings to house arming and fusing circuits.

The origins of the names *Thin Man, Fat Man,* and *Little Boy* is a matter of debate. In a history of the delivery project prepared just after the end of the war, Ramsey asserted that it was Air Force representatives who coined the names *Thin Man* and *Fat Man* to make telephone conversations sound as if aircraft were being modified to carry Roosevelt and Churchill. In a 1998 autobiography, Robert Serber claims to have coined the names, with *Thin Man* being taken from the title of a 1934 detective novel by Dashiell Hammett, and *Fat Man* referring to the role played by actor Sydney Greenstreet in the 1941 movie *The Maltese Falcon*, which starred Humphrey Bogart.

Tests of full-scale models for ballistic behavior and functioning of fusing and instrumentation circuits were begun in the spring of 1944 at Muroc Field in California. The site of a large dry lake bed, Muroc is now Edwards Air Force Base. The prototype modified B-29 arrived on February 20, and the first drop test occurred in early March. Parsons' conservatism in demanding an early start on the delivery program was well-founded. Fuses proved so unreliable that an investigation was begun into adapting radar units normally mounted on the tails of fighter aircraft as substitutes. Thin Man models proved to have very stable flight characteristics, but the *Fat Man* design wobbled violently, with its long axis departing up to 20° from the line of flight. Simply assembling the implosion bomb was arduous: one model required 1,500 bolts. (For the Nagasaki weapon, this number was cut to 90.) A release mechanism that worked properly for *Fat Man* failed completely for *Thin*

Man, with several dangerous hang-ups occurring. In what would be the last test of this series on March 16, a *Thin Man* released prematurely and fell onto the bomb-bay doors, which had to be opened to release the bomb. The doors were seriously damaged, and this accident brought testing at Muroc to an abrupt if temporary halt.

With the mid-1944 realization that the gun assembly method could not be used with plutonium, the situation for the uranium gun bomb became much simpler. The assembly speed could be reduced to about 1,000 feet per second, and the length of the bomb could be shortened to 10 feet, which meant that it could fit into a single B-29 bomb bay. The prototype bomber was reconfigured back to its original two-bay configuration, and the shortened gun bomb was dubbed *Little Boy.*

Tests resumed in June, 1944. The new radar-driven fusing units functioned satisfactorily, but *Fat Man's* wobble proved more challenging to address. Replacing its parachute-like circular-shaped tail assembly with a square one helped to suppress but did not wholly eliminate the wobble. Ramsey had steel plates added to the tail assembly at 45-degree angles; this modification resulted in stable flight and gave *Fat Man* its distinctive tail-end, which contributed 400 lb to the weight of the bomb. *Fat Man* test units were painted mustard yellow to make them easy to track, and they became known as "Pumpkin" units.

Tests of the gun-bomb firing mechanism were made at Wendover Army Air Base in Utah, one of the largest gunnery and bombing ranges in the world. In Manhattan Project lingo, Wendover was codenamed "Kingman," "Site K," and "W-47." This program was very successful from the outset: thirty-two tests involving natural uranium projectiles were conducted, and on only one occasion did the gun fail to fire, a consequence of a faulty electrical connection. These tests did result in one significant modification to the design of the breech of the gun bomb, however. The original design called for the bomb to be fully armed with conventional explosive upon aircraft take-off, but it was deemed desirable that it be possible to arm the bomb during flight lest a crash on take-off initiate a nuclear explosion. The breech was modified to permit loading or unloading powder bags in the cramped space of the bomb bay, which Parsons would do during the Hiroshima mission. No such arrangement was practical with the implosion weapon, which because of its enclosed design left the ground fully armed.

In parallel with technical refinements to bomb designs, air crews had to be selected and trained. On August 11, 1944, the Army Air Forces recommended freezing the design of the shapes of the bomb casings and starting crew training. Freezing the designs would permit modifications to a lot of B-29's to be started while crews were being assembled. In consultation with General Henry Arnold, Commander of the Army Air Forces, Groves decided for security reasons to organize a self-sustaining Air Force unit to carry out the bombing missions themselves. During the summer and fall of 1944, Air Force and Manhattan Project personnel screened possible candidates to command the new unit, settling on Lieutenant Colonel Paul W. Tibbets. A superb combat pilot, Tibbets had flown the first B-17 bomber across the English Channel on a bombing mission in World War II, and later led the first American raid on North Africa. After more than twenty-five combat missions, he returned to the United States to become involved with flight-testing the B-29 bomber. On September 1, 1944, Tibbets underwent a final security grilling at the Colorado Springs headquarters of General Uzal Ent, Commanding General of the Second Air Force. After answering questions to the satisfaction of Groves' security chief, Colonel John Lansdale, Tibbets was ushered into Ent's office. There he was introduced to William Parsons and Norman Ramsey, who briefed him on his new assignment. Tibbets was placed in command of the Army Air Forces 509th Composite Group, which would be responsible for dropping the combat bombs. The 509th trained at Wendover; Navy Commander Frederick Ashworth served as liaison between Los Alamos and Wendover, and would accompany the *Fat Man* bomb on its flight to Nagasaki.

Left: Colonel Paul Tibbets (1915–2007) waves from the cockpit of the Enola Gay shortly before takeoff for the Hiroshima mission. Right: Frederick Ashworth (1912–2005) gives a talk at Los Alamos in his later years. Sources: http://commons. wikimedia.org/wiki/File:Tibbets-wave.jpg; http://commons.wikimedia.org/wiki/File:Frederick_Ashworth.jpg

The first of 17 modified B-29's began arriving in October, 1944, and test flights began that month. Particular emphasis was put on training pilots to carry out unusual post-drop maneuvers designed to put the maximum possible distance between the aircraft and the bomb before it exploded; test bombs were filled with concrete to simulate the effect of how the plane would lurch upward after the bomb was released. This first group of modified aircraft proved to have poor flying qualities, so a new batch equipped with fuel-injected engines, variable-pitch propellers, and improved bomb-release mechanisms was obtained in the spring of 1945. Stripped of all of their guns and armor except for their tail turrets, aircraft of this second group were each 7,200 lb lighter than normal B-29's. These modifications enabled them to fly above 30,000 feet at an average speed of 260 miles per hour, while carrying a payload of 10,000 lb for almost 2,000 miles. Another modification involved the addition of a position for an electronics test officer who would monitor the bomb's electrical circuits during flight. The first test with one of the replacement B-29's occurred on March 10, 1945 at the Salton Sea, although the bomb was released early and fell near a small town.

Two of the aircraft in this second group would go down in history as the planes that carried the atomic bombs dropped on Japan. On May 9, 1945, the day after the German surrender was announced, Tibbets was at the Martin Aircraft plant in Omaha, Nebraska, to pick out the bomber that he would use for the first atomic strike. B-29 production number B29-45-MO-44-86292 would be christened as *Enola Gay*, Tibbets' mother's maiden name. It was formally delivered to the Army Air Forces on May 18, flown by pilot Robert Lewis to Wendover on June 14, and then again by Lewis to Tinian island, arriving on July 6. Serial number B29-35-MO-44-27297, *Bockscar*, was delivered on March 19, and arrived at Tinian on June 16. Named after its commander, Captain Frederick C. Bock, this aircraft is sometimes referred to with the two-word designation "*Bock's Car.*" *Enola Gay* and *Bockscar* are believed to be the only surviving Silverplate aircraft, and now reside at museums in Washington and Dayton, Ohio, respectively.

The pace of the 509th's training schedule was relentless, and went on right up to the time of the Hiroshima and Nagasaki missions. Between October, 1944, and mid-August, 1945, a total of 155 test bombs were dropped, a rate of nearly one per day. One of the most problematic issues encountered with *Fat Man* was a piece of equipment known as the X-unit, which was responsible for simultaneously triggering its network of spherically-distributed detonators. With redundant detonators for each of 32 implosion-lens segments, 64 cables were involved, all of which had to be the same length and have the same resistance. Not until late July, 1945, did a sufficient number of X-units begin to become available. The first drop test of a *Fat Man* with high explosives and an X-unit was not conducted until August 5, the day before the Hiroshima mission.

Combat models of the *Fat Man* bomb incorporated a number of safety features. Front-view photographs of the Nagasaki bomb show four cylindrical tubes about three inches in diameter protruding from the front of the casing. These were contact fuses: if the fusing circuitry failed to trigger an "airburst," these would fire the detonating system when the bomb struck the ground. *Little Boy* did not incorporate contact fuses as it would "self-assemble" upon striking the ground. *Fat Man*'s ballistic casing, plus cover plates on the *Little Boy* weapon and the rear tail covers on both bombs were made of hardened armor plate; tempered-steel casings were vulnerable to 0.50-caliber machine gun fire.

Left: assembled Fat Man bomb. Note signatures on tail. Sources: http://commons.wikimedia.org/wiki/File:Fat_Man_on_Trailer.jpg. *Right: On Tinian island, Fat Man receives a coat of sealant. Note FM stencil on nose. Source:* http://commons.wikimedia.org/wiki/File:Fat_Man_on_Tinian.jpg

Provision also had to be made for an overseas combat base at which the bombs would be assembled, checked, and loaded onto aircraft. After surveying both Guam and Tinian islands and consulting with Groves and Admiral Chester Nimitz (Commander-in-Chief of both the United States Pacific Fleet and the Pacific Ocean Areas), Ashworth selected Tinian: it was about 100 miles closer to Japan than Guam, had construction forces available, and its port facilities tended to be less overloaded than those at Guam. Tinian is a member of the Northern Mariana Islands chain, located just south of the island of Saipan. Only about 12 miles long, the island had been taken by the Marines in July, 1944, and for a time was the site of the largest airport in the world: six runways each 8,500 feet long, which served as launching points for round-the-clock bombing raids against the Japanese home islands. It was not uncommon for 400 aircraft to leave the field in less than two hours. Tinian's Manhattan codename was "Destination."

Map of Tinian and Saipan. 15 min of latitude corresponds to a distance of about 17 miles (27 km). Source: http://commons. wikimedia.org/wiki/File:Map_ Saipan_Tinian_islands_closer. jpg

Ashworth oversaw construction of 509th facilities on Tinian. Air-conditioned assembly buildings were erected, along with warehouses and shops. Special pits were constructed for loading bombs into aircraft from underneath; there was otherwise insufficient clearance between the bodies of the aircraft and the ground to accommodate the weapons. If bottlenecks in construction or transportation arose, Ashworth needed only to invoke the code word "Silverplate," which came to designate all atomic-bomb related activities within the military and which required instant cooperation from all personnel. The 509th moved its operations to Tinian in late June, 1945, to undertake practice missions in advance of their "hot runs" against Hiroshima and Nagasaki.

7.13 Trinity

Given the uncertainties with the implosion method, the idea of a full-scale test was circulating well before the spontaneous fission crisis emerged in mid-1944. No one wanted the first trial of a *Fat Man* weapon to be over enemy territory, where, if it failed, the fissile material might be recovered. Groves saw the idea of a full-scale test as a waste of fissile material, and proposed that any test device contain only enough to just start a chain reaction. Oppenheimer objected to this on the rationale that it would be practically impossible to specify the precise amount of material necessary to achieve such a circumstance. On February 16, 1944, he wrote Groves to emphasize that the "implosion gadget must be tested in a range where the energy release is comparable with that contemplated for final use." Groves relented, and preparations for a full-scale test began the next month, when Oppenheimer appointed Kenneth Bainbridge to oversee the operation. Bainbridge's subsequent report on the test, Los Alamos report LA-6300-H, is now available online.

The first issue was to locate a suitable site. Criteria included flatness in order to facilitate measurements, favorable weather, wind patterns that would not expose populated areas to excessive fallout, and proximity to Los Alamos to simplify travel. Also, Harold Ickes, the Secretary of the Interior, wanted no Indians to be displaced for the test. Four sites were considered in New Mexico, including the Jornada del Muerto ("Journey of death") desert east of the Rio Grande; one in Colorado; two in California including near the town of Rice in the Mojave desert in the eastern part of the state; and sand bars off the coast of Texas. The choice came down to the Jornada and Rice locations, with Jornada winning out. Proximity to Los Alamos was likely a factor, although it has been claimed that Groves rejected the Rice location because it was in use by General George Patton, whom Groves refused to approach regarding its use. One source quotes Groves as saying that Patton was "the most disagreeable man I ever met."

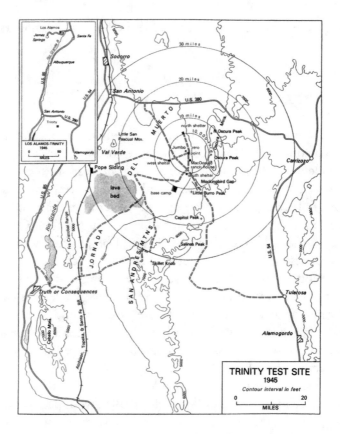

The Trinity test site. From V. C. Jones, United States Army in World War II: Special Studies—Manhattan: The Army and the Atomic Bomb. Courtesy Center of Military History, United States Army

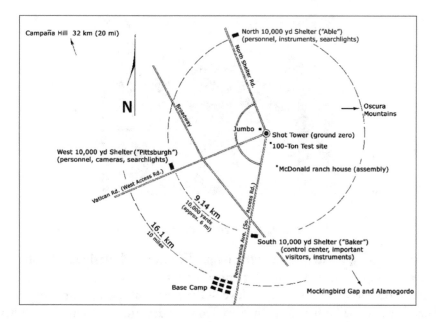

Detail map of Ground Zero area. Source: http://www.lahdra.org/pubs/reports/In%20Pieces/Chapter%2010-%20Trinity%20Test.pdf, *based on Lamont (1965)*

Located about 160 miles south of Los Alamos (250 km), the Jornada site comprised an 18 by 24-mile tract in the northern portion of the Alamogordo Army Air Field. The town of Alamogordo (present population about 32,000) is located about 60 miles southeast of the point where the bomb was detonated; Socorro (about 9,000 inhabitants) lies about 35 miles to the northwest. Summertime temperatures in the area routinely reach over 100 °F. At the time of the test, the nearest habitation was about 12 miles distant. Before the war, the land had supported some cattle grazing, but in 1942 the Army appropriated the four-room ranch house of George McDonald to serve as a part of the Alamogordo Bombing and Gunnery Range. The house was used as the assembly station for the *Trinity* bomb; while it was somewhat damaged by the explosion, it still stands about two miles southeast of ground zero. Now restored to the way it appeared in 1945, the house is accessible to tourists during the two weekends per year that the site is normally open to visitors.

The McDonald Ranch House, October 2004

An enduring mystery is how the name *Trinity*, which designated both the site and the test, was chosen. Oppenheimer claims to have suggested it, and a common speculation is that his love of the poetry of John Donne may have been involved. The first four lines of Donne's devotional poem "Batter My Heart" read

Batter my heart, three-person'd God, for you

As yet but knock, breathe, shine, and seek to mend;

That I may rise and stand, o'erthrow me, and bend

Your force to break, blow, burn, and make me new.

Donne is alluding to the Christian notion of deity as Father, Son, and Holy Ghost. Another speculation derives from Oppenheimer's interest in Hindu culture, where the concept of Trinity involves three gods: Brahma the Creator, Vishnu the Preserver, and Shiva the Destroyer. In this faith, whatever exists in the Universe is never destroyed but rather transformed, appropriate imagery for a nuclear explosion.

Except for the ranch house, the site was completely undeveloped. A Base Camp of barracks, officers quarters, warehouses, repair shops, bomb-proof structures, technical facilities, a mess hall, and other support facilities had to be constructed to serve the needs of a staff that would grow to number over 250. Over 20 miles of blacktopped roads and 200 miles of telephone lines would have to be provided, along with a fleet over 100 vehicles. Oppenheimer approved the construction plans on October 27, 1944, and the first residents, a detachment of Military Police under the command of Lieutenant Harold Bush, arrived to take up their duty in late December. The Base Camp was located about 17,000 yards (9.6 miles) south of "ground-zero," the location of the test itself. The bomb would be mounted atop a 100-foot surplus Forest Service fire-watch tower whose concrete footings were sunk some 20 feet into the earth. Groves witnessed the explosion from Base Camp along with various distinguished visitors, including Bush, Conant, and Fermi.

Trinity   Base   Camp.   Source: http://commons. wikimedia.org/wiki/File:Trinity_basecamp.jpg

Located within an area of about 100 square miles centered on ground zero were three instrument stations, roughly to the North, West, and South, all 10,000 yards from the test site. The South station also served as the control center where the final switches to activate an automatic firing sequence would be thrown; Oppenheimer witnessed the test from that point. At the time of the test, all shelters were under the supervision of a scientist until the bomb detonated, at which time command passed to a physician who was authorized to order evacuation if necessary. The scientists in charge at the North, West, and South shelters were Robert Wilson, John Manley, and Robert Oppenheimer's brother, Frank. Personnel who had participated in the development of the bomb but who were not needed at the control station witnessed the spectacle from a vantage point on Campaña Hill, some 20 miles to the northwest. This group included Hans Bethe, James Chadwick, Ernest Lawrence, Edward Teller, and Robert Serber. The contrast between the scale of the desktop-scale apparatus with which Chadwick had discovered the neutron only 13 years earlier and the *Trinity* operation could not have been more stark.

The precise number of people that witnessed the test was not documented, but film-badge counts indicate that some 350 people were at the site sometime during the day of the test, July 16. One of the major figures of the Manhattan Project who would *not* witness *Trinity* was Arthur Compton. Oppenheimer had sent him an invitation reading "Anytime after the 15th would be a good time for our fishing trip." Compton decided not to attend so as not to raise questions at the Met Lab, but after the test Oppenheimer called him to report that "You'll be interested to know that we caught a very big fish."

By early 1945, preparations for *Trinity* were becoming so complex that Oppenheimer appointed the Cowpuncher Committee to provide executive direction for the implosion program—to "ride herd" on it. Cowpuncher comprised the Laboratory's top scientific and administrative personnel: Oppenheimer, Bainbridge, Bethe, Kistiakowsky, Parsons, Bacher, Allison, and Cyril Smith. The committee first met on March 3, and assigned highest priority to initiator development, detonators, and procuring lens molds.

To test-run procedures and calibrate instruments, a rehearsal test was conducted at 4:30 a.m. on May 7, 1945. This involved detonating 108 tons of explosives mounted atop a 20-foot high tower located about 800 yards southeast of where the *Trinity* tower was erected. The height of this explosion was not arbitrary. At that time, the best prediction for the *Trinity* energy yield was about 5,000 tons TNT equivalent. Theoretical analysis indicated that for an observer at distance d from a nuclear explosion of yield E, the air pressure behind the initial shock wave would be proportional to $E^{2/3}/d^2$, so the center of gravity of the 108-ton stack was placed at 28 feet above the ground to scale to *Trinity's* planned 100-foot high detonation and anticipated yield. *Trinity's* yield would prove to be much more than 5 kilotons, however, which resulted in many instruments, set at scaled distances, being overwhelmed in the actual test. To create a low-level simulation of the fallout pattern to be expected from a nuclear explosion, the TNT was seeded with tubes containing fission products from a Hanford fuel slug sufficient to supply 1,000 Curies of beta-activity and 400 of gamma-activity. The TNT shot proved a valuable test of procedures, and revealed a number of issues that needed to be resolved before the real test. Some of these were technical, such as interference on instrument cables, while others were more prosaic, such as failure to provide enough batteries to power all of the instruments that had been deployed. Probably the most important lesson was that there should be a cutoff date beyond which no further apparatus would be introduced into the experimental area.

One of the most curious aspects of the *Trinity* test was the "Jumbo" program. When the chances for implosion looked slim, it was thought that it would be wise to set off the explosion within a vessel that could contain the force of the high explosive so that the plutonium could be recovered in the event of a nuclear fizzle. The pressure requirement was estimated to be 60,000 lb per square inch, or about 4,000 atmospheres. The only option that looked feasible was to set the bomb off within a strong containing vessel. This led to the design and procurement of Jumbo, a massive steel cylinder within which the bomb would be placed; the ends would then be closed off. As related by Bainbridge, "Jumbo represented to many of us the physical manifestation of the lowest point in the Laboratory's hopes for the success of an implosion bomb. It was a very weighty albatross around our necks."

Design of Jumbo fell to the Engineering Group of Kistiakowsky's X-Division. In its final incarnation, the container weighed in at 214 tons, was 28 feet long, 10 feet in inside diameter, had a shell 14 inches thick, and cost $12 million. Manufactured in Ohio, the giant vessel was carried 1,500 miles by rail on a special flatcar (which itself weighed 157 tons) over a circuitous route that included travel down the Mississippi river to New Orleans. Jumbo's rail journey ended at a siding 30 miles from ground zero. From there it was hauled to the test site at three miles per hour on a 64-wheel trailer. But by the time of the test, confidence in a successful implosion was much greater; the anticipated need for Jumbo had diminished. Also, experimenters were concerned that the vessel would interfere with monitoring instruments. The plan was abandoned, and Jumbo was erected on a tower some 800 yards northwest of the explosion. The tower was vaporized, but Jumbo survived; had it been used, the result would have been tons of radioactive fallout over New Mexico and chunks of shrapnel hurled to great distances. Easily large enough for more than one person to stand inside, the remaining 100-ton body of the cylinder, minus its ends (blown off, according to some sources), now lies where it was on the morning of July 16. One of the ends now serves as a tourist attraction in Socorro.

Left: Jumbo, 1945. Source: http://commons.wikimedia.org/wiki/File:Trinity_Jumbo.jpg. *Right: The author (light-colored shirt and hat) inside the 100-ton body of Jumbo, 800 yards from Trinity ground zero*

The *Trinity* test was probably the most monitored and photographed scientific experiment in history to its time. Numerous experiments were proposed, but as shop time was at a premium in the weeks leading up to the test, all proposals had to be submitted to a review committee for classification as essential (efficiency, blast pressure, detonator performance), desirable (fireball photography and analysis, motion of the surrounding earth), or unnecessary. No experiment could affect the operation of the bomb, and no experiments were allowed to be installed within four weeks of the test date in order to leave time for set-ups, rehearsals, and debugging. Proposers had to detail estimated manpower requirements, calibration procedures, signal line needs, actuation mechanisms, and shop time. Six chief groups of experiments were arranged: implosion diagnostics; energy release measurements; damage, blast, and shock; general phenomena; radiation measurements; and meteorology. Within these groups were deployed dozens of individual experiments designed to measure every conceivable aspect of the explosion. An incomplete list includes detonator simultaneity; shock wave transmission through the imploding high-explosive; fission-rate growth; gamma rays; neutrons; fission products; atmospheric pressure effects; seismic disturbances; earth displacement; and ignition of structural materials. Over 50 types of cameras were deployed, from simple pinhole models to motor-driven units capable of exposing up to 10,000 frames per second. Spectrographic cameras would record light of various wavelengths emitted by the fireball. Gold foils placed in protective tubes spread around the site would become radioactive due to neutron capture and so record the strength of the neutron flux. Fission fragments in the soil would be collected from a lead-lined tank with a trap-door in the bottom; such fragments were a valuable source of information on the efficiency of the bomb. Pressure gauges were deployed to measure the energy released in the explosion. Some pieces of equipment would be destroyed by the shock wave created by the bomb, and had to be designed to transmit their data between the time of the explosion and their destruction. In all, some 500 miles of wires and cables were installed for the test.

Groves paid particular attention to obtaining shock measurements from both airborne and ground-level sensors. Barographs were deployed at distances of 800, 1,500 and 10,000 yards, and from 50 to 100 miles from the site of the explosion. Their data would bear on setting the detonation heights of combat weapons, and he wanted evidence in the event of any damage lawsuits which might arise. To obtain radiation-exposure records, films were mailed to dummy addresses through local post offices, and picked up later by intelligence officers. Groves also deployed a security contingent of 160 men north of the test area lest it prove necessary to evacuate ranches and towns at the last moment.

Workdays at the site often stretched to 18 h. On June 9, the Cowpuncher Committee set Friday, July 13, as the earliest possible test date, with the 23rd as the probable date. On June 30, the earliest possible date was revised to Monday, July 16. For political reasons, Groves wanted the test as soon as possible. Oppenheimer thought the 14th possible, but settled on the 16th.

On July 2, the plutonium core hemispheres for the *Trinity* device were completed, and on the 4th a mockup device was assembled and checked for criticality. On the 6th, *Trinity's* uranium tamper was machined, and on the 10th the best available lens castings were selected.

Meteorological conditions were a constant headache. The Manhattan Project's meteorology supervisor, Jack Hubbard, had been obtained from the California Institute of Technology. Equipped with portable weather stations, field radar sets, sensors which gave temperature and humidity readings at different altitudes, balloons, and local and national records, one of Hubbard's first responsibilities had been to choose a date for the 100-ton test, and he identified April 27 and May 7 as the optimum dates. The latter was chosen, and his forecast proved accurate; Bainbridge described the meteorological service for the test as excellent. July weather in the southwest can be more unstable than May weather, however.

For *Trinity*, physics, meteorology and politics collided incompatibly. The demands of the various experimental groups were practically impossible to reconcile. For some groups, rain before the test might not be a concern, but for others an instrumentation cable might be useless if it had not had time to dry out. Hubbard's first choice of dates was July 18-21, with the 12th to the 14th as second choice, and the 16th as only a possible date. The 16th was favored, however, because that would be the earliest date for which the bomb would be ready, and Groves was under intense pressure to carry out the test as soon as possible. President Truman would be in Germany for the Potsdam Conference from July 16 to August 2, negotiating with Winston Churchill and Josef Stalin regarding post-war occupation arrangements in Europe and the prosecution of the war against Japan. The conference had originally been set to begin on July 6, but Truman had asked for a postponement to the 15th to give Los Alamos more time. Churchill had given his assent to the use of the bomb on July 1, and the British and Canadians were informed at a July 4 meeting of the Combined Policy Committee that America intended to use the bomb. The strategic situation was complex. Planning for a November 1 American-British invasion of the southern island of Japan was already very advanced. The Soviet Union had committed to enter the war against Japan within three months after the defeat of Germany, which had been declared on May 8. A successful test would strengthen the hand of American and British negotiators, and could be parlayed into an ultimatum to Japan to surrender. As events played out, the Soviets would honor their commitment on the last possible date, August 8, between the bombings of Hiroshima and Nagasaki. Setting the test date was out of Hubbard's hands, but he pressed on. From June 25 onwards, hourly observations were recorded by weather stations at

Base Camp and Ground Zero. On July 6, he predicted that the area would be dominated by a stagnant tropical air mass, which proved to be partly true. On learning that the test had been set for the 16th, he recorded in his diary: "Right in the middle of a period of thunderstorms, what son-of-a-bitch could have done this?"

Out of concern that radioactive fallout could be carried over populated areas, the most pressing weather considerations were wind and rain. South-southwest winds were preferred in order to blow fallout to the northeast. On the morning of the 15th, Hubbard predicted that the next day would see light and variable winds from east to west below 14,000 feet, and west-southwest winds above 15,000 feet. What he apparently did not predict were the strong localized thunderstorms that moved into the area about 2:00 a.m. on the 16th, two hours before the scheduled test time.

Groves biographer Robert Norris has presented a different view of the Hubbard story. This is that when Hubbard was acquired, the purpose of the work was not revealed, and Cal Tech assigned one of its "lesser-qualified" staff. In this version, Groves apparently began to appreciate this as the test neared, and brought in Air Force meteorologist Colonel Ben Holzman, who had participated in the selection of the date for the D-Day landings in Normandy. Groves wrote in his memoirs that Hubbard had been making accurate long-range predictions, but the only time he was not right was "on the one day that counted." Groves states that he dismissed the forecasters in the hours before the test, deciding to rely on his own predictions.

Rehearsal tests were conducted on July 8, 12, 13, and 14. *Trinity's* plutonium hemispheres were conveyed to the site by car from Los Alamos on the 11th, and initiators arrived the next day. Final assembly of the high-explosive components was carried at one of the outlying sites at Los Alamos on the 13th, and they were brought down to the site by truck later that day. The mood of the Laboratory soured when a magnetic-method test shot carried out on the 14th seemed to indicate that the bomb would not function efficiently, but Hans Bethe saved the day by demonstrating that the analysis was flawed and that acceptable detonator symmetry had in fact been achieved. A stanza of poetry caught the sense of the time:

> From this crude lab that spawned a dud
> > Their necks to Truman's axe uncurled
> > Lo, the embattled savants stood
> > And fired the flop heard round the world.

Final assembly of the *Trinity* device began at one p.m. on Friday, July 13, within a tent at the base of the 100-foot tower. The date and time were chosen by George Kistiakowsky in the hope that they would bring good luck. Just after three p.m., the core assembly was ready for insertion within the high explosive, but a hitch arose. The plutonium core, warm from its own internally-generated alpha-decay heat and the desert climate, did not fit into the cooler high-explosive assembly. Leaving them in contact for a couple minutes brought them to thermal equilibrium, and the core slipped into place. Assembly of the bomb's innards was complete by 5:45 p.m., and it was raised to the top of the tower in preparation for installation of detonators and firing circuitry the next day. As the bomb was raised, a protective bed of mattresses was placed under it. Sunday, July 15, was reserved for final inspections.

Trinity atop its test tower on July 15, 1945, with Norris Bradbury (1909–1997). The cables feeding from the box halfway up the device go to the implosion-lens detonators discussed in the text. Source: http://commons.wikimedia.org/ wiki/File:Trinity_Gadget_002.jpg

Bainbridge's report contains a copy of a detailed schedule for the test. Among minutiae regarding the precise placement and handling of components reside more prosaic matters such as "Light must be available to work in tent at night," and "Bring up G-Engineer footstool." For Sunday, July 15, the entirety of the schedule read "Look for rabbit's feet and four-leaved clovers. Should we have the chaplain down here? Period for inspection available from 0900 to 1000." The entry for July 16 read only as "Monday, 16 July, 0400 Bang!"

The last group of people to attend the bomb was an arming party headed by Bainbridge, which set out at about 10:00 p.m. on the night of the 15th to activate timing and arming switches so that the bomb could be triggered from the South-10,000 station, and also to collect Donald Hornig, who had earlier ascended the tower to switch out a practice detonating circuit for the operational one and to stand guard over the bomb. In Hornig's words:

> Oppenheimer was really terribly worried … that it would be easy to sabotage. So he thought someone had better baby sit it right up until the moment it was fired. They asked for volunteers and as the youngest guy present, I was selected. I don't know if it was that or that I was most expendable or best able to climb a 100-foot tower! By then there was a violent thunder and lightning storm. I climbed up there, took along a book, *Desert Island Decameron*, and climbed the tower on top of which there was the bomb, all wired up and ready to go. Little metal shack, open on one side, no windows on the other three, and a 60-Watt bulb and just a folding chair for me to sit beside the bomb, and there I was! All I had was a telephone. I wasn't equipped to defend myself, I don't know what I was supposed to do. There were no instructions! The possibility of lightning striking the tower was very much on my mind. But it was very wet and the odds were the tower would act like a giant lightning rod and the electricity would just go straight down to the wet desert. In that case, nothing would have happened. The other case was that it would set the bomb off. And in that case, I'd never know about it! So I read my book.

By the time of the test, Hubbard had not slept in over two days. At a weather conference held at 2:00 a.m. on the 16th, he predicted that conditions would become acceptable at dawn. Holzman agreed, and the shot was set for 5:30. Groves demanded that Hubbard sign his forecast, stating that he had better be right, "or I will hang you." Groves then placed a call to the Governor of New Mexico to inform him that it might be necessary to declare martial law throughout the central part of the state. Importantly, the winds were as desired.

At Base Camp, Enrico Fermi occupied himself by offering to take wagers on whether the bomb would ignite the atmosphere, and, if so, would it destroy only New Mexico or the entire world? He estimated that if nitrogen in the air were ignited, it would go only about 35 miles. He added that it would not make any difference whether the bomb went off or not, as it would still have been a worthwhile experiment. Groves was not amused by Fermi's diversions, but others were also in a wagering mood. A pool on the yield of the bomb was established, with an ante of $1 each. Edward Teller optimistically bet on 45 kilotons; Hans Bethe opted for 8 kilotons. Oppenheimer picked 200 tons, and had a side bet with George Kistiakowsky of $10 against a month of Kistiakowsky's salary that the bomb wouldn't work at all. The winner was I. I. Rabi, who arrived too late to choose a low number, and had to settle for 18 kilotons; he took home $102. Others had different concerns. Kenneth Bainbridge later wrote that "My personal nightmare was knowing that if the bomb didn't go off or hangfired (a delay between trigger-ing and detonation), I, as head of the test, would have to go to the tower first and seek to find out what had gone wrong." Oppenheimer was practically a nervous wreck: he had suffered a bout of chicken pox and had lost 30 lb; despite standing at over six feet, he weighed only about 115 lb.

In the control bunker at S-10,000, the tension was palpable. As described by Groves' deputy, Brigadier General Thomas Farrell:

> The scene inside the shelter was dramatic beyond words. In and around the shelter were some twenty-odd people concerned with last minute arrangements … For some hectic two hours preceding the blast, General Groves stayed with the Director, walking with him and steadying his tense excitement. Every time the Director would be about to explode because of some untoward happening, General Groves would take him off and walk with him in the rain, counseling with him and reassuring him that everything would be all right.

Groves departed for Base Camp 20 min before the detonation. He had dictated that he and Farrell were not to be together in situations where there was an element of danger, which arguably existed at both locations. The final countdown began at 5:10 a.m., and was conducted by Samuel Allison. At T-minus 45 s, arming-party physicist Joseph McKibben threw a final switch that activated a timing apparatus with a rotating drum and pin-actuated switches to trigger time-sensitive instruments. Donald Hornig manned a cutoff switch that was the only way the test could have been stopped.

The exact time of the *Trinity* detonation is only approximately known because of difficulty in picking up a national time-service radio broadcast at the shelter. Bainbridge's report gives as a best estimate 5:29:15 a.m., plus 20 s or minus 5 s. Witnesses at Base Camp were instructed to lie flat on the ground, face away from the tower, and not to rise until after the blast wave had passed. From Farrell's description:

> As the time interval grew smaller … the tension increased by leaps and bounds. Dr. Oppenheimer, on whom had rested a very heavy burden, grew tenser as the last seconds ticked off. He scarcely breathed. He held on to a post to steady himself. For the last few seconds he stared directly ahead and then when the announcer [Allison] shouted 'Now!' and there came this tremendous burst of light followed shortly thereafter by the deep growling roar of the explosion, his

face relaxed into an expression of tremendous relief. Several of the observers standing back of the shelter to watch the lighting effects were knocked flat by the blast.

The tension in the room let up and all started congratulating each other … Dr. Kistiakowsky … threw his arms around Dr. Oppenheimer and embraced him with shouts of glee.

A number of descriptions of the explosion have been published. One of the most striking was provided by Farrell, a devout Catholic, in his subsequent report to Groves:

The effects could well be called unprecedented, magnificent, beautiful, stupendous and terrifying. No man-made phenomenon of such tremendous power had ever occurred before. The lighting effects beggared description. The whole country was lighted by a searing light with the intensity many times that of the midday sun. It was golden, purple, violet, gray and blue. It lighted every peak, crevasse and ridge of the nearby mountain range with a clarity and beauty that cannot be described but must be seen to be imagined. It was that beauty the great poets dream about but describe most poorly and inadequately. Thirty seconds after the explosion came, first, the air blast pressing hard against the people and things, to be followed almost immediately by the strong, sustained, awesome roar which warned of doomsday and made us feel that we puny things were blasphemous to dare tamper with the forces heretofore reserved to The Almighty. Words are inadequate tools for the job of acquainting those not present with the physical, mental, and psychological effects. It had to be witnessed to be realized.

Farrell commented to Groves that "The war is over." "Yes," was Groves' reply, "just as soon as we drop one or two of these things on Japan."

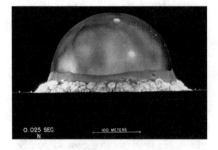 

Left: The Trinity fireball at 25 ms into the nuclear age. Right: The Trinity mushroom cloud a few seconds later. Sources: http://commons.wikimedia.org/wiki/File:Trinity_Test_Fireball_25ms.jpg. http://commons.wikimedia.org/wiki/File:Trinity_shot_color.jpg

At Base Camp, Enrico Fermi estimated the strength of the blast by an elegantly simple experiment:

> The explosion took place at about 5:30 A.M. I had my face protected by a large board in which a piece of dark welding glass had been inserted. My first impression of the explosion was the very intense flash of light, and a sensation of heat on the parts of my body that were exposed. Although I did not look directly towards the object, I had the impression that suddenly the countryside became brighter than in full daylight. I subsequently looked in the direction of the explosion through the dark glass and could see something that looked like a conglomeration of flames that promptly started rising. After a few seconds the rising flames lost their brightness and appeared as a huge pillar of smoke with an expanded head like a gigantic mushroom that rose rapidly beyond the clouds probably to a height of the order of 30,000 feet. After reaching its full height, the smoke stayed stationary for a while before the wind started dispersing it.
>
> About 40 s after the explosion the air blast reached me. I tried to estimate its strength by dropping from about six feet small pieces of paper before, during and after the passage of the blast wave. Since at the time, there was no wind I could observe very distinctly and actually measure the displacement of the pieces of paper that were in the process of falling while the blast was passing. The shift was about 2 1/2 m, which, at the time, I estimated to correspond to the blast that would be produced by ten thousand tons of T.N.T.

Kenneth Bainbridge described the test as "a foul and awesome display." After the shock wave passed, Bainbridge congratulated Oppenheimer and said to him: "Now we are all sons of bitches." In a 1975 reminiscence, Bainbridge related that in 1966, Oppenheimer told Bainbridge's daughter that her father's assessment was the best thing anyone said after the test.

> Hans Bethe on Campania Hill: "it looked like a giant magnesium flare which kept on for what seemed a whole minute but was actually only one or two seconds. The white ball grew and after a few seconds became clouded with dust whipped up by the explosion from the ground and rose and left behind a black trail of dust particles. The rise, though it seemed slow, took place at a velocity of 120 m/s. After more than half a minute the flame died down and the ball, which had been a brilliant white became a dull purple. It continued to rise and spread at the same time, and finally broke through and rose above the clouds which were 15,000 feet above the ground. It could be distinguished from the clouds by its color and could be followed to a height of 40,000 feet above the ground."
>
> James Conant at Base Camp: "Then came a burst of white light that seemed to fill the sky and seemed to last for seconds. I had expected a relatively quick

and bright flash. The enormity of the light and its length quite stunned me. My instantaneous reaction was that something had gone wrong and that the thermal nuclear transformation of the atmosphere, once discussed as a possibility and only jokingly referred to a few minutes earlier, had actually occurred."

I. I. Rabi at Base Camp: "We were lying there, very tense, in the early dawn, and there were just a few streaks of gold in the east; you could see your neighbor very dimly. Those ten seconds were the longest ten seconds that I have ever experienced. Suddenly, there was an enormous flash of light, the brightest light I have ever seen or that I think anyone has ever seen. It blasted; it pounced; it bored its way right through you. It was a vision that was seen with more than the eye. It was seen to last forever. You would wish it would stop; although it lasted about two seconds. Finally it was over, diminishing, and we looked toward the place where the bomb had been; there was an enormous ball of fire which grew and grew and it rolled as it grew; it went up into the air, in yellow flashes and into scarlet and green. It looked menacing. It seemed to come toward one. A new thing had just been born; a new control; a new understanding of man, which man had acquired over nature."

Emilio Segrè at Base Camp: "We saw the whole sky flash with unbelievable brightness in spite of the very dark glasses we wore … In a fraction of a second, at our distance, one received enough light to produce a sunburn." In his later biography of Fermi, Segrè wrote that "Even though the purpose was grim and terrifying, it was one of the greatest physics experiments of all time … The feat will stand as a great monument of human endeavor for a long time to come."

Norris Bradbury at the Control Shelter: "The shot was truly awe-inspiring. Most experiences in life can be comprehended by prior experiences but the atom bomb did not fit into any preconception possessed by anybody. The most startling feature was the intense light."

Robert Christy on Campania Hill: "It was awe-inspiring. It just grew bigger and bigger, and it turned purple … The debris was intensely radioactive, and it was sending out beta particles and gamma rays in all directions, and those ionized the air. So the air around this ball emitted a bluish glow … It was most fantastic, to see this thing going up and swirling around and eventually cooling off to the point where it was no longer visible."

Charles Thomas, also at Campania Hill: "It was awful. It looked like a giant mushroom, the stalk was thousands of tons of sand being sucked up by the explosion and the top of the mushroom looked like a flowering ball of fire…. It resembled a giant brain, the convolutions of which were constantly changing in color."

General Farrell at South-10,000: "The long-hairs have let it get away from them!

Kistiakowsky embraced Oppenheimer and exclaimed "Oppie, you owe me $10." In a 1980 reminiscence, Kistiakowsky claimed to still have the $10 bill.

Oppenheimer's reaction to the test is a matter of debate. His brother Frank, when interviewed for the 1980 documentary *The Day After Trinity*, stated that he thought all his brother had said was "It worked!" In postwar years, Oppenheimer uttered a number of dramatic, quasi-philosophical statements on his reaction to the test. A 1947 lecture on "Physics in the Contemporary World" at MIT included the following frequently-quoted passage:

> Despite the vision and the farseeing wisdom of our wartime heads of state, the physicists felt a peculiar intimate responsibility for suggesting, for supporting, and in the end, in large measure, for achieving, the realization of atomic weapons. Nor can we forget that these weapons, as they were in fact used, dramatized so mercilessly the inhumanity and evil of modem war. In some sort of crude sense which no vulgarity, no humor, no over-statement can quite extinguish, the physicists have known sin; and this is knowledge which they cannot lose.

A number of physicists were offended by this statement. Freeman Dyson, a physicist at Cornell University in the years after the war:

> Most of the Los Alamos people at Cornell repudiated Oppy's remark indignantly. They felt no sense of sin. They had done a difficult and necessary job to help win the war. They felt it was unfair of Oppy to weep in public over their guilt when anybody who built any kind of lethal weapons for use in war was equally guilty. I understood the anger of the Los Alamos people, but I agreed with Oppy. The sin of the physicists at Los Alamos did not lie in their having built a lethal weapon. To have built the bomb, when their country was engaged in a desperate war against Hitler's Germany, was morally justifiable. But they did not just build the bomb. The enjoyed building it. They had the best time of their lives while building it. That, I believe, is what Oppy had in mind when he said they had sinned. And he was right.

In a 1965 interview for a television documentary, *The Decision to Drop the Bomb*, Oppenheimer gave this reaction to *Trinity*:

> We knew the world would not be the same. Few people laughed, few people cried, most people were silent. I remembered the line from the Hindu scripture, the *Bhagavad-Gita*. Vishnu is trying to persuade the Prince that he should do his duty and to impress him takes on his multi-armed form and says, "Now I am become Death, the destroyer of worlds." I suppose we all thought that, one way or another.

Groves permitted access to the Manhattan Project to *The New York Times* science reporter William L. Laurence, whose 1940 *Saturday Evening Post* article on fission he had attempted to embargo. Laurence witnessed the *Trinity* explosion from Campañia Hill, and in the first of many articles on the Project published in the *Times*, gave a dramatic description of the explosion on the front-page of the September 26, 1945, edition (excerpted):

> At that great moment in history, ranking with the moment in the long ago when man first put fire to work for him and started on his March to civilization, the vast energy locked within the hearts of the atoms of matter was released for the first time in a burst of flame such as never before been seen on this planet, illuminating earth and sky for a brief span that seemed eternal with the light of many super-suns. ... It was like the grand finale of a mighty symphony of the elements, fascinating and terrifying, uplifting and crushing, ominous and devastating, full of great promise and great forebodings. ... And just at that instant there rose from the bowels of the earth a light not of this world, the light of many suns in one. ... On that moment hung eternity. Time stood still. Space contracted into a pinpoint. ... The thunder reverberated all through the desert, bounced back and forth from the Sierra Oscuros, echo upon echo. The ground trembled under our feet as in an earthquake.

In the same article, Laurence quoted George Kistiakowsky as saying "I am sure that at the end of the world—in the last milli-second of the earth's existence—the last man will see what we saw." The best assessment of the significance of the *Trinity* test may be that of novelist Joseph Kanon: "July 1945 at Alamagordo is the hinge of the century. Nothing after would ever be the same."

Trinity's most dramatic visual manifestation was its enormous ball of fire. Immediately following a nuclear detonation, the energy liberated is deposited in bomb debris, heating them to temperatures of millions of degrees. Much of this is promptly radiated away the form of X-rays and ultraviolet light, and since cold air is opaque to radiation at these wavelengths, the air surrounding the weapon absorbs the energy and heats up dramatically, to a temperature of about a million degrees out to a radius of a few feet. Because this bubble of hot air emits energy in the X-ray and ultraviolet regions of the electromagnetic spectrum, it will be *invisible* to an outside observer. But the bubble is surrounded by a cooler envelope, which, although incredibly hot by everyday standards, will be visible to observers at a distance. As the fireball increases in size, its total light emission increases up to a maximum due to an effect known as Stefan's law of thermal radiation, which indicates that light emission of a hot object is proportional to its surface area times the fourth power of its temperature. After this, the fireball begins cooling due to the growing mass of accreted air. Like a hot-air balloon, the fireball will also rise. The temperature within the fireball is so great that all of the weapon residues will be in the form of vapor, including the fission products. As the fireball expands and cools, these vapors condense to form a cloud of solid debris particles; the fireball also picks up water from the atmosphere. All of this material will eventually become fallout, sometimes in the form of radioactive rain. As the fireball ascends, cooling of its outside and air drag creates a doughnut-like shape; the cloud will often have a reddish appearance due to the presence of nitrogen-oxide compounds at its surface.

The air inside the fireball cools by successive radiation and re-absorption of X-rays. When the air has cooled to a temperature of about 300,000°, a "hydrodynamic shock" forms, a "front" of compressed air. This shock front travels faster than energy can be transported by successive absorption and re-emission of radiation, so it decouples from the hot sphere and moves out ahead of the latter, leaving behind a region of relatively cool air which eats into the central hot sphere. For outside observers, visible radiation comes from the shock wave. As the shock front cools, its temperature bottoms out at a minimum of about 2,000°; the time of this minimum following the detonation can be used to estimate the yield. The shock front also becomes transparent; an observer, if he or she still has eyes and sentience, can now look into higher-temperature air, which results in a second brightness maximum. This double-maximum in the time-evolution of visible radiation is uniquely characteristic of an atmospheric nuclear explosion. The second maximum actually lasts about 100 times as long as the first, and contains virtually all of the radiant energy emitted. During this time, however, the central fireball is still hot enough to be essentially opaque, and hence invisible. As the shock front progresses outwards, there soon comes a time when, for a while, the air pressure behind the front is lower than ambient atmospheric pressure, a so-called "negative pressure" region. In this phase, air rushes inward to the site of the explosion, an afterwind.

Hans Bethe and Robert Christy wrote in an undated memorandum (presumably summer 1945) that "the ball of fire will rise to the stratosphere (about 15 km height) in about two or three minutes .... The flash of light obtained in the first instant will be as bright as the sun at a distance of about 100 km from the explosion .... At a time when it reaches the stratosphere it will still appear as bright as the moon at a distance of about 250 km. The radioactive materials are expected to be near the center of the ball of fire and rise with that ball of fire to the stratosphere. Presumably the ball of fire will rise to a very considerable height (100 km or more) before its rise is stopped by either diffusion or cooling. If the radioactive material ever comes down again it will certainly be spread out over a radius of at least 100 km and probably very much more and will, therefore, be completely harmless".

It has been estimated that *Trinity* released an amount of radioactivity equivalent to an initial decay rate of nearly 14 *trillion* Curies.

The second iconic image of a nuclear detonation is the characteristic mushroom-shape cloud that forms after the explosion. This happens for air-burst weapons, that is, ones detonated above the ground. The stem of the mushroom is formed when the initial blast wave reflects from the ground. The reflected wave, however, will be traveling through air that has already been heated and compressed by the passage of the initial wave, and so moves *faster* than the initial wave. The reflected wave catches up to the initial wave, forming the stem. In technical parlance, the stem is known as a "Mach stem."

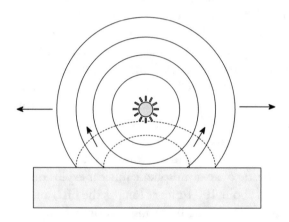

Schematic illustration of formation of a reflected shock wave. After Glasstone and Dolan (1977)

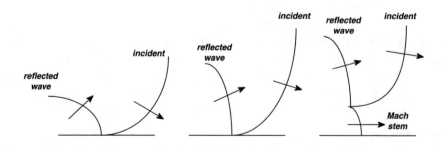

Schematic illustration of formation of the Mach stem. After Glasstone and Dolan (1977)

About one-third of the total energy liberated by a fission weapon is in the form of ultraviolet, visible, and infrared light. The rate of delivery of this energy is so prompt that combustible materials such as paper, wood, and fabrics will be charred or burst into flame out to great distances. Such materials can be ignited by the prompt delivery of 10 calories (chemical, not food) of radiant energy per square centimeter; a 20-kilton explosion delivers this much energy to a radius of 6,000 feet. At *Trinity*, some fir timbers were slightly scorched to this distance; such charring requires a temperature of about 400 °C. For human beings, moderate burns to unprotected skin can be produced by deposit of about 3 calories per square centimeter. For a 20-kiloton explosion, the radius for this effect is about 10,000 feet; at Nagasaki, skin burns were reported to 14,000 feet. *Trinity's* heat energy output alone was estimated at 3 kilotons TNT equivalent.

Some quarter-million square meters (70 acres) of surrounding desert sand was fused to a depth of about half an inch into a fragile, greenish, glassy material that came to be known as Trinitite. The greenish color is due to the presence of iron in the sand; samples are still very slightly radioactive.

Aerial view of the aftermath of the Trinity test. The 0.1 kt test crater is from the 100-ton TNT test. The area covered by the image is about 1,550 m wide by 1,400 m tall. Source: http://commons.wikimedia. org/wiki/File:Trinity_crater_(annotated)_2. jpg

Groves was anxious to get word to Henry Stimson in Potsdam, and called his secretary in Washington, Jean O'Leary, about 9:00 a.m. Washington time, some ninety minutes after the test. O'Leary proceeded to the Pentagon office of Stimson advisor George Harrison, where they drafted a coded cable:

> Operated on this morning. Diagnosis not yet complete but results are satisfactory and already exceed expectations. Local press release necessary as interest extends a great distance. Dr. Groves pleased. He returns tomorrow. I will keep you posted.

Stimson received the cable at 7:30 p.m. Potsdam time (1:30 p.m. in Washington), 6 h after the test), and immediately relayed it to President Truman. In his diary for July 18, Truman remarked that at a lunch alone with Churchill he "Discussed Manhattan (it is a success)." He also recorded that "Believe Japs will fold up before Russia comes in. I am sure they will when Manhattan appears over their homeland. I shall inform Stalin about it at an opportune time." It is at this point that Churchill truly seems to have understood the power of the bomb, remarking to Stimson that "This atomic bomb is the Second Coming in Wrath."

Groves hastened back to Washington, arriving in his office about 2:00 p.m. the day after the test. That evening, he prepared a lengthier memorandum to Stimson. Completed early in the morning of Wednesday, July 18, it was sent by courier to Potsdam, where it was handed to Stimson at 11:35 a.m. on Saturday morning, July 21. A few passages testify to the enormity of the blast:

> The light from the explosion was clearly seen at Albuquerque, Santa Fe, El Paso, and other points generally to about 180 miles away. The sound was heard … generally to 100 miles. Only a few windows were broken, although one was some 125 miles away. A crater from which all vegetation had vanished, with a diameter of 1,200 feet … in the center was a shallow bowl 130 feet in diameter and 6 feet in depth … The steel from the tower was evaporated … I no longer consider the Pentagon a safe shelter from such a bomb … Radioactive material in small quantities was located as much as 120 miles away … My liaison officer at the Alamogordo Air Base, sixty miles away [reported] a blinding flash of light that lighted the entire northwestern sky.

Upon receiving the report, Stimson took it to Marshall and Truman; Churchill was also informed. It took Stimson the better part of an hour to read the report. The version of the report reproduced in Groves' memoirs does not include a statement included in the original: "It resulted from the atomic fission of about 13½ pounds of plutonium which was compressed by the detonation of a surrounding sphere of some 5000 lb of high explosive."

On the morning of the 24th, another cable from Harrison informed Stimson that "operation may be possible any time from August 1 depending on state of preparation of patient and condition of atmosphere." Later that morning, a combined American and British Chiefs of Staff meeting convened with Churchill and Truman; Truman biographer David McCullough pinpoints this meeting as critical in the decision to use the bomb. That evening, Truman approached Josef Stalin to let him know that America had developed a new weapon "of unusual destructive force." Stalin, who was probably well-briefed on the project, apparently showed no special interest, replying only that he hoped that America would make "good use of it against the Japanese." In their analysis of Soviet nuclear espionage, *Bombshell: The Secret Story of America's Unknown Atomic Spy Conspiracy,* Joseph Albright and Marcia Kunstel write that on February 28, 1945, the State Security People's Commisariat (NKGB) in Moscow finished a comprehensive report on atomic intelligence which would go to Lavrenti Beria, the People's Commisar for Internal Affairs. The Soviets knew of the main features of the implosion weapon five months before the *Trinity* test.

Groves had prepared a number of press releases written to accommodate a range of test outcomes. The story he went with was that a remotely-located ammunition magazine containing "a considerable amount of high explosives and pyrotechnics" had exploded on the grounds of the Alamogordo Army Air Base, but that there had been no loss of life or any injuries. The story was reported in the area and along the west coast, but received no exposure on the east coast except for a few lines in the early edition of a Washington paper.

Because the *Trinity* yield was over three times greater than predicted, many instruments were overwhelmed by the explosion. No blast-measuring device within 200 feet of the tower survived, although one located at 208 feet gave a pressure reading of nearly 5 tons per square inch, almost 700 atmospheres. Gauges designed to measure the peak blast pressure gave a result of 9.9 kilotons, but radiochemical analyses of soil samples indicated nearly twice that figure, 18.6 kilotons. A number of re-evaluations of *Trinity* measurements have been carried out in light of information subsequently gathered from the Hiroshima and Nagasaki bombings, as well as postwar atomic tests. A 2006 analysis based on radiochemical and spectroscopic studies of trinitite resulted in a yield of 21.4 kilotons, with about 31% of the yield being due to fissions in the uranium tamper, and a 2016 analysis of fission products in tirinitite by a radiochemistry group at Los Alamos resulted in a figure of 22.1 kilotons. By any measure, *Trinity* was by far the largest man-made explosion in history to its time; the previous record, estimated to be 2.9 kilotons, had been set by the accidental explosion of a munitions ship in Halifax harbor in Nova Scotia in 1917.

Because wind patterns at the time of the test were favorable, there were no serious fallout issues, but there were consequences. In addition to creating direct fission (and subsequent decay) products, the explosion vaporized an estimated 100–250 tons of sand, much of which would have been rendered radioactive by neutron bombardment. The radioactive cloud split into three parts, with the majority moving northeast and dropping radioactivity over an area of about 100 miles long by 30 miles wide. Readings of about 3 rems per hour in the affected area were not uncommon; the present-day standard for maximum exposure for people who work with radioactive materials is 5 rems *per year*. The lead-lined tanks that had been prepared to retrieve soil samples could make only brief passes through the crater itself, where soil samples registered initial activities of 600–700 rems per hour. Exposure limits which would trigger evacuation of shelters and surrounding areas were not rigidly defined, although 10 rems per hour was loosely accepted as the threshold of concern, with a recommendation that no person "of his own will" receive more than 5 rems at one exposure. The North-10,000 shelter was evacuated about twenty minutes after the explosion when 10 rems per hour was recorded, but it is suspected that this may have been a mis-read.

The most seriously-affected radiation victims were likely animals, particularly grazing Hereford cattle at local ranches. A few weeks after the test, several cows began losing hair, which regrew white as opposed to its normal reddish tint; Louis Hempelmann's Health Group bought four cows and brought them to Los Alamos for study. Because the breed purity of discolored cattle would be questioned, ranchers faced a cut in price, and in December, 1945, Los Alamos bought some 75 animals that were most heavily damaged. None of them died of unexplained causes, and they reproduced normally. Some of the more seriously exposed ones did eventually develop skin cancers on their backs, but the overall conclusion was that there was no gross differences between the exposed cattle and their offspring when compared with an unexposed control group. The animals' owner, W. L. McNierney, sold 140 of them at a reduced price, and in August, 1946, was compensated $1350 for his loss.

One effect of *Trinity* fallout turned up far from the site. In the fall of 1945, the Eastman Kodak Corporation in Rochester, New York, found that several batches of industrial X-ray film were flecked with spot-like imperfections. The film itself was fine, but radioactive particles had become embedded in strawboard liners used to separate the films in their cartons. The strawboard

had been prepared by paper mills in Iowa and Indiana in the weeks following *Trinity*, and it is thought that rain washed fallout into rivers which were
used as water sources during the paper processing. One of the culprit fallout
products was Cerium-141, which has about a 32-day half-life.

In the aftermath of the bombings of Hiroshima and Nagasaki, some com-
mentators asserted that both cities were uninhabitable, an assessment which
would have been a surprise to surviving residents. To quell concern over radiation effects, Groves arranged a media day for reporters and photographers
at the *Trinity* site on September 9, with everyone wearing protective booties.
Radioactivity was measured at 12 rems per hour, and the visit was kept brief.
Ironically, because *Trinity* was detonated so close to the ground, the site was
radiologically hotter than either Hiroshima or Nagasaki. Systematic studies of
long-term effects at *Trinity* began in 1947, and were carried out periodically
thereafter. In the 1947 survey, plutonium was found in the soil and on plants
at locations up to 85 miles from the detonation site, and some birds, rodents,
and insects were malformed, had eye cataracts, or unusual spottings. Another
study a year later found no damaged birds or rodents, indicating that effects
were not genetically passed on. Trinitite proved to be insoluble in water, so it
could not easily enter plants or animals.

*Trinity ground zero, September 1945.
Oppenheimer (center, hat), Groves,
and others look at the remains of the
100-foot tower. Source:* http://commons.
wikimedia.org/wiki/File:Trinity_Test_
-_Oppenheimer_and_Groves_at_
Ground_Zero_001.jpg

In the years following the war, some efforts were launched to try to make
the *Trinity* site into a national monument. Various studies were carried out,

but competing interests of using the land for grazing and the impact of what became the White Sands Missile Range doomed these plans. In the 1950s, the Trinitite was packed into barrels and buried; a 1967 study calculated that a person would have to eat some 100,000 kg of the material to ingest the maximum permissible body burden of 4 nanoCuries of plutonium-239, although only 10 kg would have to be consumed to reach maximum permissible beta and gamma-ray exposures from fission products. In 1965, the National Park Service declared the site a National Historic Landmark, and erected a monument; in 1975, the location was designated a National Historic Site. The Army donated Jumbo to the city of Socorro, but no means could be found to remove it from the site. Tourists need have no concern about visiting the site: a 1985 Los Alamos report concluded that exposure during public visits to the ground-zero area amounts to less than 0.2% of Department of Energy Radiation Protection Standards for members of the public.

Left: The author, second from right, at the Trinity ground-zero monument, October, 2004. Right: Monument plaque

Author at the West-10,000 instrument bunker

7.14   The Combat Bombs

With *Trinity* a success, the stage was set for combat use of nuclear weapons.

The first Los Alamos bomb-preparation personnel departed for Tinian on June 18, 1945, nearly a month before bomb components began to arrive. In July, uranium for the gun bomb was delivered to Tinian in two shipments, one by sea and one by air. On Saturday, July 14, the projectile rings, encased in a lead-lined cylinder, departed Los Alamos for Kirtland Field in Albuquerque. The cylinder was attached to a parachute, loaded aboard a DC-3 transport plane, and flown to just outside San Francisco. From there it was convoyed to Hunters Point Naval Shipyard, where it resided for the 14th and the 15th until being bolted to the deck of the cruiser USS *Indianapolis*, which also carried the inert parts of *Little Boy*. *Indianapolis* departed San Francisco at 8:00 a.m. on Monday, July, 16, just three and one-half hours after the *Trinity* test, and arrived at Tinian on Saturday, July 28. The six *Little Boy* target rings, cast later, arrived by C-54 transport aircraft with two rings as the sole cargo aboard each of three planes. The C-54s departed Kirtland on the afternoon of July 26, and began arriving at Tinian on the evening of the 28th/29th; all three had arrived by 02:00 on the 29th. The projectile and target pieces and initiators were loaded inside the bomb on July 30. With the installation of radar altimeters and barometric switches on the 31st, *Little Boy* was ready for combat, awaiting only weather good enough for a visual bombing run. Back at Los Alamos, the Theoretical Division's most recent predicted yield was 13.4 kilotons, which would prove to be remarkably accurate.

On the evening of Thursday, July 26 the governments of America, China, and Great Britain (Russia was not yet at war with Japan) issued the joint Potsdam Declaration, which called on Japan to surrender unconditionally, or face "prompt and utter destruction." With summer time in effect, Potsdam was eight hours earlier than Tinian, which put the time of the declaration as the very early hours of July 27 on Tinian, the day before the arrival of *Little Boy's* projectile and target rings. Physics and politics were again crossing paths.

The declaration was broadcast to Japan by radio, and leaflets describing it were dropped from American bombers. Japan is one time zone west of Tinian and seven hours ahead of Germany; the broadcasts were picked up in Tokyo at 7:00 a.m. on the morning of Friday, July 27. Japanese government officials spent that day debating the ultimatum, but Prime Minister Kantaro Suzuki concluded that the only recourse was for Japan to fight on and treat the declaration with what historians have characterized as "silent contempt." On Saturday afternoon in Tokyo (Saturday morning in Potsdam), the same day as

Little Boy's target rings arrived at Tinian, Suzuki related the official response at a press conference. Radio Tokyo began broadcasting Suzuki's statement on Sunday afternoon in Japan. Japan's atomic fate was sealed two days before the completion of *Little Boy.*

Delivery of *Fat Man* components to Tinian went on in parallel with the preparations for *Little Boy.* At the same time as *Little Boy's* target rings departed Kirtland field, two other C-54s carrying the *Fat Man* plutonium core and initiator also departed, and arrived at Tinian on the 28th. That same morning, three B-29 bombers, each carrying a high-explosive implosion pre-assembly, departed from Kirtland. These arrived at Tinian about midday on August 2.

These various lots of components were not simply spares. In addition to bombing runs with weapons of the same shapes and weights as "active" *Little Boy* and *Fat Man* units, a number of tests were run to check various systems using both inert bombs and ones loaded with conventional explosives. *Little Boy* test bombs were known as "L" units, and *Fat Man* ones as "F" units. July 23 saw the dropping of unit L1, which was fired in the air by radar fusing.

Little Boy test units, and the Nagasaki F31 Fat Man plutonium implosion weapon shortly before its mission. Courtesy of the Los Alamos National Laboratory Archives

Units L2 and L3 followed on July 24 and 25. On July 29, unit L6 was used to test the procedure for emergency reloading of the bomb into another aircraft at Iwo Jima. The same unit was used on July 31 in a test where the bomber flew to Iwo Jima accompanied by two observation planes, rendezvoused, and returned to Tinian to complete a drop test; this was essentially a rehearsal for the Hiroshima mission. Following this test, all rehearsals preparatory to a combat delivery of a *Little Boy* with active material were complete. Unit L11 was designated for the Hiroshima bombing.

The first *Fat Man* test, with unit F13, was made on August 1. This unit was used to test fusing and detonating circuits, and was inert in that it used cast plaster blocks in place of high explosives. Unit F18 was dropped *unsuccessfully* (the firing mechanism did not operate properly) on August 5, the day before the Hiroshima mission. Unit F33, a fully-functioning model except for having an inert core, was dropped on August 8; this was a rehearsal for the Nagasaki mission the following day. The "active" Nagasaki *Fat Man* was unit F31. Another unit, F32, was held at Tinian in case a third combat drop was to be made, but its fissile material never left the United States.

Between June 30 and July 18, 509th crews flew seven training and orientation missions comprising 27 sorties (a "sortie" means the flight of an individual aircraft, whether alone or as part of a group); bombs were not carried on these missions. Between July 1 and August 2, 15 practice-bombing missions totaling 89 sorties were conducted. These used conventional 500 and 1,000-pound bombs dropped on nearby lightly-defended Japanese-held islands. Curiously, these were not considered to count as "combat" missions. What did count as combat operations were 16 Pumpkin missions (51 sorties), where *Fat Man*-shaped 10,000-pound bombs containing 6,300 lb of high-explosive were dropped from altitudes of about 30,000 feet over various cities in Japan proper. Two of these sorties had to be aborted, with the result that only 49 Pumpkins were dropped; in one case the bomb was jettisoned, and in the other it was returned safely to Tinian. Pumpkin missions extended from July 20 right up to the day of the Japanese surrender, August 14; 509th missions continued for almost a week after the Hiroshima and Nagasaki bombings.

After unloading at Tinian, the *Indianapolis* sailed to Guam, about 130 miles to the south, and then proceeded toward Leyte Island in the Philippines, where its crew of 1,196 were to join a Task Force in preparation for the November 1 invasion of Kyushu. But just before midnight on Sunday, July 29, the ship was torpedoed by the Japanese submarine I-58. The *Indianapolis* sank within 12 min; some 850 men managed to escape into the sea. A distress call was sent, but it is not clear if any transmitting power remained. Not until Thursday morning, August 2, were 316 survivors inadvertently discovered. Many of those who survived the sinking had succumbed to shark attacks. The loss of the *Indianapolis* represented the greatest single loss of life at sea in the history of the Navy, and has been called the worst naval disaster in American history. The *New York Times* reported the story at the bottom of its front page on Wednesday, August 15, the same day that the headlines reported that Japan had decided to surrender. The *Indianapolis'* Captain, Charles McVay, survived, but was court-martialed and found guilty for failing to steer an evasive zigzag course to avoid torpedoes, although he had not been explicitly ordered to do so. In recognition of McVay's bravery in combat before the sinking, however, the Secretary of the Navy lifted the sentence. McVay was promoted to Rear Admiral upon his retirement in 1949, but committed suicide in 1968. In July, 2001, the Navy announced that McVay's record had been amended to exonerate him for the loss of the *Indianapolis* and her crew. In August, 2017, the wreckage of the ship was found at a depth of 18,000 feet in the Philippine Sea.

As technical preparation of the bombs was underway, bureacratic groundwork for their use was being finalized. On July 22, General Marshall, in Potsdam, directed his acting Chief of Staff in Washington, General Thomas Handy, to prepare a directive for submission to himself and Stimson. Groves prepared the orders on the 23rd, and relayed them back to Marshall through Handy. Marshall informed Handy on the 25th that Truman and Stimson had approved them. The text of the orders read:

25 July 1945

TO: General Carl Spaatz
Commanding General
United States Army Strategic Air Forces

1. The 509 Composite Group, 20th Air Force, will deliver its first special bomb as soon as weather will permit visual bombing after about 3 August 1945 on one of the targets: Hiroshima, Kokura, Niigata and Nagasaki. To carry military and civilian scientific personnel from the War Department to observe and record the effects of the explosion of the bomb, additional aircraft will accompany the airplane carrying the bomb. The observing planes will stay several miles distant from the point of impact of the bomb.

2. Additional bombs will be delivered on the above targets as soon as made ready by the project staff. Further instructions will be issued concerning targets other than those listed above.

3. Discussion of any and all information concerning the use of the weapon against Japan is reserved to the Secretary of War and the President of the United States. No communiques on the subject or releases of information will be issued by Commanders in the field without specific prior authority. Any news stories will be sent to the War Department for specific clearance.

4. The foregoing directive is issued to you by direction and with the approval of the Secretary of War and of the Chief of Staff, USA. It is desired

that you personally deliver one copy of this directive to General MacArthur and one copy to Admiral Nimitz for their information.

(Sgd) THOS. T. HANDY

THOS. T. HANDY
General, G.S.C.
Acting Chief of Staff

These orders effectively made the decision to use the bombs the responsibility of commanders in the field; no further authorization from higher-up would be necessary. Groves also sent Marshall a memorandum describing operational plans. Attached to the memo was a small map of Japan cut out from a _National Geographic_ map, accompanied by descriptions of each of the four target cities listed in Handy's orders. All of these cities except for Nagasaki had been specifically reserved against bombing to provide virgin targets for the new weapons; the memo also included a draft of the orders necessary for the Joint Chiefs of Staff to release them to General Spaatz for attack.

An excerpt from Truman's personal diary for July 25, the day before the Potsdam Declaration was issued, offered an apocalyptic perspective:

> We have discovered the most terrible bomb in the history of the world. It may be the fire destruction prophesied in the Euphrates Valley Era, after Noah and his fabulous Ark. Anyway we think we have found the way to cause a disintegration of the atom. An experiment in the New Mexico desert was startling—to put it mildly. Thirteen pounds of the explosive caused the complete disintegration of a steel tower 60 feet high, created a crater 6 feet deep and 1,200 feet in diameter, knocked over a steel tower 1/2 mile away and knocked men down 10,000 yards away. The explosion was visible for more than 200 miles and audible for 40 miles and more.
>
> This weapon is to be used against Japan between now and August 10th. I have told the Sec. of War, Mr. Stimson, to use it so that military objectives and soldiers and sailors are the target and not women and children. Even if the Japs are savages, ruthless, merciless and fanatic, we as the leader of the world for the common welfare cannot drop that terrible bomb on the old capital or the new. He and I are in accord. The target will be a purely military one and we will issue a warning statement asking the Japs to surrender and save lives. I'm sure they will not do that, but we will have given them the chance. It is certainly a good thing for the world that Hitler's crowd or Stalin's did not discover this atomic bomb. It seems to be the most terrible thing ever discovered, but it can be made the most useful...

The "old" capital referred to by Truman is Kyoto, the historic capital of Japan. Groves wanted Kyoto on the target list, but Stimson had deleted it. Truman's belief that he had ordered the bomb to be used against purely military targets was illusory. Hiroshima and Nagasaki were indeed sites of important Japanese military bases, but the world was about to learn that nuclear weapons are of power sufficient to obliterate entire cities at one blow.

In Washington, Stimson's office was busy drafting statements and press releases in preparation for when the bombings would be reported to the public. Preparations in the Pacific proceeded so rapidly that on July 30, Stimson, by then returned to Washington, had to send an urgent cable to Truman (still in Potsdam) with proposed revisions to the statements, noting that "The time schedule on Groves' project is progressing so rapidly that it is now essential that statement for release by you be available not later than Wednesday, 1 August." Truman received the message early in the morning on the 31st, and wrote in pencil on its reverse that any release be held until at least August 2, by which time he would be at sea on his way home. Truman presumably meant August 2, Washington time (August 2 on Tinian would correspond to August 1 in America).

Also on July 30, Groves sent a follow-up memo to Marshall describing what he expected for the effects of a bomb that would be detonated at an altitude of 1,800 feet, about what was being planned for in use against Japan. His predictions were based on extrapolating from an estimated yield from the *Trinity* test of 21–24 kilotons. He expected the blast to be lethal to 1,000 feet from ground zero, with heat and flame fatal to 1,500–2,000 feet; at Nagasaki, people would suffer burns out to nearly 14,000 feet. The neutron flux was expected to be lethal to about the same distance as heat and flame, and practically all structures over an area of six to seven square miles should be largely devastated. This prediction would prove somewhat optimistic, although multistory brick buildings would be destroyed out to a radius of about a mile at Nagasaki. Since no effects from radiation were expected in view of the altitude of detonation, he expected that it would be possible to move troops through the area immediately afterwards, preferably by vehicle but on foot if necessary. Prior to the explosion, no friendly personnel should be within six miles, a distance equivalent to the nearest observers at Trinity. He then laid out a schedule for future bomb availability. In addition to three bombs expected to be available soon for combat use (Hiroshima, Nagasaki, and one which would go unused), a further three or four should be available in September, including one of the U-235 gun-type. The same was expected for October, but November would see at least five, and December seven. Beyond these, a marked increase in production was expected in early 1946. In short, Groves was expecting up to 20 bombs to be available by the end of 1945 beyond three initial combat bombs. He also anticipated that beginning in November, U-235 would be used in the more efficient implosion design.

By August 1, 1945, the use of Los Alamos' bombs was essentially a foregone conclusion, awaiting only final delivery of materials to the Pacific and acceptable weather.

8

The German Nuclear Program: The Third Reich and Atomic Energy

The success of the Manhattan Project makes it easy to overlook the fact that there *was* a German nuclear program during World War II, and that for some time it ran ahead of its Allied counterpart. For various reasons, the German effort began to lose steam in the summer of 1942, just as the Manhattan Engineer District was coming into existence. But by no means did it die out. Working feverishly in the last weeks of the war in Europe in April, 1945, German scientists came close to creating a self-sustaining chain reaction in a heavy-water-moderated pile. This was as far from a full-blown Manhattan Project as was Fermi's CP-1 pile, but, like it, was the first step on that long path.

8.1 Origins of the German Program: Competition from the Outset

Hahn and Strassmann's paper reporting the discovery of fission was published on January 6, 1939. While Lise Meitner and Otto Frisch have tended to be singled out as *the* interpreters of the discovery, they were by no means the only ones exploring the physics of the dramatic new phenomenon. On January 22, Siegfried Flügge and Gottfried von Droste submitted a report—now largely overlooked—to *Zeitschrift fur Physikalische Chemie* in which they presented a much more extensive analysis and arrived at the same conclusions regarding energy release as did Meitner and Frisch, who had submitted their paper to *Nature* on January 16.

© Springer Nature Switzerland AG 2020
B. C. Reed, *Manhattan Project*,
https://doi.org/10.1007/978-3-030-45734-1_8

Siegfried Flügge (1912–1997) in 1934.
Source: https://commons.wikimedia.org/
wiki/File:Flügge,Siegfried_1934_London.
jpg

The April 22 publication of Frédéric Joliot's measurement of 3.5 neutrons per fission appears to have been a direct impetus for the German nuclear program. Soon thereafter, a colloquium on fission was given at the University of Göttingen by Wilhelm Hanle. This caught the attention of physicist Georg Joos of the same institution, who felt it his duty to inform government authorities of the possibilities. Joos wrote a letter to the Reich Ministry of Education, which oversaw universities. His letter reached Abraham Esau, who had been an academic physicist but who had been rewarded for his support of the Nazi Party by being appointed President of the Reich Bureau of Standards and head of the physics section of the Ministry's Reich Research Council.

Esau promptly organized a conference, which was held in Berlin on April 29. Among others, this meeting was attended by Joos, Hanle, Walther Bothe, and Hans Geiger; Otto Hahn was out of town, but deputed Josef Mattauch, who had been hired to replace Lise Meitner, to attend in his place. As a result of this meeting, Esau recommended that all uranium stocks in Germany be secured. A ban was placed on export of uranium compounds, and contact was opened with the Ministry of Economics to secure radium from mines located in Joachimsthal, Czechoslovakia. Ironically, also on April 29, Niels Bohr addressed the possibility of a chain reaction in a public talk given at a meeting of the American Physical Society in Washington, but opined that isolating a large quantity of U-235 would be practically impossible.

Unknown to Esau, a second initiative was underway. On April 24, University of Hamburg physical chemist Paul Harteck and his assistant Wilhelm Groth had written a letter to the German War Office to alert them to the fact that developments in nuclear physics could lead to very powerful explosives. Harteck would become a major player in the German nuclear program.

Map of present-day Germany showing approximate locations of major nuclear project sites. Map by author based on http://www. freeusandworldmaps.com/html/Countries/Europe%20Countries/ GermanyPrint.html

Left: Abraham Esau (1884–1955) Source: https://
commons.wikimedia.org/w/index.php?search=Abraham+
Esau&title=Special:Search&go=Go&searchToken=
6t7ne3jzucy5rz9xy83wsby4q#/media/File:Esau_Abraham.
jpg. *Right: Paul Harteck (1902–1985) in 1948. Source:*
https://upload.wikimedia.org/wikipedia/commons/b/ba/
Bundesarchiv_Bild_183-2005-0331-501%2C_Paul_Harteck.
jpg

In a remarkable confluence of events, just two days later Henry Tizard
approached the Treasury and Foreign offices in London to propose that
Britain move to deny Germany access to large stocks of uranium ore held
by the Union Minière company in Belgium. In a memo to the Air Defense
Committee, Tizard estimated the odds of nuclear energy being of military
value as 100,000 to one (a number he likely pulled out of thin air), but felt
that the possibility could not be ignored. After meeting with Union Minière's
President, Edgar Sengier, on May 10, Tizard recommended against purchasing the ores, but did advise Sengier that they might become of great strategic
value. Subsequently, the British were made aware of Esau's April 29 meeting when Josef Mattauch, likely violating security protocols, mentioned it to
Paul Rosbaud, the editor of *Naturwissenschaften*. Rosbaud, who often passed
information to British contacts, related Mattauch's story to a British scientist
who happened to be visiting Berlin. Tizard's reluctance would prove a boon
to the German program: during the course of the war, Germany would seize
some 3,500 tons of uranium compounds from the Belgian stockpiles.

Harteck and Groth's letter was routed to the Army Ordnance Department
of the War Office (*Heereswaffenamt*), where it reached physicist Erich Schumann, an advisor to General Wilhelm Keitel, Chief of the Armed Forces High

Command. Schumann in turn contacted Kurt Diebner, an Army expert on nuclear physics and explosives. Diebner would become closely involved with the German nuclear program, eventually being appointed Commissioner for Norwegian Heavy-Water Production, Provisional Head of the Kaiser-Wilhelm Institute of Physics, and deputy head of the program.

Kurt Diebner (1905–1964). Source: https://upload.wikimedia.org/wikipedia/ commons/c/c4/Kurt_Diebner.jpgr

As a result of these initiatives, two rival programs were underway at the start of the war: Esau's and Diebner's. While Esau's would soon be sidelined by the much more powerful War Office bureaucracy, he would eventually return to direct the effort after the Army withdrew from the field in 1942. On September 4, 1939, just two days after Britain and France declared war on Germany, Esau met with General Karl Becker, head of the Army Ordnance Office (and also a professor at the University of Berlin) to request provision of uranium compounds before they were requisitioned by the Air Ministry for use in manufacture of luminous paint for aircraft instruments. Becker agreed, but directed Esau to see Schumann for preparation of the requisite voucher. Within days, Esau was informed that the Ordnance Department was ordering the Bureau of Standards to cease uranium research; a cache of uranium oxide that Esau had accumulated would be taken by the War Office.

In the meantime, the War Office's program was ramping up. The same week as Esau met with Becker, Erich Bagge, a physicist at the Leipzig Institute for Theoretical Physics and a student of Werner Heisenberg (of Uncertainty Principle fame), was ordered to report to Army Ordnance in Berlin. Fearing a journey to the front, Bagge must have been relieved when he was met by Diebner and Schumann, who wanted his help in arranging a conference of experimental physicists to explore the feasibility of using uranium as a source of power or explosives. This meeting, which would be held on September 16, was attended by (among others) Bothe, Geiger, Hahn, Diebner, and Flügge, a group which would come to call themselves the Uranium Club. History does not record the reactions of those participants who only months earlier had attended Esau's conference. Esau was not on the guest list.

Left: Erich Bagge (1912–1996). Source: National Archives and Records Administration, courtesy AIP Emilio Segrè Visual Archives. Right: Werner Heisenberg (1901–1976; in 1927). Source: https://upload.wikimedia.org/wikipedia/commons/b/ b0/Heisenberg_10.jpg

The September 16 meeting was held just two weeks after the publication of Bohr and Wheeler's analysis of fission. It was becoming clear that U-235 was likely the isotope that suffered slow-neutron fission, but Hahn apparently expressed skepticism regarding achieving a chain reaction in view of that isotope's low natural abundance. The possibility was not to be entirely dismissed, however, and two important results came out of this meeting. The first was that Schumann recommended to Becker that a "Nuclear Physics Research

Group" be established within the Ordnance Department. The result would be a research laboratory located in Gottow (a southern suburb of Berlin), where the Army was already conducting research on rockets and explosives. Diebner was placed in charge of this initiative. The second was Bagge's suggestion that Heisenberg be brought in to work out the theory of a chain reaction.

A second conference held ten days later saw the German program begin to move on a number of fronts. By this time, Heisenberg appreciated that two routes to utilizing fission might be possible: in reactors if a suitable neutron-moderating substance could be found, and/or as an explosive if U-235 could be separated from its sister isotope. Harteck had already begun to conceive of a reactor design wherein uranium and heavy-water would be arranged in alternating layers, and had also begun research on separating uranium isotopes using Clusius-Dickel thermal diffusion tubes, as would be used in the S-50 plant at Oak Ridge. He had already contracted with the I. G. Farben chemical cartel to produce the necessary working substance, uranium hexafluoride. Diebner and Bagge drew up a research program: Heisenberg would continue theoretical investigations of chain reactions, Bagge would undertake measurements of the neutron-collision properties of heavy-water, and Harteck would continue with his isotope-separation work. By late 1939, Harteck would have an experimental steam-heated 25-foot-long separation tube under construction.

At about this time, Schumann moved to have the War Office take over the facilities of the Kaiser-Wilhelm Institute of Physics (KWIP) as a location at which to centralize the work. The Institute, which was government-funded, could hardly refuse. However, the director of the KWIP was Peter Debye, a Dutch citizen. Debye was forced to choose between becoming a German national or being dismissed; he chose the latter and emigrated to America in January, 1940, taking up a position at Cornell University. Schumann appointed Diebner to replace Debye, but Institute staff and administrators felt that Diebner was not of the Debye's caliber, and lobbied for Heisenberg to be appointed instead. The result was a compromise: Diebner was appointed "provisional" director to serve during Debye's "absence", while Heisenberg would serve as an advisor, commuting to Berlin from Leipzig once a week. Despite considerable work on his part, Diebner's efforts to centralize the project were constantly frustrated by scientists preferring to stay at their home institutions. Lack of organization would prove a significant impediment to the German effort throughout the war.

Also about this time, the Auer chemical company was contracted to produce a few tons of uranium oxide for pile experiments; Auer's raw materials were the seized Belgian ores. The company erected a plant at Oranienburg, also near Berlin, to produce about one ton of oxide per month; the first ton was delivered to the War Office in early 1940. On March 15, 1945, this plant would be destroyed in a raid by over 600 B-17 bombers, a mission requested by General Groves in order to deny the facility to advancing Russian forces.

8.2   A Report to the War Office, and Norwegian Heavy-Water

On December 6, 1939, Heisenberg reported on the situation to the War Office, outlining the dual possibilities for power production and explosives. As a competitor to Harteck's layered design for a pile, he had conceived of his own configuration wherein 1.2 tons of uranium and a ton of heavy-water would be mixed into a paste and enclosed in a spherical chamber of radius 60 cm, which would be surrounded by a neutron-reflecting shield of water.

On February 29, 1940—just when Alfred Nier was separating his first minute samples of U-235—Heisenberg submitted a second report, which would initiate one of the enduring mysteries of the German nuclear program: what appears to be very fundamental misunderstanding on his part regarding how to calculate the critical mass. In his report, he gave an expression for the critical radius which was based on an erroneous random-walk model of neutrons as they travel through a bomb core, and which resulted in an estimate of some 600 metric tons. That Heisenberg would err so dramatically is particularly mystifying given that Perrin, Flügge, and Peierls's papers on criticality had already appeared in the open literature.

The German pile program would come to rely on heavy-water as a moderator. At the time of Heisenberg's report, the only large-scale source of heavy-water in the world was a hydrogen-electrolysis plant operated by the Norwegian national hydroelectric generating company, Norsk Hydro, in Vemork, Norway, near the town of Rjukan. The electrolysis plant was an adjunct facility to a 450-MW hydro-electric generating station, the largest in the world. The plant's primary purpose was to make hydrogen for use in fertilizers; the heavy-water was a by-product. By 1938, the plant had produced about 40 kg of heavy-water and was producing about 10 kg per month; a large-scale pile experiment would require tons. This facility played a major role in the Nazi bomb project.

The Vemork hydroelectric plant. Source: https://upload.
wikimedia.org/wikipedia/commons/8/87/Vemork_
Hydroelectric_Plant_1935.jpg

Left: Partial map of Norway, showing Vemork, Rjukan, and Oslo. Vemork is about 80 miles (straight line) from Oslo. Source: Google maps, in compliance with information on permissions page at https://www.google.com/permissions/geoguidelines.
html. *Right: Partial map of southern Scandinavia; the rectangle shows the approximate area of the map to the left. Source: From d-maps.com (*http://d-maps.
com/carte.php?num_car=5972&lang=en*). Permission for commercial use granted according as* http://d-maps.com/conditions.php?lang=en

Also in early 1940, theoretical physicists working under Carl Friedrich von Weizsäcker at the Kaiser Wilhelm Institute examined possible pile configurations, and concluded that Harteck's layered design would require about two tons of uranium and half a ton of heavy-water, whereas a spherical pile with concentric layers of uranium oxide and heavy-water would require less material but would be more difficult to construct. von Weizsäcker's position in the German nuclear program was unique: his father was the second-ranking official in Hitler's Foreign Ministry.

Carl Friedrich von Weizsäcker (1912–2007) in 1993. Source: https://commons. wikimedia.org/wiki/File:Carl_Friedrich_ von_Weizsaecker.jpg. *Photo by Ian Howard, released under the GNU Free Documentation License,* https://commons. wikimedia.org/wiki/Commons:GNU_ Free_Documentation_License

On January 15, Harteck wrote to Heisenberg to advocate large-scale production of heavy-water, apparently unaware that just nine days earlier Heisenberg had attended a meeting on this very issue at Diebner's office in Berlin. Diebner had sought Heisenberg's advice on whether a full-scale heavy-water plant should be constructed in Germany; Heisenberg suggested that they wait until he had measured its neutron-capturing properties. Diebner promised to secure about 10 L of heavy-water from Norway for the necessary experiments, and Heisenberg replied to Harteck to advise that, should a plant be built, it would be the business of physical chemists like Harteck himself. Heisenberg apparently had in mind that pile experiments would be the purview of physicists like himself. This motif of turf-protection was an element of the German project throughout the war.

Diebner's efforts to secure heavy-water came to frustration. A representative of I. G. Farben, which had a financial interest in Norsk Hydro, attempted to persuade the firm to sell its entire stock of 185 kg of the precious liquid. When the Germans refused to indicate why they wanted it, the Norwegians declined the request. The 185 kg would instead be spirited out of the country. In February, 1940, soon after the Farben approach, Frédéric Joliot approached the French minister of munitions, Raoul Dautry, to ask if heavy-water could be procured for his own pile experiments. Dautry and Joliot met with Jacques Allier, a lieutenant in the French secret service who was also involved with a French bank which held an interest in Norsk Hydro. Allier made his way through Sweden to Oslo, where on March 4 he met with the Director-General of Norsk Hydro, Axel Aubert. Allier carried a letter of credit for 1.5 million Norwegian kroner, but Aubert offered Allier all of the heavy-water at no cost. The liquid was secured into 26 stainless-steel flasks, which were flown to Scotland on March 12. Within a week they had arrived at Joliot's laboratory in Paris, but their stay there would be brief: in June they were smuggled to Britain before they could be captured by occupying German forces. Allier's initiative was just in time: On April 9, 1940, soon after Frisch and Peierls had prepared their "super bomb" memorandum, Germany invaded Norway. The Vemork area fell on May 3, and the Germans were soon developing plans to increase production of heavy-water to 1.5 tons per year. On April 10, Allier attended the first meeting of the MAUD Committee to brief the group on the heavy-water issue and chain-reaction research in France.

Undaunted by the lack of heavy water, Harteck devised an ingenious plan for a substitute: frozen carbon dioxide, known commonly as dry ice. I. G. Farben made dry ice on an industrial scale for use as a refrigerant, but there would not be much demand for it until the summer months, and offered Harteck the necessary supply free of charge. Expecting some 10 tons of dry ice to arrive within a few weeks, Harteck wrote to Diebner to ask for 100–300 kg of uranium oxide, which he believed to be all that was available. Unfortunately, Heisenberg had simultaneously put in a request for 500–1,000 kg for his own experiments. Apparently unwilling to exert authority over the matter, Diebner suggested to Heisenberg that he come to some arrangement with Harteck. Heisenberg wrote to Harteck to suggest that he content himself with 100 kg, which infuriated Harteck: he really needed some 600 kg for the scale of the experiment he had in mind. In any event, Harteck would need the oxide for only a short time before his dry ice evaporated. The oxide began arriving in late May, but he received only about 185 kg. Along with 15 tons of dry ice, he built a pile about six feet square by seven feet tall, with the uranium distributed in shafts drilled into the ice. With so little uranium, however, the experiment was hopeless, and no neutron multiplication was detected; the only results (which were of no small value) were measurements of the diffusion length of neutrons in dry ice and their capture by uranium. A startling aspect of all German wartime piles was that they incorporated no control mechanisms aside from sometimes having a surrounding shield of ordinary water. This was because Heisenberg had become convinced that as the temperature of a pile began to increase, the fission cross-section would decrease, with the result that the process would be self-regulating and stabilize at a temperature he estimated to be about 800 C. The flaw in Heisenberg's reasoning was that what matters is the rate of fissions, which is proportional to the product of the neutron speed (which will increase with temperature) and the fission cross-section. But the cross-section is essentially inversely proportional to the speed, so the two effects largely cancel each other, leaving the rate of fissions unchanged.

Despite the loss of the heavy-water, the German program did begin to achieve some results in the summer of 1940. At Heidelberg, Walther Bothe began measurements on the diffusion of neutrons through graphite, while Heisenberg and his collaborators in Leipzig made similar measurements with small amounts of heavy-water and uranium oxide. Both graphite and heavy-water looked promising as moderators, particularly the latter. In July, planning began for a building to house a subcritical pile on the grounds of the Kaiser-Wilhelm Institute of Biology and Virus Research, which was located next to the Physics Institute in Berlin. To deter the curious, the laboratory was designated "The Virus House". Construction of the laboratory proceeded under the direction of Karl Wirtz, a staff member at the KWIP.

8.3   Plutonium, Cyclotrons, and the First Berlin Pile

One of the developments in the German program that occurred in mid-1940 is an excellent illustration of how scientific ideas are not restricted by borders. In July, 1940, von Weizsäcker read in the *Physical Review* of Edwin McMillan and Philip Abelson's success in isolating element 93. Like Louis Turner, he struck on the idea that U-238 nuclei might transmute under neutron bombardment to a new fissile element. While he assumed that the decay chain would stop at element 93, he was on the right track, and wrote up his speculation in a report to the War Office.

Like their American counterparts, German scientists also invested considerable effort to investigating possible methods of enriching uranium. At Hamburg, Harteck and Groth experimented with uranium hexafluoride, determining that nickel was the only metal capable of withstanding that compound's violent corrosiveness. Clusius raised the possibility of liquid diffusion, and electromagnetic separation was also considered. However, a limiting factor in the German program was the lack of a large-scale cyclotron with which to synthesize plutonium or serve as a model for an electromagnetic separator. This issue was partly solved with the German conquest of France, which permitted the Germans to move in on the cyclotron at Joliot's laboratory in Paris, but neither country possessed anything of the scale of Ernest Lawrence's Radiation Laboratory.

The cyclotron issue brings up one of the more curious players in the German program. In early 1940, Baron Manfred von Ardenne, a largely self-educated applied physicist and inventor who had inherited sufficient wealth to establish a private research laboratory, became convinced that the electromagnetic method was the route to large-scale U-235 production. While searching for funding, he learned that the Post Office maintained a largely-unused research fund. Despite objections from von Weizsäcker, von Ardenne secured support to build an accelerator, and began work on cyclotrons. Lying as it did outside the circle of leading scientists involved with the uranium project, von Ardenne's work was essentially orphaned. By the end of the war, he had developed a method similar to that used at Oak Ridge, and which he eventually took to the Soviet Union. But with his initiative, the German nuclear program became distributed between his laboratory, the Army research site at Gottow, the KWIP, and scientists working at various universities. While these scattered efforts paralleled the early days of American fission research, by the summer of 1940 the latter was becoming more firmly coordinated under Vannevar Bush's National Defense Research Committee.

Manfred von Ardenne (1907–1997) in 1930. Source: https://upload. wikimedia.org/wikipedia/commons/ 4/4f/Bundesarchiv_Bild_183-K0917- 500%2C_Prof._Manfred_v._Ardenne. jpg. *This image is freely available for commercial use according as the terms of a Creative Commons license available at* https://creativecommons.org/licenses/by- sa/3.0/

On May 10, 1940, Germany invaded Belgium. The Belgian Army surrendered on May 28, and later that year Belgian uranium compounds began arriving in Berlin. They were soon put to use by Gottfried von Droste, who used two tons of it packed into 2,000 paper containers to build a three-foot cubical pile, an experiment which does not seem to have had any outcome beyond pointing out the need for uranium free of impurities. But a larger-scale effort was about to commence. By early October, the Virus House facility was ready. The main feature of the laboratory was a two-meter-deep brick-lined circular pit which would serve as a receptacle for a cylindrical reactor vessel. The vessel was 1.4 m tall and of equal diameter, and could be lifted into and out of the pit with a crane. In December, Wirtz, Heisenberg, and others began assembling their first pile. The cylinder was loaded with layers of uranium oxide separated by paraffin wax (as a moderator), and immersed in water in the pit. A neutron source was lowered into the pile, but no chain reaction was observed; apparently neutrons were being captured within the pile. The experiment was repeated with 6,800 kg of uranium oxide arranged in two piles within the cylinder, but again to no avail. Heisenberg concluded that a light-water or paraffin-moderated pile would not achieve criticality. Other researchers were on the same trail. At Heisenberg's home base in Leipzig, Robert Döpel, who had participated in the original April meeting in Berlin, constructed a spherical uranium/paraffin pile, but this too yielded null results. In Heidelberg, Walther Bothe and his collaborators mixed nearly four tons of uranium oxide with water in a large vat, and also came to the conclusion that heavy-water was needed as the moderator.

By the end of 1940, the War Office—presumably at the initiative of Diebner—decided that pure metallic uranium should be used as opposed to uranium oxide, and the Auer company contracted with the German Gold and Silver Exchange Corporation (German: *Degussa*) to produce it. By the end of 1940, fully two years before Enrico Fermi's CP-1 pile, some 280 kg had been produced. Degussa would produce all metallic uranium in Germany during the war, but the reduction process the firm used left the metal rife with impurities.

8.4   An Error with Graphite, Twice a Spy, and a Visit to Copenhagen

The year 1941 would prove as pivotal for the German program as for the British/American effort, but in essentially the opposite way. A devastating setback occurred in January, when Walther Bothe reported on results of measurements of neutron diffusion in a 100 cm diameter sphere of graphite, concluding that, contrary to previous expectations, that material captured too many neutrons to make it suitable as a moderator. Bothe believed his graphite to be very pure; it has subsequently been speculated that it may have been contaminated with atmospheric nitrogen, which has an appreciable neutron capture cross-section. Ironically, had Paul Harteck not been discouraged from his carbon dioxide experiments, Bothe's error might have been discovered. The only feasible moderator now looked to be heavy-water, and Karl Wirtz was dispatched to Vemork to see how the capacity of the plant there could be improved.

The other possible route to securing atomic energy, uranium enrichment, fared little better. Harteck had no success with his Clusius tubes, apparently because of the temperature at which he was operating them. Other methods of enrichment were proposed, including an ingenious rotating-shutter "sluice" system conceived by Erich Bagge. Cyclotron and centrifuge methods were also investigated, the latter especially by Groth and Harteck. No fewer than seven enrichment methods were considered, but the Germans apparently never looked seriously at the gaseous diffusion method which would prove so successful at Oak Ridge.

The emphasis on pile research and the possibility of breeding plutonium was further stimulated with the arrival at Manfred von Ardenne's laboratory of one of the most colorful if unlucky personalities of the German effort, Friedrich ("Fritz") Houtermans. A remarkably gifted researcher and seemingly inexhaustible source of jokes, Houtermans had earned his Ph.D. under Nobel Laureate James Franck in 1927. Later, he made seminal contributions to the theory of nuclear fusion as the source of energy in stars, and was involved in coining the term "thermonuclear" to describe such reactions. Part-Jewish and a dedicated communist, Houtermans left for England upon Hitler's ascendance to power, taking a job with Electrical and Musical Industries (popularly known as EMI Records) in London. England did not agree with him; in December, 1934, he left for Russia, taking up a position at the Ukrainian Physico-Technical Institute in Kharkov, where he conducted well-regarded research on thermal-neutron capture cross-sections in various materials. With the worsening political situation in Stalinist Russia, he soon came under suspicion of being a German spy, and was arrested in late 1937 by the NKVD. Thrown in prison, he was maltreated, almost starved to death, and at one point interrogated nearly continuously for 11 days. Under the threat that his wife and children would be arrested (they had escaped to America), he confessed to being a spy.

Fritz Houtermans (1903–1966) in 1927. Source: https://upload.wikimedia.org/wikipedia/commons/ 1/18/Fritz-Houtermans1927.jpg

Unlike many others caught up in Stalin's terror, Houtermans survived. In August, 1939, Russia and Germany signed a non-aggression pact, and in April, 1940, he was extradited back to Germany. The NKVD delivered him into the hands of the Gestapo, who jailed him for three months on suspicion of being a Communist agent. Upon his release, he was prohibited from working for any state agency, but Max von Laue and Carl von Weizsäcker helped him obtain a position in von Ardenne's laboratory, where he began work on New Year's Day, 1941. (Von Laue, who had been awarded the 1914 Nobel Prize for Physics for his work on X-ray diffraction, had helped many scientists escape from Nazi Germany and was bravely outspoken in his criticisms of Nazi policies; his involvement with the nuclear project was extremely limited.) Houtermans set to work, and was soon struck by the potential of using element 94 as an explosive. In August—just after the preparation of the MAUD report—he prepared an extensive report on this possibility; he also considered the magnitude of the critical mass. After the war, he became a professor at the University of Bern in Switzerland, where he remained until his death in 1966.

Intelligence contacts in Norway kept British officials apprised of the growing German interest in heavy-water. About the time of the MAUD report, indications that production was being increased at Vemork reached Reginald Jones, a physicist and scientific intelligence officer with the British Intelligence Service. Jones contacted Lieutenant-Commander Eric Welsh, who had lived in Norway and ran the Service's Norwegian section. Welsh, who would later play a significant role in organizing commando raids against the Vemork plant, began coordinating his work with one of his sources, Leif Tronstad, a Norwegian chemist living in London who had been involved in designing the plant. Tronstad became head of a section of the Norwegian High Command in London that was responsible for espionage and sabotage, and would meet his death three years later while operating in his native country.

By the end of 1941, the first 360 kg of heavy-water of a 1,500-kg contract had arrived in Germany, and over two tons of powdered uranium metal had been produced. Heisenberg and Döpel constructed a second spherical pile at Leipzig (Leipzig II, or "L-II"; other L-piles would follow until the end of the war), but it contained only 142 kg of uranium oxide and 164 kg of heavy-water. No increase in neutron production was observed, but they estimated that there would have been a slight increase if not for the capture effects of aluminum shells that separated the various layers of the device. Heisenberg has been quoted as saying that "It was from September 1941 that we saw an open road ahead of us, leading to the atomic bomb." Exactly what he had in mind is unclear; the remark may have been occasioned as much by von Weizsäcker and Houtermans speculations on using transuranic elements as explosives as much as by any success with piles.

One of the most dramatic episodes of the German nuclear program occurred in September, 1941, when Heisenberg visited Copenhagen to speak at the German Scientific Institute, a German propaganda outlet. Sometime during the week of September 15–21, Heisenberg had a private, and fateful, conversation with his old friend and mentor, Niels Bohr. No others were present; there is no record of their exchange. What we know of this meeting can only be reconstructed from letters and comments that each made after the war.

The background context of this meeting is important. Despite his position in the German program, Heisenberg was politically vulnerable. In 1936, he had been accused in an SS publication of being a "White Jew" for his advocacy of "Jewish physics", namely Einsteinian relativity and quantum physics, an accusation which cost him a distinguished position at the University of Munich. Fortunately, Heisenberg's mother and SS Reichsführer Heinrich Himmler's mother moved in the same social circles, and the former relayed a personal letter from Heisenberg to Himmler asking whether he approved of such attacks. After a year-long SS investigation, Himmler wrote back to Heisenberg to indicate that he disapproved of the attacks and that they would cease.

Heisenberg could hardly have refused the "invitation" to visit Copenhagen. This trip was but one of at least 10 such journeys he made during the Nazi regime, to places as diverse as the United States (in 1939), Hungary, Switzerland, Holland, and Poland; he might well have seen these invitations as evidence of his "rehabilitation" in Party circles following Himmler's investigation. While Heisenberg apparently personally abhorred the Nazis and hoped that Hitler would be replaced as soon as the war was won, he shocked colleagues during a visit to Holland in 1943 by stating that he felt that since Democracy could not possibly rule Europe, the continent faced a choice between Germany and the Soviets—with the former clearly to be preferred.

Heisenberg's version of his meeting with Bohr, as related in a 1948 document, was that he wanted to ask "Does one as a physicist have the moral right to work on the practical exploitation of atomic energy?" As historian David Cassidy has pointed out, however, moral issues do not seem to have been a particular concern for Heisenberg during the war. Cassidy speculates that Heisenberg, probably convinced at that time of inevitable German victory, hoped to convince Bohr to use his influence with Allied scientists to prevent them from working on a bomb which could be used against Germany. Similarly, in an article published in *Physics Today* a few years before his passing, Hans Bethe stated that he believed that Heisenberg might have been trying to tell Bohr that the Germans were working on a reactors, not bombs, so that Bohr could be a "messenger of conscience" to persuade Allied scientists to refrain from working on bombs. Bethe further remarked that he felt that Heisenberg had no interest in making an atomic bomb, and that he was sincere when he told Bethe that his rationale for working on the German uranium project was so that he could save some young physicists for the future.

Whatever Heisenberg's motivation, Bohr, who was of Jewish descent and whose country was occupied by the Nazis, must have been profoundly disturbed at Heisenberg's certainty of German victory. For Heisenberg, even having the conversation was dangerous: to hint at the existence of the German nuclear program was treasonous. Apparently Bohr became disturbed, and asked Heisenberg if an atomic weapon was truly possible. In response, Heisenberg handed him a diagram of a reactor (another treasonous act), which Bohr may have interpreted as an indication the Germans were indeed working on atomic weapons. At this point Bohr abruptly terminated the conversation; he would take the reactor diagram with him to Los Alamos in late 1944. After the war, Bohr wrote several letters to Heisenberg giving his side of the story, but never sent them. Sadly, the two men never reconciled, their friendship a casualty of the war. But even if Heisenberg's intent was sincere and he was misinterpreted by Bohr, it is hard to imagine the evolution of events on the Allied side being different from what they were.

8.5   Changes of Fortune

The perennial controlling factor in the German nuclear program was the availability of heavy-water. By the end of 1941, Norsk-Hydro production had been increased to about 140 kg per month in an effort to satisfy the War Office's contract for 1,500 kg. In early 1942, a new contract was awarded, this time for five tons. Simultaneously, consideration was given to establishing a plant in Germany which would work with a process developed by Paul Harteck and Hans Suess. I. G. Farben proposed to build a pilot plant, likely in anticipation of getting in on future energy generation.

Also by early 1942, some progress was emerging in the area of uranium enrichment. In February, Bagge operated his sluice mechanism with uranium for the first time. The first prototype centrifuge, constructed for Wilhelm Groth, was ready by April, but it exploded during a test because its steel-alloy construction was not strong enough. In other ways, however, Heisenberg's open road was encountering one detour after another. In June, 1941, Germany had invaded the Soviet Union, breaking their 1939 non-aggression pact. The eastward advance was rapid at first, but by October the German forces were bogged down in the battle of Moscow. The German economy, which had operated on the basis of lightning campaigns followed by respites during which forces and materials could be replenished, was becoming strained. This forced Hitler to decree a policy that economic needs were to be directed to military necessities, and on December 5, two days before Pearl Harbor and one day before the decisive S-1 meeting in Washington, Erich Schumann wrote to the directors of the various institutes engaged in uranium research to tell them that work on the project could only be justified "if a certainty exists of attaining an application in the foreseeable future." A conference of the directors took place on December 16, after which Schumann forwarded a report to the chief of Army Ordnance, General Emil Leeb. The result was a decision the Army should seriously reduce its funding of the project and relinquish control back to the Reich Research Council, which still lay within the Ministry of Education. The Minister of Education, Bernhard Rust, while a zealous supporter of Nazi ideals, was, however, not regarded as particularly competent. The Research Council itself was still under the directorship of Abraham Esau, who now nominally came into control of the initiative he had tried to stimulate almost three years earlier. But Rust and Esau did not enjoy the political clout of the Army, and the Army team under Diebner continued its research at Gottow.

Late February saw two further gatherings unfold. Despite the Army's desire to be out of the uranium project, Schumann convened a meeting of leading scientists to be held at the KWIP over the 26th to the 28th. An ambitious agenda listed presentations covering 25 highly-technical issues from cross-sections to isotope enrichment. The Research Council decided to hold its own meeting on the 26th, with the idea of inviting high government officials such as Albert Speer (the Minister of Munitions), General Keitel, Martin Bormann, Heinrich Himmler, and Hermann Göring to listen to more general-level talks given by the scientists before the latter went on to the KWIP meeting. Due apparently to a clerical error, the government officials were sent the agenda for the technical conference, and none of them attended. Rust was present, however, and listened to Heisenberg give an exposition on the physics of a chain reaction, the necessity for heavy-water, and the possibility of U-235 as an explosive. A translation of Heisenberg's lecture was published by David Cassidy and William Sweet in the August, 1995, edition of *Physics Today*. But for his mistaken notion that a reactor would be self-stabilizing, the lecture indicates that Heisenberg had a clear command of the issues surrounding nuclear piles, including that they could be used to breed plutonium. To add to the mystery of the confusion with critical mass arising from the February, 1940, report, Army Ordnance produced a 131-page report on their conference which included an estimate of the critical mass of from 10 to 100 kg. However, the source of this number is not indicated. Whatever its provenance, it is hard to imagine Heisenberg not having been involved with it in some way.

By May, 1942, Degussa had produced almost 3.5 tons of pure uranium in powdered form. Some of this went to Heisenberg and Döpel in Leipzig, where they were preparing pile L-IV, which they hoped would demonstrate net neutron production. But preparations were tedious because powdered uranium can catch fire when exposed to air. This had been discovered in late 1941, when a Leipzig technician had been spooning powder into pile L-III's aluminum sphere when it caught fire and a jet of flame shot out of the funnel, seriously burning the technician's hand and igniting a drum of uranium. As a result, filling of L-IV was done in a carbon dioxide atmosphere. Over three-quarters of a ton of uranium and 140 kg of heavy-water were encased within two bolted-together aluminum hemispheres about 80 cm in diameter; a radium-beryllium neutron source could be placed into the pile's center via a shaft. Heisenberg and Döpel detected about a 13% increase in neutron production, and estimated that a pile with five tons of heavy-water and ten tons of solid uranium metal in the form of slabs would be self-sustaining. In late May, Degussa began preparing a ton of uranium in slab form, but the plates tended to be plagued by impurities. From figures given in David Irving's *The Virus House*, total production of metallic uranium in Germany during the war amounted to some 13,700 kg, which would not have been enough for even a single loading of the X-10 pile.

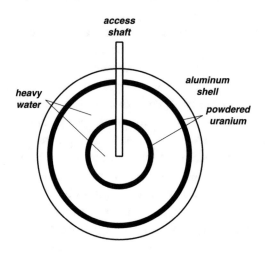

Schematic illustration (not to scale) of the Leipzig L-IV pile. Sketch by author after Irving (1967) p. 132. Right: A German spherical pile. Source: AIP Emilio Segrè Visual Archives, Goudsmit Collection

On June 6, Heisenberg traveled to Berlin for a critical meeting with Albert Speer and his staff to decide on the future of nuclear research. (Some sources give the date of this meeting as June 4. June 6 was a Saturday, which would have been convenient for travel.) Also present were Diebner, Harteck, Wirtz, Hahn, General Leeb, and Field Marshall Erhard Milch, who was in charge of German aircraft production. Heisenberg addressed the group on possible military aspects of fission, and Milch asked how large a bomb would have to be to destroy a city. Heisenberg allegedly answered that it would be about the size of a pineapple. This remark comes from a recollection of a member of the audience, but, if accurate, indicates that Heisenberg may then have had a much clearer sense of the critical mass in contrast to his earlier report. However, he apparently hastened to add that it would be impossible for Germany to produce a bomb as no large-scale method enriching uranium was in hand. Speer limited approvals to various construction projects, including a shelter equipped to house a large reactor on the grounds of the KWIP. The project was in limbo, given neither a full-scale go-ahead nor termination.

Albert Speer (1905–1981) at the Nurem-berg Trials. Source: https://commons. wikimedia.org/wiki/Albert_Speer#/ media/File:Albert-Speer-72-929.jpg

In an article published in the August 16, 1947 edition of *Nature*, Heisenberg pinpoints the meeting with Speer as a decisive turning point, claiming that from thereon, the only practical goal would be to obtain an energy-producing pile and that German physicists were "… spared the decision as to whether or not they should aim at producing atomic bombs." Historians continue to debate the extent to which this statement involved some after-the-fact self-absolution. On June 23, Speer reported briefly to Hitler on the uranium project, but the conversation was limited to one item on a long agenda. Hitler never kept himself informed on the prospects for fissions bombs as did Franklin Roosevelt.

Soon after Speer had been briefed on the project, the Reich Research Council was reorganized to better mobilize German science for the war effort. The name was retained, but the organization would now be directed by Hermann Göring. Göring subsequently appointed a civil servant with high honorary SS rank, Rudolf Mentzel, to manage the Council's affairs. Mentzel in turn delegated administration to various directors and "plenipotentiaries" for important research projects. Nuclear physics came under the purview of Abraham Esau, to whom Kurt Diebner would now report.

The same day as Speer briefed Hitler, Heisenberg's L-IV pile suffered a disaster. The pile had been immersed in its water tank for 20 days, and began to emit bubbles. Robert Döpel found that they contained hydrogen; apparently water was seeping in and reacting with the uranium. The pile was lifted out of the water, and the same technician who had previously been burned loosened an inlet cap. Air rushed into the pile, and a flame of burning uranium erupted, melting the aluminum shell and setting more uranium on fire. The fire was doused with water, and the pile lowered back into the tank. The temperature kept rising, however. About six p.m., Heisenberg and Döpel were examining the pile when it began to shudder and swell. They fled the room just seconds before the hydrogen trapped within exploded and set the building on fire. This time the fire department had to be called in, but much of the laboratory, the uranium, and the heavy-water were lost.

8.6   Freshman and Gunnerside

By June, 1942, the Vemork plant had delivered about 800 kg of the five tons of heavy-water that Heisenberg estimated would be necessary to achieve a self-sustaining chain reaction. German scientists and engineers were dispatched to Norway to improve the efficiency of the plant, and, by September, they expected production to be able to reach 400 kg per month. There was also the prospect of eventual contribution from the Farben plant being built in Germany itself.

British Intelligence continued to monitor German interest in heavy-water. By good luck, they scored a significant break in March, 1942, when a Special Operations Executive (SOE) agent in Norway and a band of volunteers captured a coastal steamer and sailed it to Aberdeen, Scotland. The SOE had been established in 1940 to conduct sabotage, reconnaissance, and espionage in occupied Europe. One of the volunteers was Einar Skinnarland, who was from the Vemork area. Given some quick training, he was parachuted back into Norway on March 29, and established contact with the chief engineer at the heavy-water plant, Jomar Brun. Brun had little sympathy for the German occupation, and arranged for photographs and drawings of the plant to be micro-photographed and smuggled to Britain through neutral Sweden in toothpaste tubes. In November, Brun escaped to Britain, where he proved an invaluable source of information.

By July, the British War Cabinet was sufficiently concerned to request its Combined Operations (CO) department to mount a ground attack on Vemork to destroy the heavy-water factory; CO had been set up to harass German forces by means of commando raids. A bombing raid was out of the question: If nearby ammonia tanks were struck, the local population could be exposed. Combined Operations coordinated with SOE, which had an advance party trained and ready to parachute into Norway at a desolate location some 30 miles northwest of Vemork. CO developed a plan where some 40 troops would land in gliders on a lake that fed the Vemork turbines, march in uniform to the plant, blow it up, and escape to Sweden. The operation was code-named *Freshman*. On the night of October 18, four Norwegians of the advance party parachuted in. Promptly hit by a snowstorm, it took them two days to gather their equipment, after which followed a grueling trek to their base. Not until November 6 could they get a brief radio message to London as to their whereabouts.

Thirty-four *Freshman* commandos began their mission on the night of November 19, (near full moon), their Horsa Mark-1 gliders towed by four-engine Halifax bombers. The flight from an airfield in northeast Scotland would be 400 miles across the North Sea, with the gliders to be dropped from an altitude of 10,000 feet; this would be the first time that gliders were used in an operation. Tragically, the mission turned into a disaster. Despite the advance team setting out lights, cloud cover made the landing area impossible to identify, and the bombers had to turn for home when they began running low on fuel. On the return journey, the tow-rope of one of the gliders snapped, and the second bomber and its glider crashed into a mountainside. Fourteen men survived, but were quickly rounded up by the Germans and shot. The glider whose rope had snapped crash-landed in southern Norway, and while some of the 17 men on that craft survived, they soon met the same fate; their bomber made it back to Scotland. Not only were many lives lost at a time when the German nuclear program was being downgraded in priority, but the Germans were now alerted to British interest in the Vemork area, and began reinforcing their garrison and laying minefields around the plant. The advance party, despite enduring conditions so miserable that they were sometimes reduced to eating moss, were ordered to wait until moonlight would be suitable for another attempt.

Following the *Freshman* disaster, the SOE volunteered to take over the mission of destroying Vemork. Jomar Brun identified an entrance to the plant in the form of an unsecured cable duct, and the War cabinet authorized another strike. A Norwegian SOE commando, Lieutenant Joachim Rönnenberg, was ordered to select five god skiers to accompany him to parachute into Norway, join up with the advance party, and blow up cells inside the plant where heavy-water was concentrated. One of these men was Knut Haukelid, who was to remain in Norway with three men from the advance party after the operation; the rest were to ski 250 miles to escape to Sweden. A mock-up of the target part of the plant was constructed, and the group was given extensive training in infantry and explosives. This new mission was code-named *Gunnerside*.

Gunnerside was scheduled to commence on January 23, 1943, with the men issued cyanide capsules to be used if they were in danger of being captured. But once again the landing area could not be identified, and their bomber tuned back for Scotland. Training continued, and the mission was rescheduled for February 16. Messages from the advance party indicated that the Vemork plant was now heavily guarded, so a new drop zone was chosen: Lake Skryken, a brutally forsaken area some 30 miles from the advance party's base, which was itself 20 miles from the target. This time the men made their drop. All survived, but a blizzard came in and they were forced to take refuge in a hunting lodge for two days. The blizzard began to abate, but continued for another three days, during which they gathered up supplies. Finally, Rönnenberg ordered that they would depart at noon on the 22nd. After skiing through the night and the next day, they met up with two men from the advance party. Everybody then settled into the advance party's hut to plan their attack.

The Vemork plant sat atop a rugged 500-foot gorge. A bridge crossed the gorge, but was guarded. The decision was made to approach the plant by scaling down the gorge on the opposite side, ford the river, and then ascend the gorge on the plant side. On the afternoon of the 26th, two men were left behind to guard the group's wireless equipment while the rest began the journey to the plant site. They were in place by the next evening, and began their descent at about 10:00 p.m. Upon approaching the plant, they split up into a covering party and a demolition party, with Rönnenberg leading the latter. After cutting a hole in a perimeter fence, Rönnenberg and another man made their way to the cable duct; others went in through a window. Charges with timed fuses were laid at the bottom of the heavy-water cells, and the men had barely escaped the building when the charges exploded, emptying the cells down drains and distributing enough shrapnel to damage other equipment. The Germans dispatched thousands of troops to search for the saboteurs, but all escaped; General Nikoluas von Falkenhorst, the German Military Governor of Norway, called the operation "the best coup I have ever seen." It has been estimated that about a ton of liquid comprising about 350 kg of pure heavy-water was lost, a serious setback to the German program. Repairs to the plant were commenced promptly, but even after it came back into operation on April 17, months would be required before heavy water could be drawn off in quantity.

8.7   Plenipotentiary for Nuclear Physics, and Vemork Bombed

Late 1942 saw further administrative changes to the German nuclear program, although it still remained divided between two factions. During the summer, von Weizsäcker and Wirtz had convinced the governors of the Kaiser-Wilhelm Foundation that Werner Heisenberg needed to be brought in as Director of the KWIP. Since Peter Debye had never formally resigned, Heisenberg was made "Director at the Kaiser-Wilhelm Institute of Physics", effective October 1; he was also appointed professor of theoretical physics at the University of Berlin, a prestigious position. Diebner retreated to the Army's research site at Gottow, where he carried on with his own Army-funded pile experiments. There, he conceived the crucial idea of distributing uranium as chunks within a moderator, as opposed to using layers of plates or shells. In the summer of 1942, he set up his first pile using uranium oxide and paraffin as a moderator inside a large cylindrical aluminum vessel. The paraffin was built up in a layered honeycomb fashion, with 6,802 cubical voids which would be filled with some 25 tons of powdered oxide. The vessel was lowered into a concrete pit which was filled with water to act as a shield and neutron reflector. No increase in the neutron flux was detected, but the idea of distributing the uranium through the moderator brought the German project closer to the successful Fermi/Szilard approach that would be used in Chicago. In the meantime, work at the Virus House in Berlin continued, and the underground bunker authorized by Albert Speer, which would hold a plate-design reactor containing three tons of uranium, was under construction.

The summer 1942 reorganization of the Reich Research Council did not particularly improve the prospects for the uranium project. Work remained fractionated between the KWIP site, various universities, the Army site at Gottow, and the Post Office's support of von Ardenne's laboratory. On November 24, Esau wrote to Rudolf Mentzel to propose centralizing the work. The issue went up to Göring, with the result that Esau was designated as "Plenipotentiary of the Reichsmarschall for Nuclear Physics" to oversee a Nuclear Physics Research Group within the Research Council. Despite this grandiose title, Speer and members of the Kaiser-Wilhelm Foundation (KWF) evidently had little confidence in Esau, who was personally more inclined to support enrichment work than the reactor program. On February 4, 1943, just two weeks after the *Gunnerside* commandos parachuted onto Lake Skryken, Albert Vögler, President of the KWF, informed Esau that he intended to apportion research between the Kaiser-Wilhelm Institute and the

Nuclear Physics Research Group; he had been promised support by Speer. Naturally, discord soon arose between the two groups over access to materials. The situation became somewhat clarified when the War Office decided to pull out of the effort altogether at the end of March, 1943, but this made Esau's no less challenging: Diebner's research group was turned over to the RRC, but was to continue research at Gottow under RRC funding. Esau drew up a two-million-Reichsmark budget for the following year, which included 600,000 Reichsmarks for the construction of 10 ultracentrifgues for Harteck and Groth.

At Gottow, Diebner carried on with his cube-based pile experiments. Theoretical considerations led him to believe that 6.5 cm cubes would be best, but he had to settle for cutting uranium plates into 5 cm cubes. In a development reminiscent of Paul Harteck's dry-ice pile, he conceived the idea of freezing his heavy-water and embedding the cubes in it to form a lattice. His first experiment along this line involved 232 kg of uranium and just over 200 kg of heavy-water ice embedded within a paraffin-wax sphere 75 cm in diameter. This was an awkward arrangement in that the lattice could not easily be reconfigured once assembled, but did show better neutron production that the Leipzig L-IV pile, and led to the idea of affixing the cubes along wires, which would be suspended vertically into a reactor vessel filled with heavy-water.

A German uranium cube being held by the
author at an undisclosed location

Aside from its internal dysfunction, the German nuclear effort was also beginning to become more and more hobbled by relentless Allied bombing raids. In July, 1943, Harteck's ultracentrifuge laboratory had to be relocated several hundred miles from Kiel to Freiburg after a series of air raids;

in November, 1944, the latter city would be devastated in another raid. Erich Bagge's isotope sluice was beginning to show promise, but in the summer of 1943 much of his time was taken up with organizing the evacuation of about a third of the KWIP from Berlin to a new headquarters in the town of Hechingen in southern Germany. Heisenberg remained in Berlin, preparing for the bunker-based pile experiment. While by late 1943 it was becoming clear that Diebner's idea of disposing uranium in a lattice of cubes was superior to Heisenberg's plate geometry, Heisenberg—who wielded immense influence by virtue of being a Nobel Laureate—was dismissive of this approach because his own plate configuration made for easier theoretical calculations. In November, Auer began casting the plates for Heisenberg's Berlin pile, even as Esau and Diebner entered into a separate contract with the firm for manufacture of cubes. Another lattice pile, this time with over 500 kg of uranium and nearly 600 kg of heavy water, gave even more promising results.

After the *Gunnerside* raid, the Vemork plant was brought back into production. Alarmed, General Groves persuaded General George Marshall to authorize a bombing attack on the hydroelectric plant that fed the heavy-water distillation plant. The attack was scheduled for between 11:30 and noon on November 16, 1943, a time when many of the plant workers would be at lunch. Nearly two hundred B-17 and B-24 bombers of the 95th and 100th Bombardment Groups of the Eight Air Force dropped over seven hundred 500-pound bombs on the plant. Many went wide, but three hit the hydro plant pipelines, causing the heavy-water plant to shut down. Twenty-one Norwegians were killed in the raid, including 16 who had taken refuge in an air-raid shelter that suffered two direct hits.

As a result of the raid, the Germans decided to relocate all heavy-water production to their home country. Esau directed that all semi-concentrated heavy-water remaining in Norway be shipped to Germany, and set aside 800,000 Reichsmarks for Farben to construct a plant to produce 1.5 tons of heavy-water per year. The most direct victim of the Vemork shutdown was Diebner, who suffered the additional setback that all of the heavy-water under his control was transferred to Heisenberg. To compound his situation, a RAF raid on Frankfurt destroyed Degussa's factories, bringing uranium metal production to a halt.

Political pressure against Abraham Esau—probably orchestrated by Speer—came to a head in late 1943. In October, Rudolf Mentzel sounded out University of Munich physics professor Walther Gerlach on the possibility of his taking over the physics section of the Reich Research Council. Curiously, Gerlach's war work mostly involved research into torpedoes; he had not been involved with the nuclear project. After consulting with Heisenberg and Hahn, he decided to accept the post, contingent on his being given absolute authority over the distribution of funds; his appointment became effective on January 1, 1944. While Gerlach devoted himself to the project, his decisions seemed to reflect a mixed attitude; he permitted the two reactor groups, Diebner's and Heisenberg's, to continue functioning separately until very near the end of the war. It has been speculated that he felt that he was contributing to saving pure research in Germany by keeping as many scientists as possible from being sent into combat, with a longer-term view to securing a place for the country in postwar nuclear energy development. For his part, Esau was reassigned to a command in high-frequency research.

Walther Gerlach (1889–1979). Source: AIP Emilio Segrè Visual Archives, Gift of Jost Lemmerich

8.8   The D/F Hydro Sinking, Alsos, and the Berlin Pile Bunkers

Early 1944 saw the final setback to the heavy-water project. After the bombing raid on Vemork, the Germans decided to ship home forty-three 400-l drums and five 50-l flasks containing the last of the partially-concentrated heavy-water drawn off from the plant; this totaled just over 600 kg of pure product. British intelligence learned of this, and instructed Knut Haukelid that the supply should be destroyed, even if it meant casualties and reprisals. The RAF dropped explosives and equipment to the saboteurs, who began considering how to execute their mission.

The drums were to be transported by rail, but part of the journey involved their being loaded onto a rail ferry, the *D/F Hydro*, for passage across Lake Tinnsjö, a 20-mile-long fjord with a maximum depth of over 1,500 feet. If the ferry could be sunk over the deepest part of the lake, the drums would be unrecoverable. The Vemork plant's new chief engineer, Alf Larsen, was sympathetic to the mission, and deliberately scheduled to drums for transport on the morning ferry of Sunday, February, 20, in the hope of minimizing casualties. After the operation, Larsen was "exported" to Sweden.

Haukelid calculated that blowing out an area of the hull of eleven square feet would cause the ferry to sink by the bow in just a few minutes, thus preventing recovery of the drums. He and his compatriots formed some 19 lb of plastic explosive into a sausage-links arrangement which would enclose the requisite area. The night before the sailing they snuck on board, telling a crew member that they were fleeing from the Gestapo. The plan worked flawlessly, with the vessel sinking in just four minutes and the rail cars carrying the drums breaking loose and tumbling into the lake. Of 53 passengers and crew aboard, 26 perished. Some of the drums did float free and were recovered, but their heavy-water content was highly diluted in ordinary water. The Germans, who anticipated needing five tons of heavy-water for their biggest reactor experiment, would have only about 2.5 tons available for the rest of the war.

In his pattern of leaving nothing to chance, General Groves remained eager to gather any intelligence he could regarding German nuclear efforts. In the autumn of 1943, he saw to the establishment of the *Alsos* mission, a collaboration of the Manhattan District, the Army's G-2 Intelligence department, the OSRD, and the Navy. Ironically, *Alsos* is the Greek word for "grove". The first *Alsos* mission would follow the American Fifth Army as it advanced through Italy, questioning government officials and scientists along the way. This group left for Naples on December 16, 1943, commanded by Lieutenant-Colonel Boris Pash, who was infamous in Manhattan Project circles for his aggressive questionings of Robert Oppenheimer. Pash had no scientific background and so the mission was not as successful as it might have been, but it did turn up some useful information. Italian scientists had themselves done no work on nuclear explosives, but an Italian officer who had been posted to Berlin for six years testified that the Germans had developed no new explosives of great violence.

Left: A formal portrait of Col. Boris Pash. Source: https://commons.wikimedia.org/wiki/File:Boris_Pash.jpg. *Middle: San Goudsmit at work during the Alsos mission. Source: AIP Emilio Segrè Visual Archives, Gift of Michaele and Terry Thurgood, Thurgood Collection. Right: Marinus Toepel (left) and Sam Goudsmit (right) in a jeep during the Alsos mission. Source:* https://commons.wikimedia.org/wiki/File:Goudsmit_Toepel.jpg

British intelligence was skeptical of rumors of German atomic-bomb development, but Groves remained cautious. In late 1943, he established a liaison office with the British Tube Alloys project in London, sending Major Horace Calvert, a very competent intelligence expert, to organize it. Calvert's staff soon set to work on developing a list of about 50 leading German nuclear scientists and establishing their whereabouts. From intelligence sources and German newspapers, their attention began to focus on the Hechingen area.

In Berlin, work on the underground pile bunker continued despite constant bombing raids. On the night of February 15, 1944, the Kaiser-Wilhelm Institute for Chemistry took a direct hit, and it was decided to relocate the work of both the Chemistry and Physics Institutes to Tailfingen, about 10 miles from Hechingen. The exception, at least for a while, was the pile bunker, which would remain under Heisenberg's supervision. Other aspects of the program suffered their own setbacks: another raid had destroyed the latest prototype of Erich Bagge's isotope sluice; he moved to the Frankfurt area to oversee construction of a replacement. The I. G. Farben works where the heavy-water plant was to be built was destroyed by a bombing raid on July 28, which ended the company's interest in the project. One area that remained promising was Harteck's ultracentrifuge program: a full-scale plant was being built near the border with Switzerland on the rationale that the Allies would not bomb so close to that country. Overall, however, by the summer of 1944 only two facets of the uranium program were enjoying top priority: Bagge's isotope-sluice, and production of corrosion-proof uranium plates by Auer. By July, the sluice produced 2.5 g of enriched uranium hexafluoride, but the following month the project was interrupted to move it to Hechingen. In the late summer, Diebner's pile-research group relocated to Stadtilm, practically in the center of Germany. It is easy to imagine this development meeting with the satisfaction of Heisenberg, who remained at the bunker facility in Berlin.

With its six-foot-thick concrete floors, walls, and ceilings, the pile bunker was immune from bombing raids, but working among constant raids must have been nearly intolerable. Compared to Enrico Fermi's CP-1, the facility was lavishly equipped. A circular pit would receive the reactor vessel, which was a 124 cm wide by 124 cm tall by 3 mm thick cylinder made of low-neutron-capturing magnesium alloy. A winch ran over the pit to raise and lower the cylinder and its lid. Special ventilation, air-conditioning, and pumping equipment would siphon off radioactive emissions; the pile could be viewed though portholes; and laboratories and workshops were available for processing uranium and heavy water. Again despite that fact that Diebner's cube geometry had been demonstrated to be superior, the first Berlin pile was constructed with alternating layers of plates and heavy water. Considerable time was spent in investigating different plate separations to see which one gave the best neutron enhancement, work that had already been done by Walther Bothe in November, 1943.

In May, 1944, Albert Vögler made it clear to Walther Gerlach that he was not satisfied with the rate of progress with the Berlin project. Gerlach's stress was compounded when in July an American bombing raid set his Munich house on fire. Weakened, it collapsed during a thunderstorm a week later, forcing him to move to Berlin. His main concern at this time, however, was to find a safer site for the Berlin pile, thinking to locate it in a narrow valley that could not easily be bombed. He chose the village of Haigerloch, which lay on a river between two sheer cliffs; it also had the advantage of being only about 10 miles from Hechingen. Conveniently, a wine cellar had been carved into the rock, and contracts were issued to local firms to enlarge it to accommodate the pile. But this would take several months of work.

In the lead-up to the June, 1944, D-Day invasion of France, the *Alsos* mission was reconstituted. The mission, now accompanied by a battalion of combat engineers, was still under the command of Colonel Pash, but this time it contained a group of scientists led by Samuel Goudsmit, a Dutch-born University of Michigan physicist. Goudsmit knew many European scientists personally, and also had the advantage that since he was not part of the Manhattan Project, he would not be a liability if captured; he was suggested to Groves by Vannevar Bush. Goudsmit was appointed to *Alsos* on May 25, 1944, and flew out for London on D-Day, June 6. There, he and his staff built up a target list of German scientists and industrial firms to be investigated. Altogether, *Alsos* comprised 55 civilian personnel, six Counter-Intelligence Corps agents, and 119 military personnel. After the war, Goudsmit wrote a memoir of his adventures, but he was not an unbiased observer: his parents had been murdered in a concentration camp.

Pash entered Paris on August 25, just behind the first column of French tanks to enter the city, and found Frédéric Joliot at his laboratory. Flown out to London for questioning on the 29th, Joliot revealed the German interest in his cyclotron, and, while maintaining that the Germans had made little headway on the uranium problem, passed on the names of Schumann, Diebner, Bothe, Bagge, and Esau. Another valuable clue came in the form of a catalog for the University of Strasbourg, which indicated that von Weizsäcker was now located there.

The city of Brussels was liberated on September 4, and Goudsmit and Major Calvert arrived a few days later to raid the offices of the Union Minière company. There they learned that the Germans had purchased over a thousand tons of uranium products and had seized much more. On the rationale that if the Germans had indeed built a pile that would need to be cooled like the Hanford piles, the *Alsos* mission even took samples of Rhine river water to be tested for radioactivity.

Strasbourg was liberated in late November. The *Alsos* mission captured seven physicists and chemists, but the real haul came when they broke into von Weizsäcker's office. von Weizsäcker had fled, but had left behind a trove of letters, files and papers which listed addresses and telephone numbers for many of the uranium project's main institutes; they also identified Gerlach as Plenipotentiary of the Reichsmarschall for Nuclear Physics. Goudsmit worked almost nonstop for four days reviewing the files.

8.9 The B-VII and B-VIII Piles

In late 1944, the last pile to be built in Berlin, B-VII, was constructed under the direction of Karl Wirtz. This pile differed in significant ways from its predecessors. A new aluminum cylinder was obtained, this one 210 cm in diameter by 210 cm tall and 5 mm thick. This would enclose the earlier magnesium-alloy vessel, with the space between the two filled with 10 tons of neutron-reflecting graphite. However, this pile still used 1.25 tons of Heisenberg's uranium plates and about a ton and a half of heavy water, not Diebner's cubes. No control rods were provided; Wirtz later claimed that the pile was intended to be subcritical. The neutron multiplication rate, while not yet-self-sustaining, was better than in previous experiments. Surprisingly, this did not raise any questions regarding Bothe's 1941 measurements of neutron capture in graphite.

Wirtz began planning another pile, B-VIII, which would use uranium cubes. But by this time the war situation was becoming dire. Berlin was under constant attack, and Russian forces were advancing rapidly from the east. Wirtz and his group were almost ready to go when Gerlach ordered, on January 30, 1945, that they had to evacuate. The uranium, heavy water, and equipment were first moved to Diebner's laboratory at Stadtilm, but Heisenberg pressured Gerlach to relocate the project to his site at Haigerloch, and the better part of a month passed before reconstruction of B-VIII got underway. This pile would use Diebner's 5 cm uranium cubes fixed on 78 wires suspended from the lid of the pile, which also contained graphite. A neutron source could be inserted through a chimney in the lid. There was little instrumentation, and the only control mechanism was a block of cadmium which could be thrown into the pile if it threatened to get out of hand.

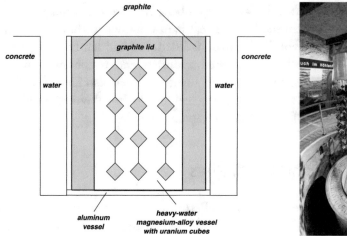 

Left: Sketch (not to scale) of the B-VIII reactor, after Irving (1967) p. 319. Right: A replica of the B-VIII pile at the Atomkeller Museum in Haigerloch. Source: https:// de.wikipedia.org/wiki/Forschungsreaktor_Haigerloch#/media/File:Haigerloch_ Atomkeller-Museum_Versuchsreaktor_2013-08-18.jpg. *This image is freely available for commercial use according as the terms of a Creative Commons license available at* https://creativecommons.org/licenses/by-sa/3.0/

The pile was reconstructed by late February. The neutron flux was monitored as heavy water was pumped in, but even when the tank was full the flux had not achieved the exponential growth characteristic of a self-sustaining reaction. It was estimated that the assembly would have to be 50% larger to obtain a chain-reaction, which would require yet more uranium and heavy water. Some of each yet remained at Stadtilm, so Diebner returned there, gathered up the supplies, and was able to set out for Haigerloch ahead of advancing American forces.

As German resistance crumbled, *Alsos* moved in to apprehend as many of the main players and as much of the material of the German nuclear project as possible. In order to beat French and Russia occupation forces before they could get in place, Pash often led daring raids into areas where fighting was still going on. Groves was particularly concerned that anything of interest in the French zone of occupation be seized as soon as possible; he did not trust them not to pass on information to the Russians.

In late March, American troops entered Heidelberg, where on the 30th Goudsmit found Walther Bothe. Bothe refused to disclose any information on military research, but did reveal that there was a second group at Stadtilm under Diebner; that Otto Hahn was in Tailfingen; that Heisenberg and Max von Laue were in Hechingen; and that the last pile had been evacuated to

Haigerloch. Stadtilm was captured on April 12 and many of Diebner's files were confiscated, but he and his convoy were yet to be seen. Another item of interest was the fate of the tons of uranium ore that the Germans had acquired from Belgium. Major Calvert traced eleven hundred tons of it to a salt mine near Strassfurt, about 50 miles northwest of Leipzig. This ore would eventually serve as feed material for Oak Ridge and Hanford.

On April 23, troops led by Pash captured Haigerloch. The next day, a group of British and American intelligence officers entered the cave and found the reactor pit. The uranium and heavy water were gone, but they dismantled the pile, seized some graphite blocks, blew up the pile's outer casing with hand grenades, and then blew up the cave.

Left: Members of the Alsos mission dismantle the Haigerloch reactor, April 1945. Source: https://commons.wikimedia.org/wiki/File:German_Experimental_Pile_-_Haigerloch_-_April_1945.jpg. Right: recovery of uranium cubes outside Haigerloch; Goudsmit is seated on the ground to the right of the stack of cubes. Source: https://commons.wikimedia.org/wiki/File:Haigerloch_uranium_cubes_uncovered.jpg

In Hechingen, von Weizsäcker, Wirtz, Erich Bagge, and an assistant of the latter, Horst Korsching, were picked up. Heisenberg had fled a few days earlier, setting out by bicycle for his summer home in Urfeld, not far from the border with Austria; his wife and six children had already been living there for some time. Today this would be about a 200-mile drive; Heisenberg made it in three days and three nights of what must have been strenuous pedaling. von Weizsäcker and Wirtz revealed that the Haigerloch heavy water was hidden in gasoline cans in a country mill, and that hundreds of uranium cubes were buried in a field outside the village; these were recovered. Most of the cubes probably ended up at Oak Ridge or Hanford; a few remain in private hands or at universities. von Weizsäcker also revealed that many documents were hidden in a canister in a cesspit at his home; Goudsmit had the unenviable task of examining these. Meanwhile, Pash had located Otto Hahn and Max von Laue in Tailfingen. Hahn, who had lost 30 lb over the preceding year, quickly turned over documents in his possession. Plenipotentiary Walther Gerlach was found at the University of Munich on May 1, the day after Adolf Hitler had committed suicide in his Berlin bunker, and Diebner was found about 20 miles southeast of Munich. Heisenberg, the big prize, was calmly waiting on the porch of his house with his bags packed when Pash found him on May 3, two days before Germany's formal surrender. The captives were sent to Paris, where they were soon joined by Paul Harteck, who had been apprehended in Hamburg. They would cool their heels in Europe for two months before being transferred to England, where they would be held for six months—and learn about the Allied nuclear effort.

In 2009, a group of Italian nuclear engineers under the direction of Giacomo Grasso published the results of an analysis of the predicted performance of the B-VIII reactor based on using software employed in the design of modern reactors. Results depended on the choice of values for various parameters, but runs with three sets of parameter choices all resulted in the reactor being subcritical, with a neutron population growth similar to what the Germans had estimated. Their results also indicated that the lack of criticality was not due to the presence of impurities in the graphite, but was rather a geometric issue: To slow neutrons to thermal energies by having them pass through heavy-water requires a path length of about 11 cm, but the distances between the surfaces of pairs of uranium cubes was about 5–8 cm, depending on the direction of neutron travel. Along a similar line, a paper published in 2015 by Klaus Mayer and his colleagues reported the results of nuclear forensic analyses of samples of metal obtained from a "Heisenberg cube" and a 1 cm thick "Wirtz plate". By examining the ratios of various isotopes, they determined that both had been manufactured from ore obtained from the Joachimsthal region of the Czech Republic. Ratios of uranium isotopes indicated that the material had undergone no enrichment, and the trace amounts of plutonium present were indicative of natural origin, that is, not as a result of any significant neutron irradiation. Using an analysis of the ratio of Thorium-230 to Uranium-234 (the former is the alpha-decay product of the latter), the group was able to estimate the date at which the uranium was last chemically treated to remove impurities and decay products during manufacture: the Heisenberg cube dated from the second half of 1943, whereas the Wirtz plate had been produced about mid-1940.

By the end of the war in Europe, the German nuclear program had failed to advance even to the point of achieving a self-sustaining chain reaction, a landmark reached in the Manhattan Project two-and-one-half years earlier. In hindsight, several contributing factors in this failure are evident: personality clashes; turf-protection; scientific missteps carried out in a culture less open to questioning an established authority figure than was the case in America (Heisenberg and the critical mass, plates versus cubes, Bothe's graphite measurements); fear of political persecution; disruption of heavy-water supplies; lack of credible scientific advice going to the highest decision-makers; and, later in the war, relentless bombing raids which would have made undertaking any large industrial effort virtually impossible. In believing that they must be ahead of their Allied counterparts, German scientists may have lacked the driving force that their adversaries might get a bomb first. Leslie Groves, while no "Plenipotentiary" in the world of nuclear physics, held the respect of his peers for his experience, leadership skills, and decision-making attributes. Once vested with the authority and priorities needed to carry out Manhattan, he never looked back.

9

Hiroshima and Nagasaki

9.1 The Pacific War and the 509th Composite Group

When Paul Tibbets was selected to command the Army Air Forces 509th Composite Group, he was given wide liberty to choose personnel, but he could tell selectees nothing of their ultimate mission. Familiar with some of the best pilots, navigators, and bombardiers of the war, Tibbets wasted no time in recruiting them. Among his earliest acquisitions were two personal friends with whom he had flown a number of missions: bombardier Major Thomas Ferebee, and navigator Theodore "Dutch" Van Kirk, veterans of 63 and 58 missions, respectively. Both would fly the Hiroshima mission with Tibbets. First Lieutenant Jacob Beser, the 509th's radar officer, would be the only crew member to fly in both the Hiroshima and Nagasaki strike aircraft, the ones that carried the bombs. Beser would be responsible for monitoring Japanese radar to determine if they were trying to jam the bomb's firing mechanisms, or perhaps even cause a premature detonation.

The 509th was unique in being "Composite". Air Force units were normally designated for single purposes: maintenance, bombardment, engineering, or transport. The 509th drew together a number of separate units to form a self-sustaining whole: the 393rd Heavy Bombardment Group (which comprised 15 bomber crews); the 320th Troop Carrier Squadron; the 390th Air Service Group; the 603rd Air Engineering Squadron; the 1027th Air Material Squadron; the 1st Special Ordnance Squadron (Aviation); the 1395th Military Police Company (which included some 50 Manhattan Project agents); and the 1st Technical Detachment, War Department Miscellaneous Group,

© Springer Nature Switzerland AG 2020
B. C. Reed, *Manhattan Project*,
https://doi.org/10.1007/978-3-030-45734-1_9

a catch-all unit of civilian and military scientists and technicians. The 509th was authorized to a complement of 225 officers and 1,542 enlisted men; the addition of some 50 members of Norman Ramsey's Project Alberta brought the total to over 1,800.

The 509th's first B-29 flight at Wendover Field occurred on October 21, 1944, with pilot Robert Lewis at the controls. In addition to test drops at Wendover, practice bombing runs involved flying from Wendover almost 600 miles due south to the Salton Sea in southern California, dropping a single "blockbuster" bomb (often filled with concrete), and then executing a 155-degree diving escape turn designed to put about eight miles between the bomber and the eventual nuclear explosion. A 20-kt bomb at eight miles would create an overpressure of about three-tenths of a pound per square inch, which the bomber was expected to have no trouble surviving. The 509th's bombers would have no fighter escorts during their missions in order to avoid drawing the attention of Japanese defenders; also, to survive the shock wave, fighters would have to have been so far from the bombers that they could provide no real protection. Tibbets knew that B-29's stripped of their guns and armor could fly as high as 34,000 feet, well out of the range of anti-aircraft guns and above the ceiling of Japanese Zero fighters. The 393rd bombardment group received its fifteenth stripped-down B-29 on November 24, 1944, bringing it to full strength. On December 17, the 41st anniversary of the Wright brothers first flight, the 509th was formally activated. Ferebee and Van Kirk were appointed as Group Bombardier and Group Navigator.

In the spring of 1945, the pace of training built toward deployment to the Pacific. On May 19, the first members of the group arrived on Tinian; others would follow until all personnel and aircraft were present by early August. Technically, the 509th lay in the theatre of operations of General Curtis LeMay, who had taken command of the Twenty-First bomber command of the Twentieth Air Force in January, 1945.

LeMay kept his headquarters on the island of Guam, about 130 miles south of Tinian. To cripple Japanese industry, he had adopted a strategy of nighttime low-level (~5,000 feet altitude) incendiary bombing. On the night of March 9–10, 1945, a fleet of nearly 300 B-29 s firebombed Tokyo, dropping some sixteen hundred tons of incendiary bombs. Individual fires coalesced into a firestorm, which burnt out 16 square miles of the city, about a quarter of its total area. Some one million people were rendered homeless, and 84,000 were killed (some estimates claim over 100,000), a toll greater than the number of immediate deaths that would occur at either Hiroshima or Nagasaki; only 14 aircraft were lost. The air was heated so much that B-29s at 6,000 feet experienced serious turbulence; crew members could smell the burning flesh of victims below. Similar raids followed against the cities of Nagoya, Osaka, and Kobe, and another raid on April 13 burnt out eleven more square miles of Tokyo.

The ferocity of World War II in the Pacific is now almost beyond comprehension. Some 5,000 Americans and many more Japanese were dying each week as American forces advanced through Japanese-held islands. At a June 18, 1945, meeting with the Joint Chiefs of Staff called to discuss the proposed invasion of Japan, Harry Truman was briefed on the sobering statistics. To take the islands of Leyte (late 1944), Luzon (early 1945), Iwo Jima (February–March 1945), and Okinawa (April–June 1945), United States casualties totaled some 110,000. An incomplete tally of Japanese killed and taken prisoner totaled over 300,000. Some 140,000 civilians on Okinawa alone are estimated to have been killed or committed suicide. If an invasion of the home islands of Japan went ahead and the Japanese kept up such a fanatic level of resistance, casualties could be astronomical. During the entire war, no Japanese unit had ever surrendered.

The proposed invasion of Japan, operation Downfall, comprised two elements. The southern island of Kyushu (home to Nagasaki) was to be the target of Operation Olympic, scheduled to begin on November 1, 1945. This would involve landing over 760,000 ground forces over 45 days; this number does not include offshore naval support personnel. The number of troops landed during the June 6, 1944, D-Day invasion of Normandy had been about 156,000. The logistics of coordinating thousands of landing craft, aircraft, and supply vessels would have been immense. The plan was for the invasion force to advance about one-third of the way along the island, setting up air bases to support an invasion of the area around Tokyo (on the main island of Honshu), Operation Coronet, which was scheduled for March 1, 1946. This would involve just over one million ground forces. In the meantime, LeMay's bombers would continue laying waste to Japanese cities; he anticipated delivery of 100,000 tons of bombs per month by the end of 1945, and 220,000 tons per month by March, 1946. General Douglas MacArthur is said to have expressed the fear that once American forces had established themselves on Kyushu, they might face a guerrilla war which could go on for 10 years. One American intelligence estimate indicated some 560,000 Japanese troops stationed in Kyushu as of August, 1945.

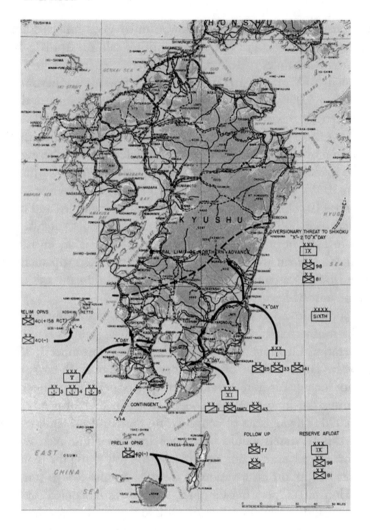

Map showing invasion locations for southern Kyushu. The scale bar at the lower right is 50 miles long. Compare Fig. 8.2. Source http://commons.wikimedia.org/wiki/File: Operation_Olympic.jpg

An issue that continues to be hotly debated among historians is that of how many casualties might have been sustained if the invasion had gone ahead. Estimates ranged from a few tens of thousands to a half-million or more. The confusion with this is due to the fact that there was no one single official estimate that went up to the Secretary of War or the President. Various estimates were developed by various components of the military bureaucracy and continuously modified as campaigns evolved. Aside from estimates that came up from combat-theatre commanders, four groups in Washington alone prepared estimates: the War Department, Navy Department, Joint Chiefs of Staff, and the Combined Chiefs of Staff, a joint American-British group. Within these, statistical data were drawn from various sub-groups: the Joint Chiefs, for example, had access to data from the Army Service Forces medical planners, the Joint War Plans Committee, the Joint Logistics Plans Committee, and the Operations Division of the War Department General Staff. Many estimates were restricted to losses anticipated in just the first thirty, sixty, or ninety days of the planned invasion, with the intent that they would be modified as the mission progressed.

Many estimates were predicated on extrapolating casualty rates from battles already fought. In the first 30 days following the invasion of Normandy, American forces suffered 42,000 casualties; by August 31, 1944, the total had climbed to over 133,000 American and over 91,000 British casualties. To the summer of 1945, American casualties amounted to about 1.25 million, a figure which does not include non-battle casualties due to causes such as disease and accidents. As for an invasion of Japan, one number that gained wide circulation beginning in the summer of 1944 was based on experience with the battle for the island of Saipan, where 3,426 Americans and 23,811 Japanese were killed: a ratio of nearly one to seven. Projecting from an anticipated Japanese home-island strength of 3.5 million defenders, this "kill ratio" would imply just over 500,000 American deaths if the entire Japanese force had to be eliminated. The pool of Japanese males in the 17–44 age bracket was estimated at over 11 million, of which some 80% were considered fit for military duty; 3.5 million may have been an underestimate. There is no reason to imagine that the Japanese were not prepared to mount a fierce defense of their homeland.

During the June 18 meeting, Truman heard a spectrum of casualty projections for Olympic. From his headquarters in the Pacific, MacArthur produced an estimate of just over 100,000 during the first 90 days, plus an additional 4,200 non-battle casualties for each thirty-day period; about a month later he would revise this to a total of about 125,000 Army and Marine battle casualties by the time Operation Coronet got underway. General Marshall estimated a number on the order of the Luzon toll, about 31,000, over the first 30 days.

Subsequent to the JCS meeting, a War Department Intelligence Division estimated that the Japanese could have about 1.1 million men ready for service by the end of 1945. Another estimate was that if it became necessary to kill 5–10 million Japanese, the American toll might be between 1.7 and 4 million casualties, including up to 800,000 killed. In late July, another estimate from MacArthur's staff revised the numbers upwards to over 200,000 casualties over all of Operation Olympic; this was in response to growing numbers of Japanese reinforcements being deployed to Kyushu. The 200,000 number was based on extrapolating from approximately 40,000 American battle casualties suffered per 2.5 Japanese divisions and the Japanese having some 13 divisions deployed. Of course, none of these figures include Japanese casualties or the toll on Japanese civilians by continuing bombing raids; Japanese leaders exhibited an astonishingly callous indifference to the suffering of their people. The Japanese had fully anticipated the landing areas, and intended to inflict 20% casualties before any GI set foot on a beach. With the Japanese culture of not surrendering, worst-case scenarios might well have become reality. Post-war investigations revealed that the Japanese had thousands of suicide aircraft and torpedoes ready for use against an invasion force.

Figures that are more secure are that the Army Service Forces estimated requiring approximately 720,000 replacements for "dead and evacuated wounded" if the war went on through the end of 1946. In the spring of 1945, the Selective Service call-up for the Army alone was raised to 100,000 per month; for the month of March, the total of Army, Navy, and Marines call-ups amounted to over 140,000 men, a number larger than the estimated number of immediate deaths at Hiroshima and Nagasaki combined. It has been estimated that in the last few months of the war, some 400,000 people were dying *each month* in East Asia and the Western Pacific.

Some historians have suggested that the bombings of Hiroshima and Nagasaki were not intended so much as inducements to the Japanese to surrender but rather to intimidate the Russians, prevent them from securing territorial gains in the Far East, and make clear that America would be the dominant political power in the post-war world. But this "revisionist" thesis ignores the fact that both Roosevelt and Truman had authorized extensive Lend-Lease supplies to the Russians, and were expecting them to enter the war against Japan. The surrender/Russia dichotomy seems a false one: Why could not Truman and his advisors have had *both* wartime and postwar strategies in mind? But in the end, debates over anticipated casualties are largely academic: it is impossible to know what might have happened between the time when the bombings did occur and the scheduled start of Olympic, even if the bombs had not been used.

In the early spring of 1945, Groves turned his attention to selecting targets. He met with General Marshall in early March, and asked him to designate a contact within the Army's Operations Planning Division. Marshall was reluctant to bring any more people into the issue than necessary, and directed Groves to see to targeting himself. For Groves, his target criteria were, as he put it, "places the bombing of which would most adversely affect the will of the Japanese people to continue the war." Beyond that, targets should be military in nature: headquarters, troop concentrations, and centers of production. Groves contacted General Lauris Norstad, Chief of Staff of the Army Strategic Air Force, to establish a committee to make target recommendations. Chosen were three staff members from General Arnold's office, plus three scientists from Los Alamos: John von Neumann, Robert Wilson, and William Penney of the British Mission; the latter had carried out extensive analyses of what levels of damage to Japanese cities might result from bombs of various yields detonated at various heights. The committee's charge was to develop a list of four previously unbombed cities, chosen such that three could be available for each mission, with weather predicted to be good enough for visual bombing.

The first of three meetings of the Target Committee took place at Norstad's office in the Pentagon on Friday, April 27. Groves opened the meeting with a short briefing, after which he left General Farrell in charge. Much of the discussion in this meeting concerned the dismal prospects for acceptable weather over Japan in the summer months. Experience indicated that June would be the worst month. In July, seven good days (defined as 3/10 or less cloud cover) could be expected, only six in August, and even fewer in September. Only once in five years had there been two successive good visual bombing days for Tokyo. January would be the best month, but there was no question of waiting that long. After lunch, the discussion turned to possible targets. Colonel William Fisher of the Air Force summarized ongoing operations. The Twenty-First bomber command of the 20th Air Force had 33 primary targets on its priority list. As the minutes of the meeting recorded, "the 20th Air Force is operating primarily to lay waste all the main Japanese cities, and they do not propose to save some important primary target for us if it interferes with the operation of the war from their point of view. Their existing procedure has been to bomb the hell out of Tokyo, bomb the aircraft manufacturing and assembly plants, engine plants and in general paralyze the aircraft industry". Then followed a list of eight cities, including Tokyo and Nagasaki, which were being bombed "with the prime purpose in mind of not leaving one stone lying on another."

As to possible Manhattan District targets, four were specifically discussed. Hiroshima was the largest untouched target not on the Twenty-First's priority list, and was the site of the Japanese Second Army Headquarters, from which the defense of Kyushu would be directed. The others were Yawata, not far from Osaka, a site of steel production; Yokohama (on Tokyo Bay, south of Tokyo); and Tokyo itself. However, Tokyo was not considered a high priority as it was "practically all rubble with only the palace grounds left standing." (The Palace and Emperor Hirohito had deliberately been spared destruction.)

The meeting adjourned at 4:00 p.m. with a list of 17 target areas identified as needing further research regarding damage already inflicted, weather data, amount of damage expected from the new weapons, and "the ultimate distance at which people will be killed." Particular consideration was to be given to large urban areas not less than three miles in diameter which were sited within larger populated areas.

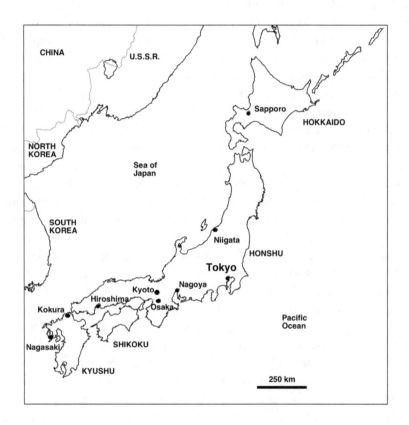

Map of Japan, showing main islands and major cities. A number of smaller islands are omitted. Adapted from http://www.hist-geo.co.uk/japan/outline/japan-cities-1.php

The committee's next meeting was held in Oppenheimer's office at Los Alamos over May 10–11, just after the 100-ton test at the *Trinity* site. The agenda included optimum bomb detonation heights; weather reports; procedures for a bomber having to jettison a bomb or return to base with a non-released one; status of targets; expected psychological and radiological effects; rehearsals; and coordination with the Twenty-First's regular bombing campaigns.

Detonation heights which would yield 5-pound-per-square-inch overpressures were desired, but corresponding damage radii were not specified in the record of the meeting. In spite of a lack of firm bomb-yield estimates, considerable latitude was available in the detonation heights; it was predicted that the bombs could be detonated as much as 40% below or 15% above optimum height with only a 25% loss in the damage area. For *Little Boy*, detonation heights of 1550 and 2400 feet were considered appropriate for yields of 5 and 15 kt, respectively. The yield outlook for *Fat Man* was still pessimistic, with heights being estimated for yields of only 0.7, 2, and 5 kt. In view of the uncertainties, it was decided that four different fuse-height settings should be available: 1000, 1400, 2000, and 2400 feet, with 1400 feet likely to be used for both bombs.

The Hiroshima and Nagasaki bombing missions. The distance from Tinian to Hiroshima is about 2740 km (1700 miles). Source http://commons.wikimedia.org/ wiki/File:Atomic_bomb_1945_mission_map.svg

By the time of the second meeting, the Air Force had relented from its position at the April 27 meeting, and was willing to reserve (leave unbombed) five targets for Manhattan consideration. First on the list was Kyoto, the historic capital and cultural center of Japan, with a population of about one million; industries were being moved there as other cities were being destroyed. It was pointed out that "Kyoto is an intellectual center for Japan and the people there are more apt to appreciate the significance of such a weapon". Second on the list was Hiroshima. Yokohama remained in third place, although considered disadvantageous in view of its heavy concentration of anti-aircraft defenses. In fourth place, and new on the list, was Kokura, the site of one of the largest arsenals in Japan. Bringing up the rear came Niigata, north of Tokyo on the western side of Honshu, a port of embarkation that was also the site of machine-tool industries and oil refining. After some discussion, the first four were recommended in the order described here for target status. Nagasaki seems not to have been discussed during this meeting.

The third and final meeting of the committee was held in the Pentagon on May 28, with Parsons, Ramsey, Ashworth, and Tibbets in attendance. After a brief discussion of revisions to detonation-height settings—now five options between 1100 and 2500 feet—Tibbets gave a detailed description of his crews' training regimens. Each of 15 bombardiers had accrued at least 50 releases at altitude, with most having performed 80 to 100. Drops conducted with radar-based bombing runs at 20,000 feet altitude were averaging within 1,000 feet of targets; visual runs were achieving 50% success within 500 feet. Round-trip flights up to 4,300 miles with 10,000 lb. bomb loads had been conducted, with plenty of fuel to spare. Parsons reported that 19 Pumpkins had been shipped from Wendover, and that it looked feasible to have 25 to 30 at Tinian by July 15, with production reaching 75 per month by mid-June. 509th ground echelons were already in place on Tinian; the entire group would arrive by mid-July. A 37-man bomb-assembly field crew comprising both civilian and military personnel had been designated. The civilians would hold assimilated military ranks; Robert Serber became an instant Colonel. The list of reserved targets had shrunk to three: Kyoto, Hiroshima, and Niigata; no reason was recorded as to why Yokohama and Kokura had been dropped. The overall conclusion was that the program was solidly on track for *Little Boy* to be ready by August 1. *Fat Man* was not discussed in detail, pending results of the *Trinity* test.

Groves' personal preferred target was Kyoto, in view of its having a large enough area to gain maximum knowledge of the bomb's effects. However, that city was spared by the personal intervention of Henry Stimson. On

May 30, two days after the target committee meeting, Groves was conferring with Stimson when the latter asked about the status of targets. Stimson, who had personally visited Kyoto on two occasions, objected to targeting the city on the grounds of its historical, cultural, and religious significance to the Japanese.

On June 14, Groves forwarded to General Marshall a revised list of Kokura, Hiroshima, and Niigata, but was not about to give up on his preference. Through June and July, he attempted, on up to perhaps a dozen or so occasions, to get Kyoto back on the list, even after Stimson had departed for the Potsdam conference. In a cable to Stimson on July 21, George Harrison stated that "All your local military advisors engaged in preparations definitely favor your pet city and would like to feel free to use it as first choice". Stimson consulted with Truman, who concurred with his Secretary of War; Stimson replied that he was aware of no factors to change his decision. Kyoto's reprieve was Nagasaki's doom: in the Groves-Handy orders of July 25, Nagasaki had replaced the historic capital, with Kokura listed afresh. In his memoirs, Groves took credit for sparing Kyoto, claiming that that he prevailed upon General Arnold to keep it on the reserved list when he realized that the Air Force might delete it from the list after Stimson's refusal to approve it. The fates of Hiroshima and Nagasaki were cast weeks before the bombing missions.

9.2   Fall 1944: Postwar Planning Begins

That the development of nuclear weapons would radically alter the balance of power in the world and could precipitate a dangerous arms race was evident to many of the leading figures of the Manhattan Project. Consideration had to be given at the highest levels to issues that would come to the fore when bombs were ready and their existence revealed. Should they be used without warning, or demonstrated first? After the war, should atomic energy come under civilian or military control? What could be revealed of the work of the Project without violating security concerns? What legislation and oversight would need to be established? What should be the role of the government in supporting research and regulating private nuclear industries? To forestall an arms race, would some form of international control be necessary, with knowledge being shared among various countries?

Some of the first to raise these issues were scientists at Arthur Compton's Metallurgical Laboratory in Chicago. By the summer of 1944, the X-10 reactor was functioning and construction at Hanford was well underway. Technical work at Chicago was beginning to wind down, and attention began to turn to questions of possible use of bombs, long-term prospects for the Laboratory, and the wider ramifications of atomic energy. Compton asked Isaiah "Zay" Jeffries, a metallurgist and General Electric executive whom he had brought into the Met Lab as a consultant, to head a committee to prepare a "Prospectus on Nucleonics," the latter being the term Met Lab scientists used for what they foresaw as a vast postwar research and industrial field. Other members of the committee included Robert Mulliken and Enrico Fermi. Their report, submitted to Compton on November 18, 1944, contained seven sections. The first five reviewed the history of nuclear physics and potential peacetime applications in areas as diverse as commercial power production, naval propulsion, medicine, agriculture, and industry. Potential research applications for radioactive tracer isotopes were numerous; one specifically mentioned was tracking metabolic pathways. It is for its last two sections, however, that the Jeffries Report is now remembered. Section six, "The Impact of Nucleonics on International Relations and the Social Order," was remarkably prophetic in its vision of possible future events.

Since the laws of physics are universal and any advanced country could harness nuclear energy, the report cautioned that America could not secure lasting security by simply attempting to stay ahead of other nations in nucleonics research and development; breakthroughs could happen anywhere. Anticipating much of the future Cold War and current-day concerns with nuclear proliferation and terrorism, the report stated that

> Nuclear weapons might be produced in small hidden locations in countries not normally associated with a large scale armament industry … A nation, or even a political group … will be able to unleash a "blitzkrieg" infinitely more terrifying than that of 1939–40 … The weight of the weapons of destruction required to deliver this blow will be infinitesimal compared to that used up in a present day heavy bombing raid, and they could easily be smuggled in by commercial aircraft or even deposited in advance by agents of the aggressor.

To forestall such a destabilizing situation, the committee advocated that a central international authority would have to be established to exercise control over nuclear power, supervise associated materials, and make available such materials for legitimate research needs. In unknowing anticipation of what would come to be called the strategy of "mutually assured destruction," the group felt that until such an authority was established, "The most that an independent American nucleonic re-armament can achieve is the certainty that a sudden total devastation of New York or Chicago can be answered the next day by an even more extensive devastation of the cities of the aggressor, and the hope that the fear of such a retaliation will paralyze the aggressor." The report also addressed the need for broad public education on nuclear issues to assure the "moral development necessary to prevent the misuse of nuclear energy".

About the same time that the Jeffries committee was formed, forward planning also began to garner more attention at the upper administrative levels of the Manhattan Project. The vast complex of production facilities and laboratories constructed for the Project would not vanish the day after the war ended. Should they come under civilian or military jurisdiction? How would they be funded and operated? In August 1944, the Military Policy Committee authorized Richard Tolman to head a Committee on Postwar Policy to study the relation of atomic energy to national security. Tolman's small group (himself, Warren Lewis, Henry Smyth of Princeton, and Rear Admiral Earle Mills of the Navy) conducted interviews with over 40 scientists and also received written submissions. Their December 28 report to Groves emphasized that nuclear power for propelling naval vessels should be developed immediately, and that, within bounds dictated by security considerations, a nucleonics industry should be strongly encouraged. Also, wide dissemination of knowledge would be essential to encourage a level of post-war progress in the field necessary to maintain national security. Perhaps most importantly, they envisioned a national authority which would distribute research and development funds among military, civilian, academic, and industrial laboratories. International relations lay outside the committee's charge, and it ventured no opinions in that area.

Vannevar Bush and James Conant had their own ideas as well. On September 19, 1944, they wrote to Henry Stimson to point out that the time would soon come to consider how to release basic scientific information and enact legislation for domestic control of nuclear power. On September 30, they followed up with a more extensive memorandum titled "Salient Points Concerning Future International Handling of Subject of Atomic Bombs." This 3-page document would prove just as prophetic as the Jeffries report. In six brief paragraphs, Bush and Conant laid out what was virtually a script for the Cold War, predicting the future development of very powerful fusion-based bombs, how any country with good technical and scientific resources could reach the U.S.-British nuclear position in three or four years, why it would be dangerous for the United States and Britain to attempt to carry on further development in complete secrecy, and how if another country were to develop fusion bombs first, the United States would find itself "in a terrifying situation." To counter this, they proposed that an international system of free exchange of all scientific information be set up, with technical staff to be given "free access in all countries not only to the scientific laboratories where such work is contained, but to the military establishments as well." While acknowledging the naivety of this idea, they closed with the warning that "the hazards to the future of the world are sufficiently great to warrant this attempt … Under these conditions there is reason to hope that the weapons would never be employed and indeed that the existence of these weapons might decrease the chance of another major war."

The arguments common to the Jeffries, Tolman, and Bush-Conant reports are striking. Their recommendations needed to be considered at the highest levels, but the day-to-day pressures of the war naturally intervened. In late October 1944, Bush suggested to Stimson that one approach might be to establish an advisory group that would report directly to the President. Stimson and Groves updated President Roosevelt on the status of the Project on December 30, but they did not discuss postwar planning; Stimson apparently felt that the time was not yet right to broach the idea of an advisory committee. As the calendar turned to 1945, planning was relegated to official limbo, where it would remain until Harry Truman assumed the Presidency in April of that year. Stimson did raise the issue with Roosevelt in their last conversation together on March 15, 1945, but nothing came of it at the time.

9.3   Truman Learns of the Manhattan Project

On the afternoon of April 12, 1945, President Franklin Roosevelt died of a cerebral hemorrhage at the age of 63. Vice-President Harry Truman, who knew of the existence of the Manhattan Project but little of its details, was sworn in that evening at the White House. Truman had officially met with Roosevelt only eight times since becoming the Vice-Presidential candidate in the 1944 election.

Left: President Harry S. Truman (1884-1972). Right: Truman, Secretary of State James Byrnes, and Ambassador to Belgium Charles Sawyer in Antwerp, Belgium, July 15, 1945 Sources: http://commons.wikimedia.org/wiki/File:Harry_S._Truman_-_NARA_-_530677.jpg;NARA-198780.tif

After a brief Cabinet meeting following his swearing-in, Truman was approached by Stimson, who related that he wished to inform the new President "about an immense project that was underway—a project looking to the development of a new explosive of almost unbelievable destructive power." The next afternoon, James Byrnes, head of the Office of War Mobilization (and soon to be Truman's Secretary of State), dramatically told Truman that "we are perfecting an explosive great enough to destroy the whole world. It might well put us in a position to dictate our own terms at the end of the war." In the pressure of adjusting to his new job, almost two weeks would elapse before Truman received a full briefing on the Project. As Truman biographer David McCullough has written, the bomb was a Roosevelt legacy inherited by Truman with no written guidance beyond Roosevelt's Quebec City agreement with Churchill. Without Roosevelt's personal backing, the project would never have obtained the priority it needed to succeed, but the results fell in Truman's lap.

At noon on Wednesday, April 25, Stimson and Groves briefed Truman. The President's schedule was crowded, but Stimson felt that a full briefing could not be put off any longer. That evening, the opening conference of the United Nations would take place in San Francisco. Truman was to address the delegates by radio, and Stimson felt that it would be inappropriate for him to do so without an appreciation of the potentialities of the new weapon. Two days earlier, Groves had given Stimson a background memorandum to be given to the President. Essentially a primer on the entire Project, this memorandum, titled "Atomic Fission Bombs," ran to only 24 double-spaced pages, but managed to cover every aspect of the Project from the idea of uranium fission up to the prospects for fusion weapons.

Stimson arrived at the Oval Office a few minutes before Groves, and had prepared a two-page covering memorandum of his own. The first sentence read: "Within four months we shall in all probability have completed the most terrible weapon even known in human history, one bomb of which could destroy a whole city." Echoing the Jeffries report, Stimson expressed the fear that the future could see a time when such weapons could be constructed in secret and used suddenly with devastating power against an unsuspecting nation or group, unless some system of control could be developed. Such a system, however, would "undoubtedly be a matter of the greatest difficulty and would involve such thorough-going rights of inspection and internal controls as we have never heretofore contemplated." The development of this weapon, he felt, "has placed a certain moral responsibility upon us which we cannot shirk without very serious responsibility for any disaster to civilization which it would further." After Stimson had finished reading his memo, Groves entered the meeting, and the three men went through his longer document in detail.

Groves' memorandum is a model of how to prepare an effective summary document. The opening sentences could not have failed to catch Truman's attention: "The successful development of the Atomic Fission Bomb will provide the United States with a weapon of tremendous power which should be a decisive factor in winning the present war more quickly with a saving in American lives and treasure. If the United States continues to lead in the development of atomic energy weapons, its future will be much safer and the chances of preserving world peace greatly increased. Each bomb is estimated to have the equivalent effect of from 5,000 to 20,000 tons of TNT now, and ultimately, possibly as much as 100,000 tons." The balance of the report includes discussions of the history of the discovery of fission; properties of uranium and plutonium; establishment of the Briggs committee; the scale of work going on at Oak Ridge, Hanford, and Los Alamos; the concept of a graphite pile; the notion of critical mass; gun and implosion bombs; anticipated operational plans; collaboration with the British; a summary of what foreign countries might be up to; and the necessity for postwar planning. Groves related that the gun bomb was expected to yield between 8 and 20 kt, and the implosion device between 4 and 6 kt. The first gun bomb, which was not expected to require a full-scale test, was predicted to be ready by August 1; a second one should be ready by the end of the year, with subsequent ones to follow at about 60-day intervals thereafter. A test of the implosion device should be possible by the early part of July. A second test of the implosion bomb, if necessary, should be ready by the first of August, with bombs themselves ready in quantity - about one every ten days - by the latter part of August. The target, wrote Groves, "is and was always expected to be Japan." Construction and operations costs to March 31, 1945, had accumulated to nearly $1.5 billion, a figure which was expected to grow to nearly $2 billion by the end of June. After outlining issues that would have to be addressed in the postwar period, Groves closed by remarking that Stimson aide George Harrison had suggested setting up a committee to develop recommendations for consideration by the executive and legislative branches of the government for the time when secrecy was no longer in effect.

In a summary of the meeting for his own files, Groves remarked that the President did not show any concern over the amount of money being spent, and that he was "in entire agreement with the necessity for the project." Truman approved the idea of a committee to begin developing policy proposals; Stimson was to recruit members. If the project succeeded, it could represent for Truman an incredible deliverance from the war that he had inherited.

9.4   Advice and Dissent

Stimson wasted no time in pulling together his committee. On May 2, he was back at the White House with a proposed list of eight members: himself, his aide George Harrison (who would serve as alternate Chair when Stimson could not attend), Undersecretary of the Navy Ralph Bard, Vannevar Bush, James Conant, Karl Compton, Assistant Secretary of State William Clayton, and, as the President's personal representative, James Byrnes. In recognition of the fact that Congress would presumably establish a permanent body to supervise and regulate atomic energy, this group was known as the Interim Committee. Their first meeting took place on May 9, with Stimson making opening remarks: "Gentlemen, it is our responsibility to recommend action that may turn the course of civilization." The committee's charge was to "to study and report on the entire problem of temporary war-time controls and later publicity, and to survey and make recommendations on post-war research, development, and control, and on legislation necessary for these purposes." For background, the group reviewed Groves' April 23 report.

When Stimson first approached Conant to serve on the committee, the latter suggested that it might be valuable to invite some of the leading scientists to present their views on international relations and the bomb. This suggestion was the first item on the agenda for the committee's second meeting, which was held on May 14. It was agreed to appoint a Scientific Panel whose members were Arthur Compton, Ernest Lawrence, Robert Oppenheimer, and Enrico Fermi. The Panel would be free to advise not only on technical matters, "but also to present to the Committee their views concerning the political aspects of the problem." The balance of the second meeting was taken up with consideration of relations with the British, and development of statements to be made public following the *Trinity* test and eventual use of the bomb; William Laurence was assigned to work up drafts. At its third meeting on May 18, the group decided to invite the Scientific Panel to meet with the committee on May 31, three days after the last meeting of the Target Committee.

Including a one-hour lunch break, the May 31 meeting ran from 10:00 a.m. to 4:15 p.m., and was pivotal in the sense of arriving at a "decision" as to how the bomb would be used. The entire committee plus the Scientific Panel was present, as were Groves and Marshall as invited guests. Stimson opened with a statement as to how he viewed the significance of the Project:

> The Secretary expressed the view, a view shared by General Marshall, that this project should not be considered simply in terms of military weapons, but as a new relationship of man to the universe. This discovery might be compared to the discoveries of the Copernican theory and of the laws of gravity, but far more important than these in its effect on the lives of men. While the advances in the field to date had been fostered by the needs of war, it was important to realize that the implications of the project went far beyond the needs of the present war. It must be controlled if possible to make it an assurance of future peace rather than a menace to civilization.

Arthur Compton reviewed the development of the Project, after which the discussion turned to domestic issues. Lawrence felt that research "had to go on unceasingly," that plant expansion had to be pursued, and that a stockpile of bombs and material needed to be built up. All of the members of the Scientific Panel spoke up on the importance of a vigorous post-war research program. As to the issue of controls and inspections, Oppenheimer felt that knowledge of the subject was so widespread that steps should be taken to make American developments known to the world, and that it might be wise for the United States to offer to the world free interchange of information with particular emphasis on the development of peace-time uses. Stimson wondered what kind of inspections might be effective, and what would be the positions of democratic governments versus those of totalitarian regimes under a program of international control coupled with scientific freedom? Vannevar Bush opined that it would be hard for America to remain permanently ahead if results of research were to be turned over to the Russians with no reciprocal exchange. Marshall cautioned against putting too much faith in the effectiveness of an inspection proposal, but did suggest that it might be desirable to invite two Russian scientists to witness the *Trinity* test; this idea was apparently not pursued.

The group broke for lunch at 1:15 p.m. No record of the lunch-time conversation was kept, but the idea of giving the Japanese a demonstration of the bomb's power before deploying it in a way that would cause any loss of life was raised; perhaps a test on a remote island might do. This idea has been attributed to both Arthur Compton and Ernest Lawrence; Byrnes apparently asked for elaboration. In the discussion that followed, it seems that nobody was able to conceive of a demonstration powerful enough to convince the Japanese that continued resistance would be pointless. Other objections were that America would look ridiculous if a demonstration proved to be a dud, and that the Japanese might bring prisoners of war into the demonstration area. In his memoirs, Arthur Compton wrote that "Throughout the morning's discussions it seemed to be a foregone conclusion that the bomb would be used." Discussion on how to use the bomb resumed after lunch; Marshall did not attend the afternoon session. As the minutes recorded (underlining as in original):

After much discussion concerning various types of targets and the effects to be produced, the Secretary expressed the conclusion, on which there was general agreement, that we could not give the Japanese any warning; that we could not concentrate on a civilian area; but that we should seek to make a profound psychological impression on as many of the inhabitants as possible. At the suggestion of Dr. Conant the Secretary agreed that the most desirable target would be a vital war plant employing a large number of workers and closely surrounded by worker's houses.

As the meeting drew to a close, Harrison remarked that the Scientific Panel was a continuing group that should feel free to present its views to the Committee at any time. In particular, the Committee wished to hear Panel member's thoughts as to what sort of controlling organization should be established. The question arose as to what Panel members were at liberty tell their subordinates about the Committee; it was agreed that they should feel free to relate that the Committee had been appointed by Stimson, and that they (the Panel) had been given complete freedom to present their views on any phase of the subject. The Scientific Panel agreed to meet again at Los Alamos on June 16. Byrnes went directly to the White House to brief Truman on the Committee's deliberations, and Stimson further discussed the matter with the President on June 6.

Arthur Compton took to heart the notion of soliciting the views of his subordinates. After returning to the Metallurgical Laboratory, he met with a group of senior scientists on June 2, and asked them for input. Various committees were established to consider issues such as research, education, and controls and organization, but it was a group headed by James Franck that was to have the most impact. Franck had shared the 1925 Nobel Prize for Physics with Gustav Hertz, and had emigrated to the United States from Germany in the mid-1930s, settling at the University of Chicago. In the summer of 1945, he was Director of the Met Lab's Chemistry Division. Franck's committee, which included Glenn Seaborg and Leo Szilard, was to prepare a report on "Political and Social Problems" associated with the bomb. Working over the week of June 4–11, they drafted a document which became known as the Franck Report, and which is now widely acknowledged to be a founding manifesto of the nuclear non-proliferation movement.

While the Franck Report echoed many of the points already made in the Jeffries Report, it added a tone of high morality. A few excerpts drawn from the Preamble give the idea:

> The scientists on this Project do not presume to speak authoritatively on the problem of national and international policy. However, we found ourselves, by force of events during the last five years, in the position of a small group of citizens cognizant of a grave danger for the safety of this country as well as for the future of all the other nations, of which the rest of mankind is unaware … All of us, familiar with the present state of nucleonics, live with the vision before our eyes of sudden destruction visited on our own country, of a Pearl Harbor disaster repeated in thousand-fold magnification in every one of our major cities.

A section titled "Prospects of Armaments Race," emphasized that America would be at a significant disadvantage if an arms race did develop, as its population centers and industries were very centralized, as opposed to those in possible enemy countries such as Russia. After arguing that nuclear weapons could not be kept secret for long, the group then proceeded to their central thesis: "From this point of view, the way in which the nuclear weapons now being secretly developed in this country are first revealed to the world appears to be of great, perhaps fateful importance." Given that the Japanese were still fighting on after many of their cities had been reduced to rubble, the authors felt it doubtful that the first available bombs would be sufficient to break Japan's will to resist. On the other hand, if one were to look forward to an international agreement on the prevention of nuclear warfare, "the military advantages and the saving of American lives achieved by the sudden use of atomic bombs against Japan may be outweighed by the ensuing loss of confidence and by a wave of horror and repulsion sweeping over the rest of the world and perhaps even dividing public opinion at home. *From this point of view, a demonstration of the new weapon might best be made, before the eyes of representatives of all the United Nations, on the desert or a barren island.* ... After such a demonstration the weapon might perhaps be used against Japan if the sanction of the United Nations (and of public opinion at home) were obtained, perhaps after a preliminary ultimatum to Japan to surrender or at least to evacuate certain regions as an alternative to their total destruction."

The Report's Summary offered a recommendation:

To sum up, we urge that the use of nuclear bombs in this war be considered as a problem of long-range national policy rather than of military expediency, and that this policy be directed primarily to the achievement of an agreement permitting an effective international control of the means of nuclear warfare.

Franck hand-delivered the report to Compton in Washington on June 12, asking that he pass it on to Stimson. The latter was not available, but Compton did pass it to Harrison. Compton added his own cover letter to the report, summarizing its essence:

The proposal is to make a technical but not military demonstration, preparing the way for a recommendation by the United States that the military use of atomic explosives be outlawed by firm international agreement. It is contended that its military use by us now will prejudice the world against accepting any future recommendations by us that its use be not permitted.

Compton did not offer his own thoughts on this position, but added that the report did not address whether failure to make a military demonstration of the new bombs might drag out the war and cost more casualties, or whether without a military demonstration it might be impossible to impress the world with the need for national sacrifices in order to gain lasting security. It is not clear if Stimson ever saw the report.

On June 15, Harrison phoned Compton in Los Alamos to ask the Scientific Panel to also consider the question of the immediate use of nuclear weapons at its meeting scheduled for the following day. The Panel's consequent one-page report made three statements. The first was a rather vague recommendation that countries such as Russia, France and China be informed of the bombs' development before they were used, and be invited to make suggestions as to how "we can cooperate in making this development contribute to improved international relations." The second and third statements get to the nub of the issue, and are worth reproducing in their entirety:

> The opinions of our scientific colleagues on the initial use of these weapons are not unanimous: they range from the proposal of a purely technical demonstration to that of the military application best designed to induce surrender. Those who advocate a purely technical demonstration would wish to outlaw the use of atomic weapons, and have feared that if we use the weapons now our position in future negotiations will be prejudiced. Others emphasize the opportunity of saving American lives by immediate military use, and believe that such use will improve the international prospects, in that they are more concerned with the prevention of war than with the elimination of this specific weapon. We find ourselves closer to these latter views; we can propose no technical demonstration likely to bring an end to the war; we see no acceptable alternative to direct military use.
>
> With regard to these general aspects of the use of atomic energy, it is clear that we, as scientific men, have no proprietary rights. It is true that we are among the few citizens who have had occasion to give thoughtful consideration to these problems during the past few years. We have, however, no claim to special competence in solving the political, social, and military problems which are presented by the advent of atomic power.

The Interim Committee met on June 21; Groves was present, but not Stimson or the members of the Scientific Panel. The morning was spent dealing with draft publicity statements and some legal issues. After lunch, the Scientific Panel's report was taken up. Discussion of future policy was left to an eventual "Post-War Control Commission," but as to use of the weapon,

> Mr. Harrison explained that he had recently received through Dr. A. H. Compton a report from a group of the scientists at Chicago recommending, among other things, that the weapon not be used in this war but that a purely technical test be conducted which would be made known to other countries. Mr. Harrison had turned this report over to the Scientific Panel for study and recommendation. Part II of the report of the Scientific Panel stated that they saw no acceptable alternative to direct military use. The Committee reaffirmed the position taken at the 31 May and 1 June meetings that the weapons be used against Japan at the earliest opportunity, that it be used without warning, and that it be used on a dual target, namely, a military installation or war plant surrounded by or adjacent to homes or other buildings most susceptible to damage.

The Interim Committee held a number of subsequent meetings, but never revisited the use-versus-demonstration issue.

Despite the reaffirmation of the May 31 decision (the June 1 meeting dealt largely with post-war industrial issues), members the Committee were not monolithic in their thinking. On June 27, Ralph Bard prepared a brief memorandum:

> Ever since I have been in touch with this program I have had a feeling that before the bomb is actually used against Japan that Japan should have some preliminary warning for say two or three days in advance of use. The position of the United States as a great humanitarian nation and the fair play attitude of our people generally is responsible in the main for this feeling. During recent weeks I have also had the feeling very definitely that the Japanese government may be searching for some opportunity which they could use as a medium of surrender … emissaries from this country could contact representatives from Japan … and … give them some information regarding the proposed use of atomic power, together with whatever assurances the President might care to make with regard to the Emperor … The stakes are so tremendous that it is my opinion very real consideration should be given to some plan of this kind …

Harrison passed Bard's memo on to Stimson and Byrnes, and Bard secured an interview with Truman, during which he tried to argue that naval blockade would make an invasion unnecessary. The President assured him that the questions of invasion and offering a warning had received careful attention.

On July 2, two weeks before the *Trinity* test, Henry Stimson sent Truman a three-page memorandum titled "Proposed Program for Japan." Since an invasion of Japan would almost certainly lead to a costly, drawn-out battle which would leave that country destroyed, Stimson raised the question of whether some alternative could be proposed that would avoid an invasion while securing the equivalent of unconditional surrender. In particular, he suggested that a warning which made clear that the Allies did not desire to destroy Japan as a nation, coupled with a policy of not excluding a constitutional monarchy, might improve the chances of success. Japan's situation was desperate: she had no allies; her Navy was effectively destroyed; she was vulnerable to air attack; the rising force of China was against her; the threat of Russia loomed; and America had the industrial capacity to continue the war and the "moral superiority through being the victim of her first sneak attack." The memorandum made no mention of atomic bombs. Many of Stimson's suggestions would appear in the Potsdam Declaration just over three weeks later, but not the suggestion regarding a constitutional monarchy. It may well be that the Japanese response would have been the same; the faction within the Japanese government that sought peace could not yet point to the specter of further atomic bombings to bolster their position.

If political statements were being formulated, Leo Szilard was certain to be involved. Convinced that Project hierarchy stifled any real avenue for making known his concern that an arms race would be inevitable if no international control agreement was reached, he decided to attempt another direct approach to the President. In early March, 1945, he drafted a memorandum titled "Atomic Bombs and the Postwar Position of the United States in the World," wherein he argued that if a control agreement with Russia could not be achieved, America would be forced to engage in a costly arms race and that the greatest danger might be the outbreak of a "preventative war." Szilard finished his memo on March 12, and decided to again enlist Albert Einstein to prepare a letter of introduction. Szilard traveled to Princeton, where Einstein obliged him with a one-page letter dated March 25. Secrecy forbade Szilard from disclosing the contents of his memorandum (Einstein knew little of the details of the Project); Einstein summarized the issue by writing that "I understand … he is now greatly concerned about the lack of adequate contact between scientists who are doing this work and those members of your cabinet who are responsible for formulating policy," and asked Roosevelt to give Szilard's presentation his personal attention.

Szilard dispatched a copy of Einstein's letter to Mrs. Roosevelt, who replied in early April with a proposal that Szilard meet with her in New York on May 8. But Roosevelt died beforehand, and Szilard found himself in limbo. Luckily, he found an employee at the Met Lab, mathematician Albert Cahn, who had some political connections in Truman's home town of Kansas City. Cahn managed to secure an appointment at the White House for Friday, May 25. Szilard traveled to Washington with Cahn and University of Chicago Dean of Science Walter Bartky, but they were redirected by the President's Appointments Secretary to meet with James Byrnes, who was then living in South Carolina. Szilard and Bartky, now accompanied by Harold Urey, proceeded by train to South Carolina (tailed by some of Groves' agents), where they met with Byrnes on May 28, the day that the last Target Committee meeting was underway in Washington. The meeting was a disaster: Byrnes was not happy with Szilard's attempt to interfere in policy-making, and Szilard felt that Byrnes completely failed to grasp the significance of atomic energy.

Not to be deterred, Szilard moved on to his next tactic: a direct petition to the new President. The first version of his petition, dated July 3 and signed by Szilard and 58 others, expressed the opinion that atomic bombing of Japan could not be justified in the present circumstances, and that atomic bombs were primarily a means for the "ruthless annihilation" of cities. The signers reminded the President that in his hands lay the fateful decision of whether or not to use these bombs, and argued that "Thus a nation which sets the precedent of using these newly liberated forces of nature for purposes of destruction may have to bear the responsibility of opening the door to an era of devastation on an unimaginable scale." The text closed with a plea that the President exercise his power as Commander-in-Chief to rule that the United States not, "in the present phase of the war," resort to the use of atomic bombs.

Perhaps through Compton, word of Szilard's activity reached Oak Ridge. Kenneth Nichols asked Compton to poll his colleagues' attitudes on use of the bomb. Compton delegated the task to Farrington Daniels, formal Director of the Met Lab. Five options were offered (paraphrased):

(i) Use the weapons in the most effective military manner;
(ii) Give a demonstration in Japan followed by an opportunity to surrender before full use of the weapon is employed;
(iii) Perform a demonstration within the United States with Japanese representatives present;
(iv) Withhold military use of the weapons but make public experimental demonstration of their effectiveness;
(v) Maintain as secret as possible all developments of the new weapons and refrain from using them in the present war.

Responses were received from 150 of approximately 250 employees; Daniels reported the results on July 13. The distribution of votes was 23, 69, 39, 16, and 3 (15%, 46%, 26%, 11%, 2%). At the level of destruction caused by a nuclear weapon, the distinction between options (i) and (ii) is not clear, but it is evident that over half of the respondents felt that some direct use of the bomb against Japan was appropriate. In the meantime, Szilard redrafted his petition, producing a second version on July 17—the day after *Trinity* - which garnered 69 co-signers. This version dropped the "ruthless annihilation" phrase of the original, but added a moral dimension:

> The added material strength which this lead gives to the United States brings with it the obligation of restraint and if we are to violate this obligation our moral position would be weakened in the eyes of the world and in our own eyes. It would then be more difficult for us to live up to our responsibility of bringing the unloosened forces of destruction under control.

Szilard handed the petition to Compton on July 19 with a request that it be forwarded to the President. Compton instead sent the petition and the results of the poll to Nichols, who passed them on to Groves. Groves held on to them until an August 1 meeting with Stimson, after which George Harrison filed them with his papers; the President apparently never saw the petition. Groves' action may seem high-handed, but the scientists had their chance for input through the Scientific Panel of the Interim Committee. By the beginning of August, the 509th Composite Group's orders had been approved by the President, and the full machinery of preparations for the bombing missions was in motion.

The question of whether a demonstration shot should have been carried out continues to be debated. Rudolf Peierls offered an assessment in his memoirs:

> To me the obvious answer would have been to drop a bomb on a sparsely populated area to show its effects, coupled with an ultimatum to the Japanese government to avoid a large-scale nuclear attack. This would have involved killing some people and destroying some buildings, since otherwise the power of the bomb would not have been obvious; the effects visible after the Alamogordo test were frightening to the expert but not impressive to the layman. Of course such an ultimatum might have failed, but at least it would have been an attempt to avoid unnecessary casualties. … My regrets are that we did not insist on more dialogue with the military and political leaders, based on full and clear scientific discussions of the consequences of possible courses of action. It is not clear, of course, that such discussions would have made any differences in the end.

An aspect of the situation that advocates of a demonstration shot seem not to have considered is that if the bombs *not* been used and the consequences of nuclear combat so starkly demonstrated to the world, what much worse horrors might have unfolded in a subsequent war? But for practical purposes, the momentum that the Project had acquired by the summer of 1945 was practically unstoppable. President Truman did not make a "hard decision" to use the bomb so much as he elected not to alter a chain of events that was already far along when he inherited the Presidency. The decision of when to end the war rather lay largely in the hands of the Japanese cabinet.

9.5   The Bombings

When Truman approved the Handy/Groves orders of July 25 and replied to Henry Stimson's request for permission to prepare public statements for release, the last formal high-level authorizations for deployment of atomic bombs against Japan were completed. The intricate program that General Groves had developed to design, develop, and deliver a revolutionary new weapon was about to come to fruition.

In the Pacific, August 1, 1945, saw various organizational changes come into effect. General LeMay moved up to become Chief of Staff to General Spaatz, and the Twenty-First Bomber Command and the Twentieth Air Force came under the command of Lieutenant General Nathan Twining. Thus it came to be that Twentieth Air Force Field Order number 13, issued on August 2, was over Twining's signature. The orders specified Hiroshima, Kokura Arsenal, and Nagasaki as the primary, secondary, and tertiary targets. Niigata had been scratched for being too far away from the other targets. Hiroshima had been bombed on May 7 and June 2, but the bombs had fallen ineffectively in the Ota river. Nagasaki had been the target of two bombing raids, on July 22 and August 1.

The weather for the first few days of August was overcast and rainy, but on Saturday, August 4, Commander Parsons was informed that the forecast was improving. At 4:00 p.m. that afternoon, 509th flight crews were given their first briefing. Tibbets told his men that what they had trained for was at hand, but he did not reveal the nature of their payload. He then introduced Parsons, who attempted to show a film of the *Trinity* test. The projector jammed and chewed up the film, so Parsons could give only a verbal description of the test. He began his comments with "The bomb you are going to drop is something new in the history of warfare. It is the most destructive weapon ever produced. We think it will knock out everything within a three mile area."

LeMay authorized the mission order at 2:00 p.m. on August 5. At the local level, this took the form of 509th Operations Order number 35, dated the same day. The mission called for sorties by seven aircraft, identified by their "Victor" numbers. V-82, the *Enola Gay*, to be piloted by Paul Tibbets, was the strike plane—the one which carried the bomb. Victors 83, 71, and 85 were weather planes, directed toward Nagasaki, Kokura, and Hiroshima, respectively, and which were to depart an hour before the strike planes. Victors 89 and 91 carried blast-measurement instruments and high-speed cameras. Victor 90 was deployed to Iwo Jima as a backup for the *Enola Gay*.

Hiroshima is located on the delta of the Ota river in the southern part of Honshu, the main island of Japan. The river breaks into channels which divide the city into islands, giving it a distinctive fingered appearance as seen from above. Before the war, Hiroshima was the seventh-largest city in Japan, with a population of about 340,000. Its population in August, 1945, has been estimated at about 280,000 civilians plus approximately 43,000 soldiers, although some estimates have put the number somewhat lower; many civilians had been evacuated, but a number of troops and workers had been brought into the city. Flat and unbroken by hills, the city was a perfect target for determining the effects of the new weapon.

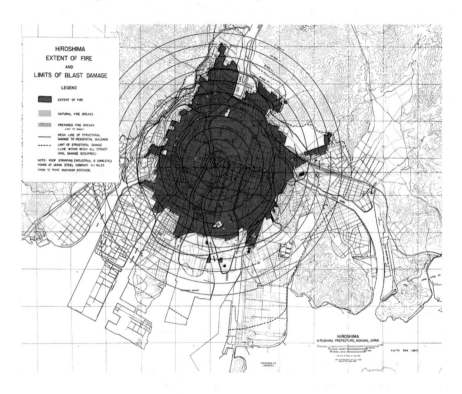

United States Strategic Bombing Survey map of Hiroshima atomic bomb damage. The darkened area shows the extent of fire damage. The curved solid line is the mean line of structural damage to residential buildings, and the dashed line is the limit of structural damage. The circles are in 1000-foot increments from ground zero out to 11,000 feet. Source http://commons.wikimedia.org/wiki/File:Hiroshima_Damage_Map.gif

Little Boy was wheeled out of its assembly building at 2:00 p.m. on Sunday afternoon. By 2:30 it had arrived at the loading pit, into which it was lowered so that the *Enola Gay* could be backed over it. The plane was in position by 3:00, and loading was complete by 3:45. Fusing checks were completed by 5:45, and a final inspection was made at 6:45. Tibbets had *Enola Gay* painted on the left-side nose of the airplane, and guards were posted to prevent any tampering.

On Saturday, General Farrell had informed Groves by cable that the *Enola Gay* should take off at approximately noon on Sunday, Washington time (Washington was 14 h behind Tinian; this would be 2:00 a.m. Monday, Tinian time.) Far from Groves' reach, Parsons decided that he would arm the bomb in flight, and spent Sunday afternoon practicing the procedure. After the bomb had been loaded, he practiced again in the cramped confines of the bomb bay. Not until Sunday evening Tinian time did Farrell cable Groves with the change in plan, too late for Groves to interfere. Final briefings began at 11:00 p.m.

The three weather planes began departing at 1:37 a.m., about an hour before the strike and observation planes. The weather crews missed the show back at Tinian that began at 2:00 as *Enola Gay* was floodlit and camera crews began filming; Groves wanted the mission recorded for posterity. Norman Ramsey compared the scene to a Hollywood premiere; one scientist allegedly compared it to the opening of a drugstore. Harlow Russ, who had helped engineer the implosion mechanism, estimated the crowd at about 350. Tibbets began the *Enola Gay's* takeoff roll at 2:45 a.m. Tinian time, Monday, August 6, using practically every yard of the two-mile runway to get airborne. The instrument, photo, and backup planes followed at two minute intervals. In Washington, it was 12:45 p.m. on Sunday afternoon.

Hiroshima and Nagasaki Strike Crews

Position	Hiroshima	Nagasaki
Commander	Paul Tibbets 1915–2007	Charles Sweeney 1919–2004
Pilot	Robert Lewis 1917–1983	Don Albury 1920–2009
Co-Pilot		Fred Olivi 1922–2004
Navigator	Theodore Van Kirk 1921–2014	James Van Pelt 1918–1994
Bombardier	Thomas Ferebee 1918–2000	Kermit Beahan 1918–1989
Bomb commander	William Parsons 1901–1953	Frederick Ashworth 1912–2005
Electronic countermeasures	Jacob Beser 1921–1992	Jacob Beser 1921–1992
Electronics test officer	Morris Jeppson 1922–2010	Philip Barnes 1917–1998
Flight engineer	Wyatt Duzenbury 1913–1992	John Kuharek 1914–2001

(continued)

(continued)

Position	Hiroshima	Nagasaki
Assistant engineer	Robert Shumard 1920–1967	Ray Gallagher 1921–1999
Radio operator	Richard Nelson 1925–2003	Abe Spitzer 1912–1984
Radar operator	Joseph Stiborik 1914–1984	Edward Buckley 1913–1981
Tail gunner	George Robert Caron 1919–1995	Albert Dehart 1915–1976

Source Campbell 30, 32

Some parameters of the Hiroshima and Nagasaki missions

	Hiroshima	Nagasaki
Strike aircraft	*Enola Gay*	*Bockscar*
Takeoff (Tinian time)	02:45 Aug 6	03:48 Aug 9
Takeoff (Washington time)	12:45 Aug 5	13:48 Aug 8
Bombing (Japan time)	08:15 Aug 6	11:08 Aug 9
Bombing (Washington time)	19:15 Aug 5	22:08 Aug 8
Landing (Tinian time)	14:58 Aug 6	23:06 Aug 9
Landing (Washington time)	00:58 Aug 6	09:06 Aug 9
Mission duration	12 h 13 min	19 h 18 min
Drop height (ft/m)	31,600/9,630	28,900/8,810
Bomb detonation height (ft/m)	1,900/580	1,650/503
Bomb yield (kt)	~15	~21

Sources Coster-Mullen, 39, 326; Campbell 31-34; Los Alamos report LA-8819. Mission time for Bockscar includes three-hour stop at Okinawa

Fifteen minutes after takeoff, Parsons and Second Lieutenant Morris Jeppson crawled into the bomb bay to begin the arming procedure. Jeppson held a flashlight and handed Parsons tools as the latter worked through his 10-step checklist:

1. Check that green plugs are installed.
2. Remove rear plate.
3. Remove armor plate.
4. Insert breech wrench in breech plug.
5. Unscrew breech plug, place on rubber pad.
6. Insert charge, 4 sections, red end to breech.
7. Insert breech plug and tighten home.

8. Connect firing line.
9. Install armor plate.
10. Remove and secure catwalk and tools.

In step 1, the "green plugs" were three "safing" plugs that isolated the firing system of the bomb from its batteries; Jeppson would later replace these with red-colored "live" plugs. The entire procedure took about 20 min.

At some point not long into the flight, Tibbets went on the plane's intercom system to inform his men that they were carrying the world's first combat atomic bomb. At the request of William Laurence, who was disappointed that he was not allowed to fly as an observer, co-pilot Robert Lewis kept a journal, which, in 1971, would be auctioned for $37,000. Laurence got his wish on the Nagasaki mission, when he flew on the instrument plane.

Little Boy in its loading pit. Source http://commons. wikimedia.org/wiki/File:Atombombe_Little_Boy_2. jpg

Enola Gay on Tinian. *Source* http://commons.wikimedia.org/wiki/File:050607-F-1234P-090.jpg

About three hours after takeoff, *Enola Gay* rendezvoused at Iwo Jima with the camera and instrument planes, *Number 91* and *The Great Artiste.* (After the atomic missions, *Number 91* would be renamed *Necessary Evil*). One of the crew members on *The Great Artiste*, Lawrence Johnston, is believed to have witnessed all three of the *Trinity*, Hiroshima, and Nagasaki explosions; he had been a student of Luis Alvarez, and had joined Los Alamos in May, 1944, to work on detonators for the implosion device.

Left: Partial crew of the Enola Gay: Standing (l-r): John Porter (ground mainte-nance officer), Theodore Van Kirk, Thomas Ferebee, Paul Tibbets, Robert Lewis, Jacob Beser; kneeling (l-r): Joseph Stiborik, George Robert Caron, Richard Nelson, Robert Shumard, Wyatt Duzenbury. Not present: William Parsons, Morris Jeppson. Photo courtesy John Coster-Mullen. Right: Morris Jeppson. Source http://commons.wikimedia.org/wiki/File:Morris_Jeppson.jpg

Parsons kept a log of the mission. In terse, unadorned words, it narrated the progress of what would prove to be a textbook operation. (Events in brackets were not in Parson's original log, but have been added here for completeness. All times are Tinian time; subtract one hour for Japan time, and subtract 14 h for Washington time. All events occurred on August 5, Washington time):

02:45   Take off
03:00   Started final loading of gun
03:15   Finished loading
05:52   (Approach Iwo Jima. Begin climb to 9,300 feet)
06:05   Headed for Empire from Iwo
07:30   Red plugs in.

After Jeppson had installed the red arming plugs, the bomb was live. In his journal, Robert Lewis wrote "The bomb is now independent of the plane. It was a peculiar sensation. I had a feeling the bomb had a life of its own now that had nothing to do with us." Jeppson kept one of the green safing plugs and a spare red plug as souvenirs; they sold at auction in 2002 for $167,000.

As the *Enola Gay* approached Hiroshima, Lewis added to his journal: "There'll be a short intermission while we bomb our target." Resuming with Parsons' log:

07:41   Started climb. Weather report received that weather over primary and tertiary targets was good but not over secondary target
08:25   (Weather plane – cloud cover less than 3/10 at all altitudes Advice: bomb primary)
08:38   Leveled off at 32,700 feet
08:47   All Archies tested to be OK
09:04   Course west
09:09   Target (Hiroshima) in sight
09:12   (Initial point)
09:14   (Glasses on)
09:15 ½   Dropped bomb. Flash followed by two slaps on plane. Huge cloud
10:00   Still in sight of cloud which must be over 40,000 feet high
10:03   Fighter reported
10:41   Lost sight of cloud 363 miles from Hiroshima with aircraft being 26,000 feet high
14:58   Landed at Tinian

Little Boy free-fell for about 43 s before detonating. Bombardier Thomas Ferebee's aiming point was the distinctive T-shaped Aioi bridge in the heart of the city; he missed by only a few hundred feet. Van Kirk's navigation had been flawless. The scheduled time for the drop was 09:15; after a flight of eight and one-half hours, *Enola Gay* arrived at its target only seconds behind schedule. In post-strike photos, the Aioi bridge, which survived, is clearly visible.

Hiroshima   mushroom   cloud.   Source http://commons.wikimedia.org/wiki/File: Atomic_cloud_over_Hiroshima.jpg

Left: Aerial view of Hiroshima, post-bombing. The Aioi bridge is in the center of the image. Source http://commons.wikimedia.org/wiki/File:AtomicEffects-p7a.jpg. *Right: General view of damage at Hiroshima. Source* http://commons.wikimedia.org/wiki/File:AtomicEffects-Hiroshima.jpg

Tibbets executed his escape maneuver, and then turned south to permit the crew to observe the city for a couple minutes before setting course back to for Tinian. As thousands suffered below, Robert Lewis wrote "My God, what have we done?" He was later quoted as saying "If I live a hundred years, I'll never quite get these few minutes out of my mind." Crew members in *Enola Gay* and the observation planes reported that, five minutes after the drop, a low grey cloud three miles in diameter hung over the city, out of the center of which rose a column of white smoke to a height of 35,000 feet with the top of the cloud considerably enlarged.

In Washington, Groves expected to hear by about 2:00 p.m. that the *Enola Gay* had taken off, but communications were delayed. To work off his nervous energy he went for a game of tennis, and then had dinner with his family. Finally at about 6:45 p.m. the call came through that the plane had taken off; by that time *Enola Gay* had climbed to her bombing altitude and was approaching Hiroshima. Groves returned to his office, where he intended to spend the night. In his memoirs, he described how he abandoned his usual formality: "In order to ease the growing tension in the office, I made a point of taking off my tie, opening up my collar and rolling up my sleeves."

Immediately after the drop, Parsons sent Groves a brief coded message, which finally arrived about 11:30 p.m. Washington time, more than four hours after the bombing:

Results clearcut, successful in all respects. Visible effects greater than New Mexico test. Conditions normal in airplane following delivery. Target at Hiroshima attacked visually. One-tenth cloud at 052315Z. No fighters and no flak.

By the time Groves received Parsons' message, *Enola Gay* was only ninety minutes from returning to Tinian. The 052315Z in Parsons' message means August 5, 23:15 Greenwich time, or 7:15 p.m. Sunday evening in Washington. Groves promptly informed General Marshall of the message, and before going to bed on a cot in his office prepared a draft report to be delivered to Marshall in the morning.

Enola Gay landed at Tinian at about 1:00 a.m., Washington time. Tibbets was immediately decorated with a Distinguished Service Cross by General Spaatz; Parsons was later awarded a Silver Star. Farrell sent Groves a lengthier cable:

Following additional information furnished by Parsons, crews, and observers on return to Tinian at 060500Z. Report delayed until information could be assembled at interrogation of crews and observers. Present at interrogation were Spaatz, Giles, Twining, and Davies.

Confirmed neither fighter or flak attack and one tenth cloud cover with large open hole directly over target. High speed camera reports excellent record obtained. Other observing aircraft also anticipates good records although films not yet processed. Reconnaissance aircraft taking post-strike photographs have not yet returned.

Sound—None appreciable observed.

Flash—Not so blinding as New Mexico test because of bright sunlight. First there was a ball of fire changing in a few seconds to purple clouds and flames boiling and swirling upward. Flash observed just after airplane rolled out of turn. All agreed light was intensely bright and white cloud rose faster than New Mexico test, reaching thirty thousand feet in minutes it was one-third greater in diameter.

It mushroomed at the top, broke away from column and the column mushroomed again. Cloud was most turbulent. It went at least to forty thousand feet. Flattening across its top at this level. It was observed from combat airplanes three hundred sixty-three nautical miles away with airplane at twenty-five thousand feet. Observation was then limited by haze and not curvature of the earth.

Blast—There were two distinct shocks felt in combat airplane similar in intensity to close flak bursts. Entire city except outermost ends of dock areas was covered with a dark grey dust layer which joined the cloud column. It was extremely turbulent with flashes of fire visible in the dust. Estimated diameter of this dust layer is at least three miles. One observer stated it looked as though whole town was being torn apart with columns of dust rising out of valleys approaching the town. Due to dust visual observation of structural damage could not be made.

Parsons and other observers felt this strike was tremendous and awesome even in comparison with New Mexico test. Its effects may be attributed by the Japanese to a huge meteor.

Farrell's message reached Groves about 4:30 a.m. The two shocks felt in the plane were due to the direct shock wave of the explosion, and the reflection of the shock wave from the ground. Groves revised his report, and was at Marshall's office by 7:00 a.m.

When film from the camera plane came back from being developed, half of the emulsion was gone; it was never determined whether any images had been recorded. The brief films available of the bombings were taken by crew members with hand-held cameras; in the case of the Hiroshima mission, Los Alamos scientist Harold Agnew, riding aboard *The Great Artiste*, filmed the explosion. Reconnaissance planes found that Hiroshima was still mostly covered by the cloud created by the explosion, although fires could be seen around the edges; clearer images would have to wait until the next day.

The world learned of the bombing when Truman's pre-authorized statement was released in Washington at 11:00 a.m.; Truman was still at sea on his way home from Potsdam, and would not arrive until the evening of the seventh. The text of the release read in part as

> Sixteen hours ago an American airplane dropped one bomb on Hiroshima, an important Japanese Army base. That bomb had more power than 20,000 tons of T.N.T. It had more than two thousand times the blast power of the British "Grand Slam" which is the largest bomb ever yet used in the history of warfare.
>
> The Japanese began the war from the air at Pearl Harbor. They have been repaid many fold. And the end is not yet. With this bomb we have now added a new and revolutionary increase in destruction to supplement the growing power of our armed forces. In their present form these bombs are now in production and even more powerful forms are in development.
>
> It is an atomic bomb. It is a harnessing of the basic power of the universe. The force from which the sun draws its power has been loosed against those who brought war to the Far East.
>
> ...
>
> The battle of the laboratories held fateful risks for us as well as the battles of the air, land and sea, and we have now won the battle of the laboratories as we have won the other battles.
>
> ...

The United States had available the large number of scientists of distinction in the many needed areas of knowledge. It had the tremendous industrial and financial resources necessary for the project and they could be devoted to it without undue impairment of other vital war work. In the United States the laboratory work and the production plants, on which a substantial start had already been made, would be out of reach of enemy bombing, while at that time Britain was exposed to constant air attack and was still threatened with the possibility of invasion. For these reasons Prime Minister Churchill and President Roosevelt agreed that it was wise to carry on the project here. We now have two great plants and many lesser works devoted to the production of atomic power. Employment during peak construction numbered 125,000 and over 65,000 individuals are even now engaged in operating the plants. Many have worked there for two and a half years. Few know what they have been producing. They see great quantities of material going in and they see nothing coming out of these plants, for the physical size of the explosive charge is exceedingly small. We have spent two billion dollars on the greatest scientific gamble in history - and won.

But the greatest marvel is not the size of the enterprise, its secrecy, nor its cost, but the achievement of scientific brains in putting together infinitely complex pieces of knowledge held by many men in different fields of science into a workable plan. And hardly less marvelous has been the capacity of industry to design, and of labor to operate, the machines and methods to do things never done before so that the brain child of many minds came forth in physical shape and performed as it was supposed to do. Both science and industry worked under the direction of the United States Army, which achieved a unique success in managing so diverse a problem in the advancement of knowledge in an amazingly short time. It is doubtful if such another combination could be got together in the world. What has been done is the greatest achievement of organized science in history. It was done under high pressure and without failure.

We are now prepared to obliterate more rapidly and completely every productive enterprise the Japanese have above ground in any city. We shall destroy their docks, their factories, and their communications. Let there be no mistake; we shall completely destroy Japan's power to make war.

It was to spare the Japanese people from utter destruction that the ultimatum of July 26 was issued at Potsdam. Their leaders promptly rejected that ultimatum. If they do not now accept our terms they may expect a rain of ruin from the air, the like of which has never been seen on this earth. Behind this air attack will follow sea and land forces in such numbers and power as they have not yet seen and with the fighting skill of which they are already well aware.

...

The fact that we can release atomic energy ushers in a new era in man's understanding of nature's forces. Atomic energy may in the future supplement the power that now comes from coal, oil, and falling water, but at present it cannot be produced on a basis to compete with them commercially. Before that comes there must be a long period of intensive research.

It has never been the habit of the scientists of this country or the policy of this Government to withhold from the world scientific knowledge. Normally, therefore, everything about the work with atomic energy would be made public.

But under present circumstances it is not intended to divulge the technical processes of production or all the military applications, pending further examination of possible methods of protecting us and the rest of the world from the danger of sudden destruction.

I shall recommend that the Congress of the United States consider promptly the establishment of an appropriate commission to control the production and use of atomic power within the United States. I shall give further consideration and make further recommendations to the Congress as to how atomic power can become a powerful and forceful influence towards the maintenance of world peace.

Truman's 20,000 tons was an overestimate, probably caused by confusing *Little Boy* with the *Trinity* test. The War Department released a longer statement which included details regarding the manufacturing plants, some of the contractors and universities involved, the cost of the project, and the existence of the Interim Committee.

Henry Stimson dispatched a message to Truman, who received it while he was having lunch aboard the *USS Augusta*.

WHITE HOUSE

MAP ROOM

6 August 1945

```
FROM:  THE SECRETARY OF WAR
TO  :  THE PRESIDENT

NR  :  335
```

Big bomb dropped on Hiroshima 5 August at 7:15 p.m., Washington time. First reports indicate complete success which was even more conspicuous than earlier test.

```
RECD:  061510Z
```

Truman is informed of the Hiroshima bombing. Source https://www.trumanlibrary.org/whistlestop/study_collections/naval/berlin/index.php?documentid=hst-naval_naid1701772-13

At 2:00 p.m. Washington time, Groves telephoned Oppenheimer to extend his congratulations. A partial transcript of their conversation:

Groves:	I'm very proud of you and all of your people.
Oppenheimer:	It went alright?
Groves:	Apparently it went with a tremendous bang.
Oppenheimer:	When was this, was it after sundown?
Groves:	No, unfortunately it had to be in the daytime on account of security of the plane and that was left in the hands of the Commanding General over there and he knew what the advantages were of doing it after sundown and he was told just all about that and I said it was up to him; that it was not paramount but that it was very desirable.
Oppenheimer:	Right. Everybody is feeling reasonably good about it and I extend my heartiest congratulations. It's been a long road.

Groves:	Yes, it has been a long road and I think one of the wisest things I ever did was when I selected the director of Los Alamos.
Oppenheimer:	Well, I have my doubts, General Groves.
Groves:	Well, you know I've never concurred with those doubts at any time.

At Los Alamos that evening, a crowd gathered in the auditorium. As related by physicist Sam Cohen:

> That evening we gathered, long before the appointed time of Oppenheimer's appearance … . Normally at one of these colloquia Oppenheimer, more or less punctual, would walk unobtrusively onstage from a wing, quiet down the audience, make a few remarks in his low-key manner and introduce the speaker. But that was not to be the case on this heroic day: He was late, very late. He did not casually slip onstage from a wing. He came in from the rear of the theatre, strode down the aisle and up the stairs onto the stage, and he made no effort to quiet a yelling, clapping, foot-stomping bunch of scientists who began to cheer him when he entered and continued to do so long after he got onstage.

> Now, keep in mind that while this pandemonium was going on, about seventy thousand Japanese civilians lay dead in Hiroshima, with an equal number injured. About 30% of the victims had received lethal or injurious does of nuclear radiation … . Most of the scientists were, or should have been, very much aware that radiation would take a terrible toll, but at this moment of triumph they couldn't have cared less about any particular moral transgression associated with it. They were flushed with their success and they showed it. And I was one of them.

> Finally Oppenheimer was able to quiet the howling crowd and he began to speak, hardly in low key. It was too early to determine what the results of the bombing might have been, but he was sure that the Japanese didn't like it. More cheering. He was proud, and he showed it, of what we had accomplished. Even more cheering. And his only regret was that we hadn't developed the bomb in time to have used it against the Germans. This practically raised the roof.

As the implications seeped in over the following hours and days, the reaction at Los Alamos was by no means one of unrestrained celebration. Alice Smith, wife of metallurgist Cyril Smith, described the atmosphere:

> As the days passed the revulsion grew, bringing with it – even for those who believed that the end of the war justified the bombing – an intensely personal experience of the reality of evil. It was this, and not a feeling of guilt in the

ordinary sense, that Oppenheimer meant by his much quoted, and often misunderstood, remark that scientists had known sin.

McAllister Hull, who cast implosion lenses:

> I do not fault Truman's decision to use the bombs, for he was accountable for every Allied casualty he had a means to prevent. I had no such responsibility. I just wish he – or we – had found a way to use them to stop the war immediately without making those of us who had worked on them accessory to several hundred thousand deaths – and scarring wounds to thousands more – in Hiroshima and Nagasaki. I do not know about my friends, but I have never for a moment forgotten that responsibility.

Hans Bethe:

> You can no longer use atomic bombs for saving lives. Hiroshima saved lives, lots of them, lots of Japanese and many Americans. If there were a nuclear war today, it would be a destruction of both countries, so in that sense it cannot be repeated. But I think the realization that it cannot and must not be repeated was very much facilitated by Hiroshima. If we hadn't had these two atomic bombings, people would not have realized what a terrible thing this is.

Following the bombing, some six million leaflets were dropped over 47 Japanese cities, encouraging ordinary citizens to pressure the Emperor and ruling militarists to end the war. Ironically, Nagasaki did not receive its quota of leaflets until after it was bombed. The text read

> To the Japanese People: America asks that you take immediate heed of what we say on this leaflet.

> We are in possession of the most destructive explosive ever devised by man. A single one of our newly developed atomic bombs is actually the equivalent in explosive power to what 2000 of our giant B-29s can carry on a single mission. This awful fact is one for you to ponder and we solemnly assure you it is grimly accurate.

> We have just begun to use this weapon against your homeland. If you still have any doubt, make inquiry as to what happened to Hiroshima when just one atomic bomb fell on that city.

> Before using this bomb to destroy every resource of the military by which they are prolonging this useless war, we ask that you now petition the Emperor to end the war. Our president has outlined for you the thirteen consequences of

an honorable surrender. We urge that you accept these consequences and begin the work of building a new, better and peace-loving Japan.

You should take steps now to cease military resistance. Otherwise, we shall resolutely employ this bomb and all our other superior weapons to promptly and forcefully end the war.

The Japanese government was not yet ready to quit, but its situation was becoming more perilous by the hour. At 5:00 p.m. local time on the afternoon of August 8, the Japanese ambassador in Moscow was informed that the Soviet Union would consider itself in a state of war with Japan as of August 9. Five time zones to the east, it was already 10:00 p.m., and Russian forces were advancing in Manchuria. The Japanese government had been hoping to use Russia as a go-between in surrender negotiations, but their proposals had been vague, and that hope was in any event now dashed. When Truman announced the Russian declaration of war at 3:00 p.m. in Washington, *Fat Man* was already airborne over the Pacific.

The second nuclear strike was originally scheduled for August 20, but by late July enough time had been made up to permit advancing the date to the 11th. By the 7th, the day after the Hiroshima mission, it appeared that the schedule could be further tightened to the 10th. Good weather was forecast for the 9th, but bad weather for the five days thereafter. Groves wanted the second atomic blow to follow the first as quickly as possible, and Project Alberta personnel set to work to try to have the first live *Fat Man* ready by the evening of the August 8. From its start, however, the Nagasaki mission suffered almost every possible misfortune that the Hiroshima mission had avoided. The front and rear halves of F31's armor-plate ballistic casing were out of round, with the result that bolt holes for attaching the casing segments to a flange on the high-explosive casing did not align properly. No other armor-plate casings were available, so an attempt was made to hammer the parts into shape. When that failed, an effort was made to enlarge the bolt-holes with a two-man drill, but it jammed and gashed the leg of one of the workers. Desperate and running short of time, the assemblers substituted an ordinary steel casing; *Fat Man* would have to take its chances against Japanese machine-gun fire. After receiving a coat of pumpkin-colored paint and sealant to close off cracks which might result in erroneous barometric readings, the assembly crew made a small profile-view stencil of *Fat Man* and applied it to the nose of the bomb, along with the letters JANCFU. The first four stood for "Joint Army Navy Civilian"; the meaning of the last two can be extrapolated from popular vernacular. Before it was rolled out for loading, a number of people autographed the bomb, including Purnell, Farrell, Parsons, and Ramsey; there were some 60 signatures in total.

The casing was not the only problem. On the night of August 7, Bernard O'Keefe, one of the members of the assembly team, was responsible for carrying out a last check of *Fat Man's* firing unit before it was encased:

> By ten o'clock on the night of August 7, the sphere was complete, the radars installed, and the firing set bolted onto the front end of the sphere. I broke out for some sleep while others did final checkup and the mechanical assembly crew put the final touches on the casing. I was to come back at midnight for final checkout and to connect the two ends of the cable between the firing set and the radars; the cable had been installed the day before. Then I would turn the device over to the mechanical crew for installation of the fin and the nose cap.

> When I returned at midnight, the others in my group left to get some sleep; I was alone in the assembly room with a single Army technician to make the final connection …

> I did my final checkout and reached for the cable to plug it into the firing set. It wouldn't fit!

> "I must be doing something wrong," I thought. "Go slowly; you're tired and not thinking straight."

> I looked again. To my horror, there was a female plug on the firing set and a female plug on the cable. I walked around the weapon and looked at the radars and the other end of the cable. Two male plugs. The cable had been put in backward. I checked and double-checked. I had the technician check; he verified my findings. I felt a chill and started to sweat in the air-conditioned room.

> What had happened was obvious. In the rush to take advantage of good weather, someone had gotten careless and put the cable in backward. Worse still, the checklist had been bypassed so that it was not double-checked before assembling the casing.

Fixing the problem would mean unsoldering the connectors from the two ends of the cable and reversing them. But to follow orders that no source of heat was allowed in the explosives assembly room would mean partially disassembling the bomb, which would take time. O'Keefe decided to proceed on his own:

> My mind was made up. I was going to change the plugs without talking to anyone, rules or no rules. I called in the technician. There were no electrical outlets in the assembly room. We went out to the electronics lab and found two long extension cords and a soldering iron. We … propped the door open (another safety violation) so it wouldn't pinch the extension cords. I carefully removed the backs of the connectors and unsoldered the wires. I resoldered the plugs onto the other ends of the cable, keeping as much distance between the soldering iron and the detonators as I could as I walked around the weapon … We must have checked the cable continuity five times before plugging the connectors into the radars and the firing set and tightening up the joints.

Field Order number 17 and Operations order number 39 detailed primary and secondary targets: Kokura Arsenal and City, and the Nagasaki Urban Area; there was no tertiary target for this mission. Located about 100 miles apart on the southernmost main island of Kyushu, both areas were rich in targets. Kokura, a city of about 168,000, was home to Kokura Arsenal, a large armaments complex where vehicles, machine guns, and anti-aircraft guns were manufactured. Nagasaki, with a population estimated to be about 250,000 at the time, is located at what has been described as Kyushu's best natural harbor. A shipbuilding center and military port, major targets there included the Mitsubishi Heavy Industries shipbuilding complex and the adjacent Mitsubishi Steel and Arms Works. The latter was where torpedoes used at Pearl Harbor had been manufactured. Unlike Hiroshima and Kokura, Nagasaki is a somewhat constricted city, surrounded by hills.

Partial Bockscar crew. Standing (l-r): Kermit Beahan, James Van Pelt, Don Albury, Fred Olivi, Charles Sweeney; kneeling (l-r): Edward Buckley, John Kuharek, Ray Gallagher, Albert Dehart, Abe Spitzer. Not present: Frederick Ashworth, Philip Barnes. Photo courtesy John Coster-Mullen

Fat Man was ready by 10:00 p.m. on the evening of August 8, and loaded into *Bockscar*. Major Charles Sweeney was assigned to pilot the strike plane; its usual commander, Captain Frederick Bock, would pilot *The Great Artiste*. The final crew briefing took place at 00:30 on the 9th.

As *Bockscar* was prepared for takeoff, another problem arose. As ballast to compensate for the weight of the bomb, the rear bomb-bay had been fitted with two 320-gallon fuel tanks. Flight Engineer John Kuharek discovered that a pump for transferring fuel from the tanks appeared to be malfunctioning. The fuel would not only be inaccessible, but at about six pounds per gallon would represent almost two tons of dead weight to be carried through the mission. To empty the tanks, replace the pump, or transfer the bomb to another plane would be too time-consuming; the window of good weather was narrowing. Sweeney decided to proceed with the mission. Bockscar departed at 03:48 Tinian time, Thursday, August 9; in Washington, it was 1:48 p.m. on Wednesday afternoon, August 8.

The rendezvous point for *Bockscar* and the camera and instrument planes was at the island of Yakushima, immediately off the southern coast of Kyushu. After flying through a storm, *Bockscar* arrived at about 09:00 and was promptly joined by *The Great Artiste*, but the camera plane, *Big Stink*, piloted by Captain James Hopkins, was nowhere to be seen. Hopkins was there, but for some reason was flying at 39,000 feet versus *Bockscar's* 30,000. In his memoirs, Sweeney claims that he was told later that Hopkins began making 50-mile dog-leg sweeps in the area of Yakushima, as opposed to circling as he should have been. Although Tibbets had instructed Sweeney to wait for no more than 15 min at the rendezvous point, he waited about 45 before deciding to strike out for Kokura.

Another element of confusion seems to have been that Commander Ashworth, who was overseeing the bomb, wanted to be sure that at least the instrument plane accompanied *Bockscar* on the strike mission. Ashworth claims that Sweeney never informed him which other plane they had rendezvoused with, and *The Great Artiste* remained too distant for Ashworth to get a visual identification. Sweeney did not address this issue in his own memoirs except to say that he felt that it was vital to have the photographic plane along to fulfill the mission plan. Ashworth claims to have stuck his head up into the flight deck to recommended that they proceed to their primary target; Sweeney implied that it was his decision to do so. The positive news was the both weather planes were reporting good conditions at the targets.

Hopkins' incorrect altitude was not *Big Stink's* only problem. Robert Serber was to fly on Hopkins' plane for the specific purpose of operating a high-speed camera to record the explosion. As Hopkins taxied to the end of the runway at Tinian in preparation for take-off, he called for a parachute check. Serber had not been issued one, and was forced off the plane, which then departed without him. After walking back to base (and fearing the presence of Japanese snipers), he was authorized to break radio silence in an attempt to transmit instructions to the plane, but this proved to be for naught. At one point, Hopkins, speaking in the clear, radioed "Has Sweeney aborted?" At Tinian this was heard as "Sweeney aborted," which caused General Farrell to run outside and throw up.

Bockscar's flight to Kokura from the rendezvous point took about 50 min, but by the time it arrived at its aiming point at about 10:44 (Tinian time), the city was obscured by smoke and industrial haze. The nearby city of Yawata had been firebombed the previous day, and smoke was drifting over Kokura. The Japanese started sending up flak, so Sweeney rose to 31,000 feet. The smoke and haze made visual bombing runs impossible; after three attempts from different directions at different altitudes, Sweeney decided to head for

Nagasaki. By this time, *Bockscar's* fuel supply was getting low. Sweeney estimated that they would have enough fuel for one run over Nagasaki, but that they would likely have to ditch in the ocean some fifty miles from Okinawa, the nearest friendly base. *Bockscar* departed Kokura about 11:30 a.m. (10:30 Japan time). The term "Kokura luck" is sometimes used by Japanese as a euphemism for the unknown avoidance of a horrible misfortune.

Left: Bockscar nose art, added after the Nagasaki mission. Source http://commons. wikimedia.org/wiki/File:Bockscar.jpg. *Right: The Nagasaki mushroom cloud. Source* http://commons.wikimedia.org/wiki/File:Atomic_cloud_over_Nagasaki_ from_B-29.jpg

The flight to Nagasaki from Kokura took only about 20 min; *Bockscar* arrived at about 11:50 a.m., Tinian time. But the weather had changed there as well, with the city now obscured by 80–90% cumulus clouds between 6,000 and 8,000 feet. The fuel situation was becoming critical. Some accounts have Ashworth directing bombardier Kermit Beahan to make a radar-based bomb run, for which Ashworth would take responsibility. Sweeney claims in his memoirs that he gave the same order. But about 30–45 s before the drop, a hole opened in the clouds, and Beahan shouted "I see it! I see it! I've got it!" They had already passed the original aiming point in the dock area of the city, so Beahan chose a new one in the industrial area. Control of the aircraft was relinquished to him, and he released *Fat Man* from an altitude of about 29,000 feet at 11:08 a.m. Nagasaki time (10:08 p.m. Washington time, August 8). The bomb detonated over the Mitsubishi complex; because of the reflective hilly geography, the crew felt five shock waves.

Sweeney ordered radio operator Abe Spitzer to transmit a strike report:

> Bombed Nagasaki 090158Z visually. No opposition. Results technically successful. Visible effects about equal to Hiroshima. Proceeding to Okinawa. Fuel problem.

The time given in Spitzer's report differs by 10 min from that listed in the accompanying Table; slightly different times have been reported by various sources. Eighty miles away, the crew of *Big Stink* noticed the explosion. As related by Group Captain Leonard Cheshire, a British observer aboard Hopkins' plane:

> We reached the target some 10 min after the explosion at a height of 39,000 feet. At this time the cloud had become detached from the column and extended up to a height of approximately 60,000 feet. From the bomb aimer's compartment I had an excellent view of the ground and could see that the center of the impact was some four miles north-east of the aiming point and that the city proper was untouched. Fortunately however the bomb had accidentally hit the industrial center north of town and had caused considerable damage.

After lingering only briefly to view the results of his work, Sweeney set course for Okinawa. Spitzer sent a Mayday call, but received no reply. "Fuel problem" was an understatement; Sweeney estimated that they had one hour of flying time available, but Okinawa was about seventy-five minutes away. By utilizing a technique known as "flying on the step" where he would leave power settings steady but put the plane into a very gradual descent, Sweeney was able to pick up a bit more airspeed without using additional fuel. Alternating descents and level-offs allowed him to stretch the fuel supply to Okinawa.

But *Bockscar* was not yet out of the woods. As they approached Yontan Field on Okinawa, Spitzer was unable to raise the tower. The nearest American base to Japan, Okinawa was always busy with incoming and outgoing traffic. Sweeney ordered Fred Olivi to fire emergency flares. Different-colored flares were used to indicate different emergency conditions, such as low fuel, damage, prepare for crash, dead and wounded aboard, or fire on board. Olivi fired all of them, and the field began to clear of aircraft and vehicles. Cutting into the active traffic pattern, Sweeney came in directly behind a B-24 that was taking off. *Bockscar* bounced into the air and slammed back down just as its left inboard engine cut out; only by using the reversible propellers were Sweeney and co-pilot Don Albury able to bring the craft to a stop before running out of runway. As Sweeney described it, "I was so mentally and physically exhausted at that point that I just let the airplane roll to the side of the runway and onto a taxiway. Another engine quit." According to various accounts, they arrived with only 7 or 35 gallons of fuel remaining—exclusive of the trapped fuel. After the crew had a meal and *Bockscar* was refueled, they made their way back to Tinian, arriving about 11:00 p.m. to no fanfare after a mission of over 19 h. Some sources state that *Bockscar* spent more time over enemy territory than any other plane on a single mission in all of World War II. Because of bad weather, reconnaissance photos of Nagasaki could not be obtained until after a week following the mission.

Rumors that the crew members of *Enola Gay* and *Bockscar* suffered debilitating illnesses due to radiation exposure or became mentally disturbed over their participation in the bombings are simply untrue. These twenty-four men lived to an average age of 76 years, with three of them (Tibbets, Van Kirk, and Ashworth) surviving into their nineties. Their causes of death included heart attacks, cancers, respiratory illnesses, and an automobile accident. The longest-lived was Van Kirk, who lived for 93 years and five months before passing away in July, 2014, of what his family described as natural causes. Between them, these men fathered over 50 children, including 10 by Charles Sweeney alone.

Wars are full of indiscriminate cruelties, but random occurrences of astonishing survivals also occur. In the history of Hiroshima and Nagasaki, one of these improbable stories involves what the Japanese came to call the "nijyuu hibakusha," or "twice bombed." After the bombing of Hiroshima, a number of survivors were relocated or moved of their own accord to Nagasaki, where, three days later, they experienced the *Fat Man* explosion. While it is estimated that some 165 people survived both bombings, the Japanese government officially recognized only one: Mr. Tsutomu Yamaguchi. A Mitsubishi engineer, Yamaguchi was in Hiroshima on a business trip on the morning of August 6, and was stepping off a streetcar less than two miles from ground zero when *Little Boy* detonated. His eardrums were ruptured and he sustained some burns, but was able to return to Nagasaki after spending the night in a bomb shelter. On the morning of the 9th, he was in his office telling his boss about what he had witnessed at Hiroshima, when "suddenly the same white light filled the room." Mr. Yamaguchi died of stomach cancer in early 2010 at the age of 93; his daughter has been reported as stating that he remained in good health for most of his life.

In the weeks and days before the bombings, American intelligence services had been intercepting and decrypting Japanese messages; it was known that many elements in the Japanese government wished to find a way toward what they considered to be an honorable surrender. The sticking point was the fate of Emperor Hirohito in the context of the "unconditional surrender" sought by the Allies. In Tokyo on August 9, high-level conferences ran on through the day. At a morning meeting of the Supreme War Council, it was decided that an absolute condition of accepting the Potsdam terms would have to be retention of the imperial house. A militarist faction demanded that if occupation of Japan could not be avoided, then the Japanese should at least be responsible for their own disarmament and trials of any war criminals. As the meeting progressed, word was received of the strike on Nagasaki. The meeting continued into the late evening with no consensus being reached. At about midnight, the Council met with the Emperor himself, who made it known that he was in favor of ending the war.

At 8:47 a.m. Tokyo time on the 10th (7:47 p.m. on the 9th in Washington), a deliberately low-security message went out from the Foreign Ministry to legations in Switzerland and Sweden. The text included a statement that the Japanese were ready to accept the Potsdam conditions so long as they were understood to not include "any demand for modification of the prerogatives of His Majesty as a sovereign ruler." Intercepted and decrypted, the message was on President Truman's desk early on the morning of the 10th. By noon, a response had been drafted that stipulated that "the authority of the Emperor and the Japanese Government to rule the state shall be subject to the Supreme Commander of the Allied Powers who will take such steps as he deems proper to effectuate the surrender terms".

Also on the 10th, Groves informed General Marshall as to the delivery schedule of the next bomb, writing

> The next bomb of the implosion type had been scheduled to be ready for delivery on the target on the first good weather after 24 August 1945. We have gained 4 days in manufacture and expect to ship from New Mexico on 12 or 13 August the final components. Providing there are no unforeseen difficulties in manufacture, in transportation to the theatre or after arrival in the theatre, the bomb should be ready for delivery on the first suitable weather after 17 or 18 August.

But Truman had exercised his prerogative as Commander-in-Chief, ordering a halt to any more atomic strikes. Henry Wallace, who had preceded Truman as Vice-President and was serving as Secretary of Commerce, recorded in his diary that afternoon that

> The President, who usually comes to cabinet not later than 2:05, came in about 2:25 saying he was sorry to be late but that he and Jimmie [Byrnes] had been busy working on a reply to Japanese proposals … Truman said he had given orders to stop atomic bombing. He said the thought of wiping out another 100,000 people was too horrible. He didn't like the idea of killing, as he said, "all those kids."

Truman's decision overrode the July 25 orders which authorized use of bombs as they became available. In Washington, Marshall's thinking was already moving to use of further bombs in tactical support of an invasion, a strategy not particularly contemplated during meetings of the Target and Interim Committees. In anticipation that the invasion would involve three corps of troops, Marshall was considering using one or two bombs for each corps' landing area before their landings, and reserving another for each to eliminate Japanese replacements that might come up. Historian Barton Bernstein has pointed out that Marshall's thoughts on tactical use speak against the revisionist thesis that the bombs were used primarily to intimidate the Soviet Union.

The Allied reply to the Japanese proposal began to be picked up by radio intercepts in Tokyo in early hours of August 12. Japanese officials debated through the day and into the evening. The continue-the-war faction favored holding out, with some speaking of mounting a coup. On the morning of the 14th, the Emperor himself called for an Imperial Conference at 10:30 a.m. (9:30 p.m. on the 13th in Washington). Again making clear to the gathered ministers his desire for peace, Hirohito directed that a public statement, an Imperial Rescript, be prepared, which he would record for broadcast over national radio; this would be the first time many Japanese would hear their Emperor's voice. That evening, a formal statement accepting the proposed compromise on the status of the Emperor was drafted. But the national Japanese news agency was already broadcasting a message indicating that an Imperial message accepting the Potsdam conditions was expected soon. At 11:48 p.m. (10:48 a.m. on the 14th in Washington), the Foreign Ministry began sending the appropriate coded messages to Switzerland and Sweden.

With negotiations dragging on, Henry Arnold felt that the Japanese needed more motivation, and decided to mount one last punch: 449 B-29's carried out daylight strikes on the 14th. Raids continued into the night, with the last bombs of the war falling on the city of Tsuchizaki at 3:39 a.m. on the 15th, Japan time (2:39 p.m. on the 14th in Washington). The official surrender note was received at the State Department at 6:10 p.m., three and a half hours later. President Truman announced the surrender to reporters in the Oval Office at 7:00 p.m., and then publicly from the portico of the White

House. In Tokyo, Hirohito's statement was broadcast at noon on the 15th, just four hours after Truman's. Hirohito's public statement did not include the word "surrender," referring instead to effecting "a settlement of the present situation by resorting to an extraordinary measure. We have ordered our Government to communicate … that our Empire accepts the provisions of the Joint Declaration. … Our one hundred million people, the war situation has developed not necessarily to Japan's advantage, while the general trends of the world have all turned against her interest. Moreover, the enemy has begun to employ a new and most cruel bomb". Formal surrender documents would be signed aboard the battleship *USS Missouri* in Tokyo Bay on September 2.

With bombs delivered and surrender in the offing, Groves moved to his next task: assessing the effects of his creations. On August 11, he directed Colonel Nichols to begin organizing teams to carry out on-site investigations in Japan; General Farrell would be in charge of organization in the Pacific. The resulting Manhattan Project Atomic Bomb Investigating Group consisted of three teams: one for Hiroshima, one for Nagasaki, and one to investigate Japanese activities in the field of atomic bombs. Nichols brought together a group of 27, including physicists Robert Serber, Philip Morrison, and William Penney.

The results of the surveys were published in June, 1946, in a Manhattan Engineer District report titled "The Atomic Bombings of Hiroshima and Nagasaki." The group carried out preliminary inspections in Hiroshima on September 8 and 9, and in Nagasaki on September 13 and 14; these were to ensure that occupying forces would not be exposed to any excessive lingering radiation. In total, the Manhattan teams spent sixteen days in Nagasaki and four in Hiroshima. At the same time, the United States Strategic Bombing Survey (USSBS) also conducted its own analysis of the bombings, with a particular emphasis on surveying their effects on Japanese morale. A selection of statistics drawn from the two reports testify to the power of the bombs. "Point X" is ground zero, the location on the ground below the point of explosion of the bomb:

At Hiroshima:

- Estimated 66,000 dead and 69,000 injured of estimated pre-raid population of 255,000; a Japanese survey indicated some 71,000 dead and 68,000 injured. 60% of deaths were attributed to burns, and 30% to falling debris.
- Of over 200 doctors in the city before the attack, over 90% were casualties, with only about 30 able to perform their normal duties a month after the bombing.
- Of 1,780 nurses, 1,654 were killed or injured.

- Only three of 45 civilian hospitals could be used after the bombing.
- 60,000 of 90,000 buildings destroyed or severely damaged.
- 70,000 breaks in water pipes.
- Heavy fire damage in a circular area of about 6,000 feet radius and a maximum radius of about 11,000 feet.
- Almost everything up to about one mile from X was completely destroyed except for about 50 heavily-reinforced concrete buildings, most of which had been designed to withstand earthquakes. Multistory brick buildings were completely demolished to 4,400 feet from X, and suffered structural damage to 6,600 feet. Steel-framed buildings destroyed to 4,200 feet, and suffered severe structural damage to 5,700 feet. Light concrete buildings in both cities collapsed out to 4,700 feet.
- Firestorm burnt out about 4.4 square miles around X.
- People suffer burns to 7,500 feet.
- Roof tiles were melted out to 4,000 feet.
- In both cities, trolley cars were destroyed up to 5,500 feet and damaged to 10,500 feet.
- Flash ignition of dry combustible material observed to 6,400 feet.
- All homes seriously damaged to 6,500 feet; most to 8,000 feet.
- Flash charring of telephone poles to 9,500 feet.
- Fires started by primary heat radiation in both cities to about 15,000 feet.

At Nagasaki:

- Estimated 39,000 dead and 25,000 injured of estimated pre-raid population of 195,000.
- 95% of deaths attributed to burns.
- About 20,000 of 50,000 buildings and houses destroyed. Total destruction area about 3 square miles.
- Nearly everything was destroyed within 0.5 miles of X, including heavy structures.
- At 1,500 feet from X, high-quality steel buildings were not collapsed, but suffered mass distortion, and all panels and roofs were blown in. At 2,000 feet, reinforced concrete buildings with 10-inch walls were collapsed; buildings with 4-inch walls were badly damaged. At 3,500 feet, church buildings with 18-inch walls were completely destroyed. Multistory brick buildings were destroyed to 5,300 feet, and suffered structural damage to 6,500 feet. Steel-framed buildings destroyed to 4,800 feet and suffered severe structural damage to 6,000 feet. The extreme range of building collapse was 23,000 feet.

- Twelve-inch brick walls were severely cracked as far as 5,000 feet.
- Roof tiles were melted out to 6,500 feet.
- People suffered burns to almost 14,000 feet.
- Flash ignition of dry combustible material observed to 10,000 feet.
- About 27% of 52,000 residential units completely destroyed, and a further 10% half-burned or destroyed. All homes seriously damaged to 8,000 feet; most to 10,500 feet.
- Hillsides scorched to 8,000 feet.
- Foliage turned yellow to about 1.5 miles.
- Flash charring of telephone poles to 11,000 feet.
- Heavy fire damage south of X up to 10,000 feet, stopped by a river.

At Nagasaki, mortality was estimated at 93% within 1,000 feet of X, falling to 49% at 5,000 feet. By far, blast and burn effects were the greatest causes of mortality and injury. The Manhattan Project's medical director, Dr. Stafford Warren, estimated that some 7% of deaths resulted primarily from radiation, although some estimates of radiation-caused deaths ran as high as 15–20%. Radiation effects included depressed blood counts; loss of hair; bleeding into the skin; inflammations of the mouth and throat; vomiting; diarrhea; and fever. Deaths from radiation began about a week after exposure, peaked in about 3–4 weeks, and ceased by 7–8 weeks. Individuals who survived and remained in the cities were estimated to have received dosages estimated at 6–25 rems (Hiroshima) or 30–110 rems (Nagasaki), with the latter figure referring to a localized area. The USSBS report states that of women in Hiroshima in various stages of pregnancy who were known to be within 3,000 feet of ground zero, all suffered miscarriages, and some miscarriages and premature births where the infant died shortly after birth were recorded up to 6,500 feet. Two months after the bombing, the city's total incidence of miscarriages, abortions, and premature births ran to 27%, as opposed to a normal rate of 6%.

The USSBS report offered a comparison of the atomic bombings with the March 9/10 firebombing raid on Tokyo:

	Hiroshima	Nagasaki	Tokyo
Planes	1	1	279
Bombs	1 atomic	1 atomic	1,667 tons
Population per square mile	46,000	65,000	130,000
Square miles destroyed	4.7	1.8	15.8
Killed and missing (thousands)	70–80	35–40	83.6
Injured (thousands)	70	40	102
Mortality (thousands/square mile)	15	20	5.3

The report estimated that by November 1, the population of Hiroshima was back to 137,000, although the city required complete rebuilding. The population of Nagasaki had come back to 143,000.

The survey teams used a number of methods to determine parameters such as blast pressure and the detonation heights of the bombs. Concrete from the remains of buildings could be tested for breaking strength. William Penney sought out gas cans at various distances that had been crushed. After taking them back to England, he had similar cans made up and measured the pressure necessary to crush them. At the Post Office Building in Hiroshima just a mile from Ground Zero, Robert Serber found a room facing the explosion where the glass had been blown out of a large window, but the frames of the windowpanes had remained intact and had cast shadows on an adjacent wall. By measuring the angles of the shadows, he determined that the bomb had detonated at an altitude of 1,900 feet, and by measuring the penumbra of the shadow he could get an idea of how big the fireball had been. In a more humorous vein, William Penney found an unusual situation in Nagasaki: a door with paper panels where half were broken and half were intact. On asking the woman who lived in the house "Atomic bomb?", her reply was "No. Small boy." In 1970, Penney and some collaborators published an extensive paper on results of measurements of the yields of the explosions, determining 11–13 kt for Hiroshima and 20–24 kt for Nagasaki.

Scores of accounts of the horrifying deaths and injuries suffered by the people of Hiroshima and Nagasaki have been published. Some of the most disturbing were published by psychiatrist and writer Robert Jay Lifton, who interviewed a number of Hiroshima survivors, in his book *Death in Life: Survivors of Hiroshima.*

A grocer who was severely burned:

> The appearance of people was … well, they all had skin blackened by burns. … They had no hair because their hair was burned, and at a glance you couldn't tell whether you were looking at them from in front or back. … They held their arms bent … and their skin – not only on their hands, but on their faces and bodies too – hung down. … If there had been only one or two such people … perhaps I would not have had such a strong impression. But wherever I walked I met these people. … many of them died along the road – I can still picture them in my mind …

A sociologist at twenty-five at hundred meters from ground zero:

> Everything I saw made a deep impression – a park nearby covered with dead bodies waiting to be cremated … The most impressive thing I saw was some

girls, very young girls, not only with their clothes torn off but their skin peeled off as well … I thought that should there be a hell, this was it … And I imagined that all of these people I was seeing were in the hell I had read about.

A thirteen-year-old trying to save his mother from the debris of their house:

The fire was all around us so I thought I had to hurry. … I was suffocating from the smoke and I thought if we stayed like this, then both of us would be killed. I thought if I could reach the wider road, I could get some help, so I left my mother there and went off. … I was later told by a neighbor that my mother had been found dead, face down in a water tank. … If I had been a little older or stronger I could have rescued her. … Even now I still hear my mother's voice calling me to help her …

A seventeen-year-old, looking for her parents:

I walked past Hiroshima station … and saw people with their bowels and brains coming out. … I saw an old lady carrying a suckling infant in her arms. … I just cannot put into words the horror I felt …

A professional cremator who suffered radiation sickness:

I was all right for three days … but then I became sick with fever and bloody diarrhea. … After a few days I vomited blood also. … There was a very bad burn on my hand, and when I put my hand in water something strange and bluish came out if it, like smoke. After that my body swelled up and worms crawled on the outside of my body.

In 1946, President Truman directed the National Academy of Sciences to conduct investigations of the effects of radiation among survivors of Hiroshima and Nagasaki. The resulting Atomic Bomb Casualty Commission (ABCC) functioned until 1975, when it was replaced by the Radiation Effects Research Foundation, a nonprofit Japanese foundation binationally managed and supported with equal funding by the governments of Japan and the United States. Most notable of the Commission's work was a long-term genetic study on the effects of ionizing radiation and its effects on pregnant women and their children. No widespread evidence of genetic damage was found, although some instances of microcephaly and mental retardation in children exposed in utero did turn up.

As to the effects of the bombs on the Japanese decision to surrender, the USSBS report came to mixed conclusions. As far as public morale went, it was apparent that there was a substantial effect only within about 40 miles of Hiroshima and Nagasaki, likely a result of censorship and lack of mass communication. While the bombs had more effect on the thinking of government leaders, the report concluded that (excerpted).

> It cannot be said, however, that the atomic bomb convinced the leaders who effected the peace of the necessity of surrender. The decision to surrender, influenced in part by knowledge of the low state of popular morale, had been taken at least as early as 26 June at a meeting of the Supreme War Guidance Council in the presence of the Emperor. … The atomic bombings considerably speeded up these political maneuverings within the government. … The bombs did not convince the military that defense of the home islands was impossible, if their behavior in government councils is adequate testimony. It did permit the government to say, however, that no army without the weapon could possibly resist an enemy who had it, thus saving "face" for the Army leaders … There seems little doubt, however, that the bombing of Hiroshima and Nagasaki weakened their inclination to oppose the peace group. … It is apparent that in the atomic bomb the Japanese found the opportunity which they had been seeking, to break the existing deadlock within the government over acceptance of the Potsdam terms.

9.6   The Aftermath

In the United States, demand for information on the Manhattan Project by media outlets and the public following the bombings was voracious. Not surprisingly, Groves had anticipated this, and had been laying groundwork to deal with the onslaught. In early 1944, he had discussed with James Conant the necessity of having some account of the Project ready for release upon the successful use of an atomic bomb, and in April of that year he asked Henry Smyth of Princeton University to take on the task of preparing a report. The purpose of the report was not only to satisfy the public's demand for information, but also to make clear what information Project employees could disclose. Groves exempted Smyth from his usual compartmentalization rules in order that he could gather information from all parts of the Project, and Richard Tolman was appointed to review the report to ensure that no security concerns were breached. Smyth completed the report on July 28, 1945. Before the Hiroshima bombing, Groves had a thousand copies printed up using top-secret reproduction facilities at the Pentagon. Despite some misgivings that it might help the Russians, Stimson recommended release of the report on August 2, and President Truman gave his own clearance on the

9th. The report was released for use by radio broadcasters after 9:00 p.m. on August 11, and for the Sunday-morning newspapers of August 12.

The formal title of Smyth's report is *Atomic Energy for Military Purposes: The Official Report on the Development of the Atomic Bomb under the Auspices of the United States Government, 1940–1945*. The original public version was published by Princeton University Press; it is now available online and has come to be known as the *Smyth Report*. While the report does not reveal any information regarding the actual construction of a nuclear weapon, what it did disclose was remarkable given the secrecy with which the Project was pursued. After chapters dealing with background physics, readers were informed of general ideas of critical size and the use of a tamper, how to separate isotopes and produce plutonium, and the idea of assembly via a target/projectile arrangement. Implosion was not discussed. Smyth's report was not intended for broad public consumption, but rather, as stated in its Preface, "to be intelligible to scientists and engineers generally and to other college graduates with a good grounding in physics and chemistry". In a summary section which alludes to the political and social questions raised by the development of the bomb, its appeal to public education is still worth contemplating:

In a free country like ours, such questions should be debated by the people and decisions must be made by the people through their representatives. This is one reason for the release of this report. It is a semi-technical report which it is hoped men of science in this country can use to help their fellow citizens in reaching wise decisions. The people of the country must be informed if they are to discharge their responsibilities wisely.

Groves' own attitude, as expressed in a memo he later wrote for his files, was surprisingly liberal:

Maintaining security is always a losing battle in the end. ... No one can predict exactly the scientific developments of the next decade or two, but it can be assumed that most of them will come from the minds of young men working untrammeled and undirected, with full access to information, in an atmosphere of freedom. ... America's capacity to win wars with new weapons ... depends on the general scientific, technical, and industrial strength of the country, not on secret researches in either private or government laboratories. ... Therefore we should put our trust in continued scientific progress rather than solely in the keeping of a secret already attained.

9.7   Farm Hall: The German Reaction

After being shuttled between various holding areas in Europe, the ten lead-
ing German scientists rounded up at the end of the war in Europe (Erich
Bagge, Kurt Diebner, Walther Gerlach, Otto Hahn, Paul Harteck, Werner
Heisenberg, Horst Korsching, Max von Laue, Carl von Weizsäcker, and Karl
Wirtz) were flown to England on July 3, and held incommunicado for six
months at Farm Hall, a country estate near Cambridge used as a safe house
by British intelligence. Formally, this was dubbed *Operation Epsilon*. Under
British law, six months was the longest a person could be "detained at His
Majesty's pleasure" without charge; the two months in Europe did not count
under British law. Before their arrival, Reginald Jones had the rooms bugged
with hidden microphones to record the internees' conversations. The recordings were made on shellacked metal disks, which were translated and tran-
scribed by a team of eight listeners before the disks were recycled for further
use. Transcripts of sensitive material were sent directly to Groves. In all, 153
pages of transcripts were produced; one listener estimated that only about
10% of conversations were recorded. The Farm Hall transcripts were declassified in 1992, and have since been analyzed extensively (see in particular
Jeremy Bernstein's *Hitler's Uranium Club* and David Cassidy's *Farm Hall and
the German Atomic Project of World War II*), and have formed the basis of at
least two plays.

Farm Hall, date unknown. Source https://
upload.wikimedia.org/wikipedia/commons/c/
c0/FarmHallLarge.jpg

On the afternoon of August 6, the internees' handler, Major T. H. Rittner, informed Otto Hahn about the bombing of Hiroshima. Hahn was shattered by the news, feeling responsible for the deaths of tens of thousands of people. Rittner calmed Hahn with "considerable alcoholic stimulant", after which Hahn went down to dinner and announced the news to his companions. The resulting conversation reflected the German scientists' growing realization of how far behind the Allies they in fact were, revisited Heisenberg's muddled conception of critical mass, and, most strikingly, revealed the development of a self-serving rationale for the failure of their own program:

Gerlach:	Would it be possible that they have got an engine running fairly well, that they have had it long enough to separate 93?
Hahn:	I don't believe it.
Heisenberg:	All I can suggest is that some dilettante in America who knows very little about it has bluffed them by saying "If you drop this it has the equivalent of 20,000 tons of high explosive," and in reality doesn't work at all.
Hahn:	At any rate, Heisenberg, your just second-raters and you might as well pack up.

...

von Weizsäcker:	I think it's dreadful of the Americans to have done it. I think it is madness on their part.
Heisenberg:	One can't say that. One could equally well say "That's the quickest way of ending the war."

Hahn: That's what consoles me.

Heisenberg: I still don't believe a word about the bomb, but I may be wrong. I consider it perfectly possible that they have about ten tons of enriched uranium, but not that they can have ten tons of pure U-235.

Hahn: I thought one needed only very little 235.

Heisenberg: If they only enrich it slightly, they can build an engine which will go but with that they can't make an explosive which will –

Hahn: But if they have, let us say, 30 kg of pure 235, couldn't they make a bomb with it?

Heisenberg: But it still wouldn't go off, as the mean free path is still too big.

Hahn: But tell me why you used to tell me that one needed 50 kg of 235 in order to do anything. Now you say one needs two tons.

Heisenberg: I wouldn't like to commit myself for the moment, but it is certainly a fact that the mean free paths are pretty big....

Wirtz: I would bet that it is a separation by diffusion with recycling.

...

Later that evening, the group listened to an official announcement of the bombing on the BBC. The conversation resumed:

Harteck: It is a fact that an explosive can be produced either by means of the mass spectograph - we would never have done it, as we could never have employed 56,000 workmen ...

...

Heisenberg: We wouldn't have had the moral courage to recommend to the government in the spring of 1942 that they should employ 120,000 men just for building the thing up.

von Weizsäcker: I believe the reason we didn't do it was because all the physicists didn't want to do it, on principle. If we had all wanted Germany to win the war we would have succeeded.

Hahn: I don't believe that, but I am thankful we didn't succeed.

von Weizsäcker's argument was latter dubbed by von Laue as the scientists' *Lesart*, or "version": that they knew how to make a bomb, but did not do so on principle. In a letter written in 1959, von Laue related that (translated) "Later, during the table conversation, the version was developed that the German atomic physicists really had not wanted the atomic bomb, either because it was impossible to achieve it during the expected duration of the war or because they simply did not want to have it at all. The leader in these discussions was von Weizsäcker. I did not hear the mention of any ethical point of view. Heisenberg was mostly silent."

Later during the night of August 6, Heisenberg and Hahn were speaking privately, and the former made the stunning claim that he had never bothered to work out the critical mass for U-235:

Hahn: They can't make a bomb like that once a week.

Heisenberg: No, I rather think Harteck was right and that they just put up a hundred thousand mass spectographs or something like that. If each spectograph can make one milligram a day, they have got a hundred grams a day ... That would give them 30 kg a year.

Hahn: Do you think they would need as much as that?

Heisenberg: I think so certainly, but quite honestly I have never worked it out, as I never believed one could get pure 235 …

Not until August 14 did Heisenberg produce a calculation that reproduced the essence of a diffusion analysis for estimating the critical mass.

In the days immediately following Hiroshima and Nagasaki, public opinion in the United States was strongly in favor of the bombings. A Gallup poll taken between August 10 and 15, 1945, showed 85% of respondents approving use of the bomb, 10% disapproving, and 5% having no opinion. The next Gallup poll on the issue, taken in 1990 (there were apparently none conducted between 1946 and 1989—the period of the Cold War), had approval at 53% and disapproval at 41%; in 2005 the numbers were 57% and 38%. The 2005 poll indicated that 80% of respondents felt that dropping the bombs saved American lives by shortening the war, but, curiously, 47% felt that dropping the bombs ultimately cost *more* Japanese lives than would have been lost had the war continued.

As the immediacy of the war faded and the implications of the bombs began to become more deeply appreciated, second-guessing as to the necessity of using them began to arise. A significant factor in this evolution was the publication of an article titled *Hiroshima* in *The New Yorker* magazine in August, 1946, by journalist John Hersey; it soon became a best-selling book. In direct, understated prose, Hersey described the stories of six survivors of the bombing of that city. For many Americans, this was their first exposure to the human costs of nuclear warfare.

In response to concern that the United States had callously deployed an inhumane weapon, individuals involved in the Manhattan Project soon began telling their side of the story. In the December, 1946, edition of *The Atlantic Monthly*, Karl Compton published a three-page article aimed at refuting what he described as the "wishful thinking among those after-the-event strategists who now deplore the use of the atomic bomb on the ground that its use was inhuman or that it was unnecessary because Japan was already beaten." Some excerpts:

> It is easy now, after the event, to look back and say that Japan was already a beaten nation, and to ask what therefore was the justification for the use of the atomic bomb to kill so many thousands of helpless Japanese in this inhuman way; furthermore, should we not better have kept it to ourselves as a secret weapon for future use, if necessary? This argument has been advanced often, but it seems to me utterly fallacious. … I believe, with complete conviction,

that the use of the atomic bomb saved hundreds of thousands—perhaps several millions—of lives, both American and Japanese; that without its use the war would have continued for many months; that no one of good conscience knowing, as Secretary Stimson and the Chiefs of Staff did, what was probably ahead and what the atomic bomb might accomplish could have made any different decision.

Compton offered arguments as to the role of the bomb in accelerating the Japanese surrender:

(1) Some of the more informed and intelligent elements in Japanese official circles realized that they were fighting a losing battle … These elements, however, were not powerful enough to sway the situation against the dominating Army organization … (2) The atomic bomb introduced a dramatic new element into the situation, which strengthened the hands of those who sought peace … (3) When the second atomic bomb was dropped, it became clear that this was not an isolated weapon, but that there were others to follow. With dread prospect of a deluge of these terrible bombs and no possibility of preventing them, the argument for surrender was convincing.

By far the most influential such article was one which appeared in the February, 1947, edition of *Harper's Magazine* under Henry Stimson's name, although it was actually written by Stimson and a number of others. Stimson opened by describing his April 25 meeting with Truman and Groves, the work of the Interim Committee and the Scientific Panel, estimates of Japanese force levels in the summer of 1945, his July 2 "Proposed Program for Japan," and, like Compton, details of the surrender process which were theretofore largely unknown to the public. He then offered some reflections (excerpted):

But the atomic bomb was more than a weapon of terrible destruction; it was a psychological weapon. … The bomb thus served exactly the purpose we intended. The peace party was able to take the path of surrender, and the whole weight of the Emperor's prestige was exerted in favor of peace. … I cannot see how any person vested with such responsibilities as mine could have taken any other course or given any other advice to his chiefs. … My chief purpose was to end the war in victory with the least possible cost in the lives of men in the armies which I had helped to raise. In light of the alternatives which, on a fair estimate, were open to us I believe that no man, in our position and subject to our responsibilities, holding in his hands a weapon of such possibilities for accomplishing this purpose and saving those lives, could have failed to use it and afterwards looked his countrymen in the face.

Stimson came in for no small amount of criticism over the fact that a number of American government officials felt that a surrender might have been possible as early as June had America been willing to clarify its position on the fate of the Emperor. Groves' opinion, expressed shortly after the end of the war, was not surprising:

> I have no qualms of conscience about the making or using of it. It has been responsible for saving perhaps thousands of lives. ... From an official standpoint I knew its success would be greatly to our advantage and from a personal standpoint it might save my own son.

In the words of Groves' biographer, Robert Norris:

> The bomb was not necessary to end the war, but it was critical in ending it when it did. Had the bombs taken longer to prepare, history might have turned out quite differently. ... What we do know is that Groves succeeded in building atomic bombs by July 1945; that the two dropped on Japan concentrated certain tendencies and forces at work within the ruling circles of Japan; and that the war ended on August 14. All the rest is speculation.

On October 16, Robert Oppenheimer resigned as Director of Los Alamos. Groves presented the Laboratory with a Certificate of Appreciation from the Secretary of War, and Oppenheimer's addressed the assembled crowd:

> It is with appreciation and gratitude that I accept from you this scroll for the Los Alamos Laboratory, for the men and women whose work and whose hearts have made it. It is our hope that in years to come we may look at this scroll, and all that it signifies, with pride.

> Today that pride must be tempered with a profound concern. If atomic bombs are to be added as new weapons to the arsenals of a warring world, or to the arsenals of nations preparing for war, then the time will come when mankind will curse the names of Los Alamos and Hiroshima.

> The peoples of this world must unite or they will perish. This war, that has ravaged so much of the earth, has written these words. The atomic bomb has spelled them out for all men to understand. Other men have spoken them, in other times, of other wars, of other weapons. They have not prevailed. There are some, misled by a false sense of human history, who hold that they will not prevail today. It is not for us to believe that. By our works we are committed to a world united, before this common peril, in law, and in humanity.

Epilogue

The Manhattan Project fulfilled its goal of helping America and its allies win the war and avoid an invasion of Japan. Democracy and capitalism emerged as the dominant political system of the world for the remainder of the twentieth century. But the bombs also triggered the Cold War and the nuclear proliferation predicted in the Jeffries, Tolman, and Franck reports. The nuclear genie had emerged from its bottle, likely never to be re-contained. The Cold war would probably have arisen in any event, but the presence of nuclear weapons surely raised the threat to a truly existential level.

Despite efforts by the newly-formed United Nations to control the availability of fissile materials and the development of nuclear weapons, the United States and Russia—and eventually other countries—had no intention of allowing their nuclear autonomy to be restricted. In America, planning got underway soon after the end of the war for a series of tests to determine the effects of atomic bombs on naval vessels; the resulting *Crossroads* tests were conducted on the Pacific island of Bikini on July 1 and 25, 1946, even while the first phase of UN arms-control negotiations were underway. The Soviet Union's first graphite reactor, code-named F-1 (Physics-1), essentially a copy of a reactor built at Hanford for testing fuel slugs, went critical on the evening of Christmas Day, 1946, at a power of 10 W. Between the first criticality of Enrico Fermi's CP-1 reactor and the *Trinity* test, 938 days elapsed. The Russians essentially duplicated this feat, detonating their first atomic bomb on August 29, 1949, exactly 978 days after the startup of F-1.

© Springer Nature Switzerland AG 2020
B. C. Reed, *Manhattan Project*,
https://doi.org/10.1007/978-3-030-45734-1

Known to the Soviets as RDS-1 (Joe-1 in the West), this device was a plutonium bomb identical to *Fat Man*; the design was based on information transmitted by Klaus Fuchs. Fission products from the test were picked up by B-29 bombers equipped with air-sampling devices flying weather reconnaissance missions over Japan, Alaska, and the North Pole; they were also detected in rainwater collected in Alaska. President Truman announced the test publicly on September 23.

In both the United States and the Soviet Union, attention also turned to the possibility of the more powerful "thermonuclear" fusion-based weapons which had so captivated the attention of Edward Teller since 1942. With the Joe-1 test and the 1948/49 Berlin blockade, political pressure on Truman was intense, and despite concerns of many Manhattan Project scientists that a weapon 1,000 times as powerful as a fission bomb would have no realistic military mission—even if it could be built at all—Truman announced on January 31, 1950, that he was ordering the Atomic Energy Commission (which had come into existence in 1947) to "continue work on all forms of atomic weapons, including the so-called hydrogen or super bomb." Seemingly insurmountable technical difficulties were overcome, and on October 31, 1952, the *Ivy Mike* test saw the detonation of America's first full-scale fusion weapon. This device achieved a yield of 10.4 megatons (MT), over 400 times as powerful as *Fat Man*. With a weight of 60 tons, however, this was not a deliverable weapon. In this sense the Soviet Union had leapt ahead of the United States when it tested a deliverable fusion weapon on August 12, 1953. The first deliverable American thermonuclear device was detonated in the *Castle Bravo* test of February 28, 1954, and yielded 15 MT, about three times what was expected. Fallout from *Castle Bravo* covered some 7,000 square miles, and contaminated the crew of a Japanese fishing vessel, the *Lucky Dragon 5*. *Castle Bravo* paled in comparison to the Soviet Union's 60-megaton "Tsar Bomba," detonated in October, 1961.

Ivy Mike test, October 31, 1952. Source: http://commons. wikimedia.org/wiki/File:Ivy_Mike_-_ mushroom_cloud.jpg

Between 1949 and 1964, Britain, France, and China also joined the nuclear club. The first British test was conducted on the Montebello Islands off Western Australia, and the first French test was conducted in the Sahara Desert in Algeria. Britain tested its first thermonuclear device in 1957; the Chinese followed in 1967, and the French in 1968. The United States, Russia, the United Kingdom, France, and China are now known as the "primary five," or "P5" nuclear weapons states.

Some nuclear milestones for the P-5 Nuclear States

Parameter	United States	USSR/Russia	Britain	France	China
Date of first test	16-Jul-45	29-Aug-49	3-Oct-52	13-Feb-60	16-Oct-64
First test yield (kt)	21	22	25	60–70	20–22
First test name	Trinity	RDS-1/Joe-1	Hurricane	Gerboise Bleue	596
Peak number warheads/year attained	31,255 1967	45,000 1986	520 1975–80	540 1993	240 2012?
Number of tests/Detonations[a]	1,030/1,125	715/969	45/?	210/?	45/?
Total warheads built	66,500	55,000	850	1,260	750
Warheads in stockpile[b]	3,800	4,500	225	300	290
Date of last test	23-Sep-92	24-Oct-90	26-Nov-91	27-Jan-96	29-Jul-96
Largest test, megatons[c]	15/5	50/2.8–4	3/<150 kt	2.6/120 kt	4/420 kt
Total megatonnage expended[c]	141/38	247/38	8/0.9	10/4	21.3/1.3

Sources: Robert S. Norris and Hans M. Kristensen, Nuclear pursuits, 2012. Bulletin of the Atomic Scientists **68**(1), 94–98 (2012); John R. Walker, British Nuclear Weapons Stockpiles, 1953–78. RUSI Journal **156**(5), 66–72 (2011); See also sources listed in Table below
[a]Some tests involved simultaneous detonation of more than one warhead
[b]Stockpile numbers include deployed and reserve units, but not those awaiting dismantelment
[c]Atmospheric/underground

A decade after China tested its first bomb, India did so (1974); this prompted Pakistan to follow suit in 1998. North Korea detonated its first nuclear device in 2006, although this test may have been a fizzle. Israel, which benefitted from close ties to France, is widely regarded to have acquired nuclear weapons in the 1960s, but maintains a policy of official ambiguity regarding whether it does or does not possess such weapons. While some other countries (Iraq, Libya) had nuclear weapons development programs but abandoned them for various reasons, South Africa is the only country known to have developed nuclear weapons only to later dismantle them. That country had constructed six gun-type bombs with another in preparation in 1990 when President Frederik de Klerk ordered the end of the weapons program in advance of the country's shift from the Apartheid government to control by the African National Congress. At this writing, the situation in Iran remains fluid.

As advances in weapons physics led to the development of lighter and more compact designs with a wide range of yields, the spectrum of missions to which nuclear weapons could be applied grew rapidly. In America, each branch of the armed forces wanted its own piece of the nuclear action. In a 2009 article, Robert Norris and Hans Kristensen estimated that between 1945 and 2009, the United States produced over 66,500 nuclear warheads of 100 different basic types and variants of types: On average, about 1,000 warheads per year, or almost three per day over seven decades. These included weapons to be carried on bombers; mounted on land, surface, and submarine-based ballistic missiles; in landmines; on short-range artillery rockets; on ground, air, and submarine-launched cruise missiles; on anti-submarine rockets; in torpedoes; and on air-to-air, air-to-ground, and earth-penetrating missiles. Some battlefield-scale tactical nuclear devices were small enough to be carried by an individual. Such a plethora of designs naturally demanded an extensive testing program. Between 1945 and 1992, the United States conducted 1,030 nuclear tests, plus an additional 24 in conjunction with the United Kingdom. The United States ceased nuclear testing in 1992; the Soviet Union had done so in 1990 after conducting over 700 tests. Depending on a warhead's anticipated mission, tests were configured to subject a variety of structures, vehicles, vessels, and environments to the effects of nuclear explosions. Detonations were conducted at surface level, underground, underwater, and at high altitudes via platforms such as airdrops, balloons, barges, and rockets. The 1.4-megaton *Starfish Prime* test of July, 1962, was detonated at an altitude of 400 km, and resulted in the discovery of the electromagnetic pulse phenomenon, which caused electrical damage some 900 miles away in Hawaii. Of the United States' 1,030 tests, 210 were atmospheric, 815 were underground, and 5 were underwater. The most frequently-used test location was the Nevada Test Site, which saw 928 tests involving 1,021 detonations. As of 2019, individual-country test scores stood at (US, USSR, France, UK, China, India, Pakistan, North Korea) = (1,030, 715, 210, 45, 45, 4, 2, 6).

The global inventory of deployed and readily-deployable nuclear weapons began to grow dramatically in the 1950s. The number peaked in 1986, when just over 69,000 were available for use, of which over 98% were in American and Russian hands. Since that time, reductions in numbers due to various arms-control treaties (particularly the Strategic Arms Reductions Treaties) and unilateral withdrawals from various venues on the parts of both America and Russia have brought the inventory down to about 9,500 weapons, not including many awaiting dismantlement. Current United States Inter-Continental Ballistic Missile (ICBM) warheads have yields of 300 and 335 kt, while Submarine-Launched Ballistic Missile (SLBM) warheads have yields of 100 and 455 kt. Warheads carried on smaller aircraft (so-called "tactical" or "non-strategic" weapons) have yields that can vary from a few tenths of a kiloton up to about 170 kt. Current Russian maximum ICBM and SLBM yields are estimated at 800 and 100 kt, respectively. Pakistan is thought to have the world's fastest-growing nuclear arsenal, which could exceed that of Britain by the early 2020s. China is believed to be the only P-5 state that is increasing its arsenal.

The cost of these programs has been staggering. A 1995 study by the Brookings Institution of Washington estimated costs associated with the U. S. nuclear weapons program from 1940 onwards at $5.8 trillion in 1996 dollars. This included research, development, testing, deployment, command and control, defense and dismantling of weapons systems, waste cleanup, compensation for persons harmed by the production and testing of nuclear weapons, estimated future costs for storing and disposing of waste, and dismantling and disposing of surplus materials. The final figure represented 29% of all military spending between 1940 and 1996, and 11% of all government expenditures during that period. Spending on nuclear weapons exceeded all other government spending except for non-nuclear defense and social security. New warhead production ceased in the United States in 1992, but existing devices regularly undergo modifications and refurbishments, and America and Russia (at least) are planning extensive modernization programs. The prospect of a renewed arms race and the possibility that one of the major nuclear powers could resume testing is very real. Nuclear weapons will be a factor in the global power balance for decades to come.'

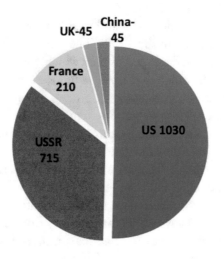

Distribution of 2045 "P-5" postwar nuclear tests 1946–1996. The UK figure includes 24 tests conducted underground in the United States. Not included here are one Indian test in 1974, three Indian tests in 1998 comprising five claimed detonantions, two Pakistani tests in 1998, and six North Korean tests (2006, 2009, 2013, 2016(2), and 2017). Data from Natural Resources Defense Council

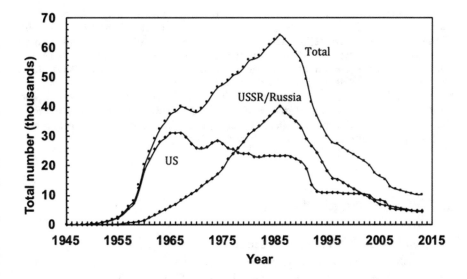

*Estimated global nuclear weapons inventories, 1945–2013. Included in the Total curve are the "smaller" nuclear powers: United Kingdom, France, China, Israel, India, and Pakistan. Data from Hans M. Kristensen and Robert S. Norris, "Global nuclear weapons inventories, 1945–2013" [Bulletin of the Atomic Scientists **69**(5), 75–81 (September 2013)]. US and Russian curves do not include "retired" warheads awaiting dismantlement*

Estimated nuclear weapons deployments as of 2020

Country	Deployed warheads	Stockpiled warheads	Retired warheads	Total inventory
United States	1,750	3,800	2,000	5,800
Russia	1,600	4,490	2,000	6,490
United Kingdom	160	225	–	225
France	290	300	–	300
China	–	290	Few	290
India	–	120	–	120
Pakistan	–	~140	–	~140
Israel	–	80	–	80
North Korea	–	~60	–	~60
Total	3,800	9,505	4,000	13,505

For the United States, deployments comprise approximatey 1,300 warheads on ballistic missiles, 300 at heavy-bomber bases, and 150 tactical weapons deployed in Europe. The stockpile figure for Russia includes an estimated 1,850 tactical weapons. Sources: Hans M. Kristensen and Robert S. Norris, "Global nuclear weapons inventories, 1945–2013" [Bulletin of the Atomic Scientists **69**(5), 75–81 (2013)]; "Worldwide deployments of nuclear weapons, 2014" [Bulletin of the Atomic Scientists **70**(5), 96–108 (2014)]; "Israeli nuclear weapons, 2014," [Bulletin of the

Atomic Scientists **70**(6), 97–115 (2014)]; "Indian nuclear forces, 2017," [Bulletin of the Atomic Scientists **73**(4), 205–209 (2017)]; "Pakistani nuclear forces, 2016," [Bulletin of the Atomic Scientists **72**(6), 368–376 (2016)]. Hans M. Kristensen and Matt Korda, "Russian nuclear forces, 2019" [Bulletin of the Atomic Scientists **75**(2), 73–84 (2019)]; "United States nuclear forces, 2020 [Bulletin of the Atomic Scientists **76**(1), 46–60 (2020)]; "Chinese nuclear forces, 2019" [Bulletin of the Atomic Scientists **75**(4), 171–178 (2019)]; The figure of ~60 for North Korea is from a Defense Intelligence Agency estimate of August, 2017. In March, 2018, the US Department of Defense announced that the US stockpile as of September 30, 2017, was 3,822 warheads. See http://open.defense.gov/Portals/23/Documents/frddwg/2017_Tables_UNCLASS.pdf

A Technological Miracle

The Manhattan Project changed the course of history. For many of the scientists involved, their wartime work represented the most exciting and dramatic time of their careers. As chemist Joseph Hirschfelder put it:

> I believe in scientific-technological miracles since I saw one performed at Los Alamos during World War II. The very best scientists and engineers were enlisted in the Manhattan Project. They were given overriding priorities. They got everything which they deemed essential to their program; the cost was unimportant. They had the full cooperation of everyone and they, themselves, devoted long hours in mixing together their ingenuity and technical skills. … In a period of two-and-a-half years, they produced a miracle—an atomic bomb which creates temperatures of the order of 50,000,000 °C … pressures of the order of 20,000,000 atmospheres … while unleashing the tremendous energy stored in the atomic nuclei. … At Los Alamos during World War II there was no moral issue with respect to working on the atom bomb. … The whole fate of the civilized world depended upon our succeeding before the Germans! … It is an open question as to whether the world is better or worse for our having made the atom bomb. … After Otto Hahn's and Fritz Strassmann's discovery it became evident that sooner or later some country would make an atom bomb. If an atom bomb had not been made and detonated in World War II, the world would be unprepared to cope with the tremendous threat of nuclear warfare. … warfare is no longer a rational means of settling differences between nations.

In late 2014, the Manhattan Project National Historical Park Act was signed it into law by President Barack Obama. The legislation provides an inventory of properties and historic districts to be included in the Park, including the site where the *Trinity* bomb was assembled, the building where Louis Slotin received his fatal dose of radiation, the houses where Robert

Oppenheimer and Hans Bethe lived, the X-10 reactor, calutrons in the Y-12 pilot plant building, B Reactor, and the 221-T Queen Mary building. Future generations will be able to view, touch, and reflect upon artifacts from one of the most pivotal undertakings in human history.

Brief Biographies

Abelson, Philip (1913–2004) American physical chemist; co-discoverer of neptunium with Edwin McMillan. While working at the Carnegie Institution of Washington and the Naval Research Laboratory during the war, Abelson developed a method of producing uranium hexafluoride ("hex") on a large scale, and pioneered the development of the liquid thermal diffusion technique for enriching uranium that was used in the S-50 plant at Oak Ridge.

Akers, Wallace (1888–1954) British chemist and industrialist. As Director of Research at Imperial Chemical Industries, Akers was selected to head the British nuclear research project, code-named "Tube Alloys." In 1942, Akers proposed that the British project be merged with the rapidly-expanding American effort, and argued vigorously but largely unsuccessfully that all information developed in America be shared with Britain. Akers was to head the British Mission to the Manhattan Project but was rejected for this role by the Americans because of his ties to Imperial Chemical; James Chadwick took on this role instead.

Alvarez, Luis (1911–1988) American physicist; Nobel Prize for Physics 1968 for contributions to elementary particle physics. During the war, Alvarez worked on ground-controlled-approach radar, reactors, and on developing detonators for the implosion bomb.

Anderson, Herbert (1914–1988) American nuclear physicist. As a graduate student at Columbia University in 1939, Anderson set up the first

© Springer Nature Switzerland AG 2020
B. C. Reed, *Manhattan Project*,
https://doi.org/10.1007/978-3-030-45734-1

demonstration of fission in the United States. His death at Los Alamos in 1988 occurred on the 43rd anniversary of the *Trinity* test.

Aston, Francis (1877–1945) British chemist/physicist; Nobel Prize for Chemistry 1922 for the discovery of isotopes by mass spectroscopy. Developed the concept of nuclear mass defect.

Becquerel, Antoine Henri (1852–1908) French physicist; discovered radioactivity early 1896, for which he shared the 1903 Nobel Prize for Physics with Marie Curie.

Bethe, Hans (1906–2005) German-American nuclear physicist; Nobel Prize for Physics 1967 for his contributions to the theory of nuclear reactions, particularly his discoveries concerning energy production in stars. At Los Alamos during the war, Bethe headed the laboratory's Theoretical Division, and made significant contributions to the study of implosion physics and initiator design. Bethe witnessed the *Trinity* test from Campañia Hill.

Bohr, Niels (1885–1962) Danish physicist; Nobel Prize for Physics 1922 for his contributions to the theory of atomic structure, especially the development of the first sound mathematical model of the structure of hydrogen. In 1939, Bohr collaborated with American physicist John Wheeler to develop understanding of the mechanism of nuclear fission.

Bothe, Walther (1891–1957) German physicist; Nobel Prize for Physics 1954 for developing the method of "coincidence counting" for studying nuclear reactions. In 1930, Bothe and his student Herbert Becker undertook light-element alpha-bombardment experiments which led to the discovery of the neutron in 1932.

Breit, Gregory (1899–1981) Russian-American theoretical physicist. In 1940, Breit proposed that American scientists censor publications on work which might have military applications. Breit had been in charge of fast-neutron studies under Arthur Compton, bur resigned from the Manhattan Project in 1942 over what he viewed as lax security; he was succeeded by Robert Oppenheimer.

Briggs, Lyman J (1874–1963) As Director of the National Bureau of Standards, Briggs, who had a background in mechanical engineering and physics, was directed by President Roosevelt to Chair the Advisory Committee on Uranium, the first formal United States government body to consider the prospects for nuclear fission as an energy source. Briggs' contributions to the Manhattan Project have tended to be overshadowed by more forceful project administrators such as Vannevar Bush and James Conant.

Bush, Vannevar (1890–1974) American engineer, academic, and government science administrator. Director of the wartime National Defense Research Committee and the Office of Scientific Research and development. Witnessed the *Trinity* test.

Chadwick, James (1891–1974) British physicist; Nobel Prize for Physics 1935 for his discovery of the neutron in 1932. Chadwick played a deep role of the administration of the British atomic program of World War II, and witnessed the *Trinity* test in 1945.

Churchill, Winston (1874–1965) British politician; Prime Minister 1940–45 and 1951–55. As Prime Minister during the war, Churchill, influenced by the advice of Sir Frederick Lindemann, mishandled an invitation from U. S. President Franklin Roosevelt to participate as an essentially equal partner in the development of the atomic bomb. By the time Churchill realized the significance of the work, the American program was far ahead of the British program, and he had to settle for a much more junior-partner status. In his postwar Prime Ministership, Churchill became a strong advocate of Britain developing its own nuclear deterrent as a counterbalance to American and Russian dominance.

Compton, Arthur (1892–1962). American physicist; Nobel Prize for Physics 1927 for his discovery of the Compton effect, wherein X-ray photons are scattered by electrons in accordance with the predictions of relativistic physics. Compton chaired the National Academy of Science Committee on Uranium, whose three reports during 1941 were strongly influential in convincing President Roosevelt to proceed with an atomic bomb program. During the war, Compton directed the Manhattan Project's "Metallurgical Laboratory" at the University of Chicago, where research on chain reactions was carried out.

Conant, James B (1893–1978) American chemist, academic administrator, and government science administrator. As deputy director and later director of the National Defense Research Committee, Conant was deeply involved with the Manhattan Project from its earliest days; he witnessed the *Trinity* test. Conant also served as President of Harvard University (1933–1953), and as the first United States Ambassador to West Germany (1955–1957).

Curie, Irene (1897–1956) French physicist; daughter of Marie and Pierre Curie. Nobel Prize in Chemistry 1935 for co-discovery with her husband Frederick Joliot of induced radioactivity caused by alpha-particle bombardment. These experiments directly stimulated Enrico Fermi to search for induced radioactivity caused by neutron bombardment, which led to the discovery of fission.

Curie, Marie (1867–1934) Polish-French physicist/chemist noted for pioneering studies in radioactivity (which term she coined); discovered radium and polonium. Curie was the first woman to be awarded a Nobel Prize, and was also the first person and only woman to be awarded two Nobel Prizes (Physics 1903; Chemistry 1911).

Einstein, Albert (1879–1955) German-American physicist; Nobel Prize for Physics 1921. Contrary to popular opinion, Einstein's Nobel Prize was not awarded for his controversial theory of relativity, but rather for his contribution to understanding a phenomenon known as the photoelectric effect. He also made significant contributions in the areas of quantum and statistical physics. His involvement in the Manhattan Project was largely limited to signing a letter to President Roosevelt in August, 1939, to alert him to the potentialities of nuclear fission.

Fermi, Enrico (1901–1954) Italian-American physicist; Nobel Prize for Physics 1938 for the creation of new elements by neutron bombardment. Fermi was legendary in the physics community for his command of both theory and experiment: He developed a theory of beta-decay; discovered neutron-induced radioactivity; narrowly missed discovering fission; built the world's first reactor at the University of Chicago; was deeply involved in the design of the of the Hanford plutonium-production reactors; and, in postwar years, contributed to the development of fusion weapons. He witnessed the *Trinity* test, famously estimating the yield of the explosion by observing how the accompanying shock wave distributed small pieces of paper he had thrown into the air.

Frisch, Otto (1904–1979) Austrian-British nuclear physicist. With his aunt Lise Meitner, Frisch correctly interpreted the phenomenon of nuclear fission, and was the first person to set up an experiment to deliberately demonstrate it. Initially skeptical of the prospect of nuclear weapons, he co-authored, with Rudolf Peierls, the "Frisch-Peierls Memorandum" of March, 1940, which initiated the British atomic program. At Los Alamos, Frisch was a member of the British Mission, and invented the "Dragon" experiment for creating a short-lived supercritical chain reaction. Witnessed the *Trinity* test.

Gamow, George (1904–1968) Soviet-American theoretical physicist and cosmologist noted for contributions to theory of alpha-decay, the liquid-drop model of nuclei, and big-bang cosmology. Gamow was one of the organizers of the 1939 Fifth Washington Conference on Theoretical Physics, at which Niels Bohr announced the discovery of fission.

Geiger, Hans (1882–1945) German physicist; co-inventor with Walther Müller of the Geiger–Müller counter, and was involved in the Geiger–Marsden gold-foil scattering experiment which led to Ernest Rutherford's discovery of the nuclear nature of atoms. Geiger died less than two months after the atomic bombings of Hiroshima and Nagasaki.

Greenewalt, Crawford (1902–1993) American chemical engineer. While working for the DuPont corporation, Greenewalt was closely involved with the design of the plutonium production reactors built at Hanford, and witnessed the startup of Enrico Fermi's CP-1 reactor in Chicago on December 2, 1942. Greenewalt served as President of DuPont from 1948 to 1962.

Groves, Leslie Richard (1896–1970) American military engineer; commanding officer of the Manhattan Engineer District. Groves graduated from the United States Military Academy in 1918, and was commissioned into the Army Corps of Engineers. By 1942, he was in charge of all military construction within the United States, most notably the Pentagon building. He was appointed to command the "Development of Substitute Materials" project—the early code name for what became the Manhattan Engineer District—in September, 1942.

Hahn, Otto (1879–1968) German chemist who specialized in the analysis of radioactive elements. Nobel Prize for Chemistry 1944 for the discovery of nuclear fission, which many historians argue should have been shared with Lise Meitner.

Heisenberg, Werner (1901–1976) German theoretical physicist; awarded 1932 Nobel Prize for Physics for contributions to the creation of quantum mechanics. Heisenberg was one of the key personalities of the wartime German nuclear program; historians of physics still debate the extent to which he understood the concept of a critical mass of a nuclear weapon.

Jewett, Frank (1879–1949) American engineer/physicist; first President of Bell Telephone Laboratories (1940–1944). As President of the National Academy of Sciences from 1939 to 1947, Jewett was involved in the creation of the Office of Scientific Research and Development in 1940. In April, 1941, OSRD director Vannevar Bush asked Jewett to facilitate a review of the military prospects of uranium fission, which led to the creation of the (Arthur) Compton Committee.

Lawrence, Ernest O (1901–1958) American physicist. Nobel Prize for Physics 1939 for the invention of the cyclotron. The cyclotron concept was put to use in large-scale uranium-isotope-separating "calutrons" during the Manhattan Project at Oak Ridge, Tennessee. Witnessed the *Trinity* test.

Lindemann, Frederick (1886–1957) German-born British physicist. As personal scientific advisor to Winston Churchill during World War II, Lindemann provided Churchill with valuable statistical advice on the progress of the war, but his lack of familiarity with contemporary scientific developments caused him to sideline advice from more informed sources such as Sir Henry Tizard and to be skeptical of the prospects for nuclear weapons.

Marsden, Ernest (1889–1970) British-born physicist who later spent much of his career in New Zealand. While an undergraduate student at Cambridge University in 1909, Marsden participated in the Geiger–Marsden experiment, which led to Rutherford's 1911 theory of the nuclear structure of atoms.

Marshall, George C (1880–1959) American Army officer and statesman. As Chief of Staff of the Army during all of World War II, Marshall was familiar with the Manhattan Project from its earliest days of Army involvement. As Secretary of State from January, 1947, to January, 1949, Marshall was instrumental in developing the "Marshall plan" to help rebuild western Europe and check the spread of Communism. Marshall also served one year as Secretary of Defense (September, 1950–September 1951), during which he helped rebuild United States armed forces at the time of the outbreak of the Korean War (June, 1950–July, 1953). Marshall was regarded as a brilliant administrator and military planner of great personal integrity.

Marshall, James (1897–1977). American military engineer who was the first Army officer to command what became the Manhattan Engineer District. Marshall was appointed to command the "Development of Substitute Materials" project in June, 1942, but was replaced by Leslie Groves in September of that year. Groves retained Marshall as District Engineer of the project until August, 1943, when he replaced Marshall with Col. Kenneth Nichols.

McMillan, Edwin (1907–1991). American physicist; Nobel Prize for Chemistry 1951 (shared with Glenn Seaborg) for discoveries in the chemistry of transuranium elements. McMillan was particularly involved in the discovery of neptunium. During the war, he worked on radar and sonar before becoming one the founding members of the Los Alamos Laboratory, where he worked on both the gun and implosion-type bombs. McMillan witnessed the *Trinity* test from Campaña Hill.

Meitner, Lise (1878–1968) Austrian-Swedish physicist. Meitner worked closely with Otto Hahn and Fritz Strassmann on experiments which would lead to the discovery of nuclear fission, but was forced to flee Germany in July, 1938, on account of her Jewish heritage. With her nephew Otto Frisch, Meitner correctly interpreted the phenomenon of fission; it is widely felt that she should have shared the Nobel Prize for the discovery with Hahn.

Nichols, Kenneth (1907–2000) American military engineer. Nichols' association with the Manhattan project began in June, 1942, when he was brought into the project by his superior officer, James Marshall. Nichols became Manhattan District Engineer in August, 1943, in which position he was responsible for the development of the Clinton and Hanford Engineer Works. After the war he became the military liaison officer with the Atomic Energy Commission and later general manager of the same organization. In this latter capacity he initiated a security hearing on the loyalty of Robert Oppenheimer, considering him to be a "Communist in every sense except that he did not carry a party card." The hearing resulted in revocation of Oppenheimer's security clearance.

Nier, Alfred (1911–1994) American mass spectroscopist. Nier isolated the first sample of uranium-235, with which that isotope's slow-neutron fissility was verified.

Noddack, Ida (1896–1978) German physicist/chemist; co-discoverer of element 75 (rhenium). Conceived idea of nuclear fission in 1934, but the concept was not taken seriously at the time.

Oliphant, Marcus (1901–2000) Australian-British physicist. As department head at the University of Birmingham, Oliphant received the Frisch-Peierls memorandum which led to the creation of the MAUD committee (of which he was a member) and the British atomic program. In August–September, 1941, Oliphant visited America and prodded scientists there to speed up work on their atomic program.

Oppenheimer, Julius Robert (1904–1967) American theoretical physicist. As director of the Los Alamos Laboratory from 1943 to 1945, Oppenheimer led the effort which produced the *Little Boy* and *Fat Man* nuclear weapons. After the war, Oppenheimer became the Chairman of the General Advisory Committee of the Atomic Energy Commission. Oppenheimer's opposition to the development of fusion weapons garnered intense political opposition, which resulted in his security clearance being revoked in 1954 following a hearing in which his loyalty was brought into question. No evidence has ever surfaced to indicate that Oppenheimer ever acted disloyally in any way.

Peierls, Rudolf (1907–1995) German-British theoretical physicist and co-author of the "Frisch-Peierls Memorandum" of March, 1940, which initiated the British atomic program. Peierls was head of the British Mission at Los Alamos; he witnessed the *Trinity* test.

Perrin, Francis (1901–1992) French physicist. Published the first analysis of criticality conditions in neutron multiplication for a chain reaction in uranium in May, 1939. In 1951, Perrin was appointed High Commissioner of the French Commissariat à l'énergie atomique (Atomic Energy Commission; CEA), and became a strong advocate for an independent French nuclear force.

Placzek, George (1905–1955) Czech physicist; contributed ideas to Niels Bohr's understanding of the mechanism of fission, and was member of the British Mission to the Manhattan Project.

Röntgen, Wilhelm (1845–1923) German physicist; discovered X-rays in November, 1895, for which he was awarded the first Nobel Prize for Physics in 1901.

Roosevelt, Franklin (1882–1945) Thirty-second President of the United States, 1933–1945. The Manhattan Project was initiated under Roosevelt as the Advisory Committee on Uranium in the fall of 1939. Roosevelt was kept briefed on the project during the war by Vannevar Bush, and at the time of his death in April, 1945, was aware that bombs would be ready in a few months. The decision of how the new weapons would be used fell to his successor, Harry Truman.

Rutherford, Ernest (1871–1937) New Zealand-British physicist; Nobel Prize for Chemistry 1908 for research into the disintegration and chemistry of radioactive elements. Regarded as the father of nuclear physics, Rutherford discovered radioactive half-life; coined the terms "alpha" and "beta" radiation; determined that alpha particles are helium nuclei; discovered artificial transmutation; developed the nuclear model of the atom; and speculated on the existence of neutrons years before their discovery. Element 104, rutherfordium, is named in his honor.

Sachs, Alexander (1893–1973) American economist and banker. As an advisor to President Franklin Roosevelt, Sachs delivered to Roosevelt the August, 1939, Einstein-Szilard letter advising that uranium fission could be of military importance. His briefing to Roosevelt led to the formation of the Uranium Committee and eventually the Manhattan Project.

Seaborg, Glenn (1912–1999) American chemist; Nobel Prize for Chemistry 1951 (shared with Edwin McMillan) for discoveries in the chemistry of transuranium elements. Seaborg was involved in the discovery of no less than 10 transuranic elements. During the Manhattan Project, Seaborg was based largely at the University of Chicago, where he researched techniques for isolating plutonium on kilogram scales. After the war, he served as Chairman of the Atomic Energy Commission (1961–1971). Element 106, seaborgium, is named in his honor.

Segrè, Emilio (1905–1989) Italian-American physicist; Nobel Prize for Physics 1959 for the discovery of the anti-proton (shared with Owen Chamberlain). Segrè was one of Enrico Fermi's students involved with the discovery of neutron-induced radioactivity; at Los Alamos in 1944, his discovery of the high spontaneous-fission rate of reactor-produced plutonium prompted the development of the implosion method for the *Fat Man* plutonium bomb.

Sengier, Edgar (1879–1963) Belgian business executive. As director of the Union Minière du Haut-Katanga (Belgian Congo) mining company, Sengier gave the United States access to uranium ore that he had stockpiled in New York in anticipation of its becoming a vital military commodity. Union Minière ore would supply some 70% of the Manhattan Project's uranium.

Serber, Robert (1909–1997) American physicist. Serber gave a series of orientation lectures to scientists arriving at Los Alamos in April, 1943, which were later declassified and published as *The Los Alamos Primer*.

Soddy, Frederick (1877–1956) British radiochemist who in collaboration with Ernest Rutherford determined that naturally radioactive elements transmute to other elements. In 1913, Soddy coined the term "isotopes" to describe forms of elements with the same number of protons but different masses, later understood to be due to differing numbers of neutrons in their nuclei.

Somervell, Brehon (1892–1955) Commanding General of the Army Service Forces during World War II. Regarded as a supremely competent logistics officer, Somervell and his Chief of Staff, Wilhelm Styer, were responsible for appointing Leslie Groves to command what became the Manhattan Engineer District.

Stimson, Henry (1867–1950) American statesman and political figure. Trained as a lawyer, Stimson held a number of government positions, including two terms as Secretary of War (1911–13, 1940–45). During the latter he oversaw the Manhattan Project, personally briefing Harry Truman when he became President upon Franklin Roosevelt's sudden death in April, 1945.

Strassmann, Fritz (1902–1980) German chemist; co-discoverer with Otto Hahn of nuclear fission.

Szilard, Leo (1898–1964) Hungarian-American physicist-inventor. Szilard is best known for his participation in drafting a letter to President Franklin Roosevelt to warn him of the military possibilities of nuclear fission. Szilard convinced Albert Einstein to sign the letter, knowing that Roosevelt would recognize Einstein's name.

Teller, Edward (1908–2003) Hungarian-American theoretical physicist. Teller was one of the organizers of the 1939 Fifth Washington Conference on Theoretical Physics at which Niels Bohr announced the discovery of fission. For a while he headed the Theoretical Division at Los Alamos, but was replaced with Hans Bethe. After the war, Teller became a vigorous advocate of further nuclear weapons development, particularly fusion-based hydrogen bombs. Following his testimony against Robert Oppenheimer in the latter's 1954 security hearing, Teller was ostracized by much of the scientific community but retained powerful allies in government defense-policy circles.

Thomson, George P (1892–1975) British physicist; Nobel Prize for Physics 1937 (shared with Clinton Davisson) for his experimental discovery of electron diffraction. Thomson chaired the 1941 MAUD committee, which concluded that a fission bomb was feasible and could be constructed in time for use in the war.

Tizard, Henry (1885–1959) British chemist and inventor. Tizard was involved in the development of radar, and as Chairman of the government's Committee on the Scientific Survey of Air Warfare, received the Frisch-Peierls memorandum in the Spring of 1940. Tizard asked George P. Thomson to convene a committee to consider the possibility of fission bombs; this group became the MAUD committee. Tizard favored linking British nuclear weapon research efforts to those in America, but was resisted by Frederick Lindemann; Tizard resigned from his Air Ministry post in frustration later in 1940.

Truman, Harry (1884–1972) Thirty-third President of the United States, 1945–1953. Truman had been Vice President for only a few months when President Franklin Roosevelt died suddenly in April, 1945. Largely unaware of the Manhattan Project, Truman faced the decision of whether or not to use atomic bombs against Japan in the hope of inducing the Japanese to surrender and thereby avoid what would likely have been a costly and drawn-out invasion of the Japanese home islands.

Turner, Louis (1898–1977) American physicist. Turner prepared an extensive review article on nuclear fission which was published in the *Physical Review* in 1940, and conceived the idea that element 94 (plutonium) could serve as a bomb material. He developed this concept just as plutonium was first being synthesized; his paper was withheld from publication until 1946.

Urey, Harold (1893–1981) American physicist; Nobel Prize for Chemistry 1934 for his discovery of deuterium. Urey played a significant role in the development of the gaseous diffusion method of uranium enrichment during the Manhattan Project.

von Halban, Hans (1908–1964) French physicist. In early 1939, von Halban and co-workers published one of the first measurements of the number of secondary neutrons released in uranium fission. von Halban initially headed a reactor research group in Montreal during the war, but was replaced by John Cockcroft.

Wheeler, John (1911–2008) American theoretical physicist. In 1939, Wheeler collaborated with Niels Bohr on an extensive analysis of the physics of fission. During the war, Wheeler was deeply involved with the design of the Hanford plutonium-production reactors. In the postwar era, Wheeler was involved with the development of fusion weapons. His personal research interests involved particle physics and cosmology; he is credited with coining the term "black hole".

Wideröe, Rolf (1902–1996) Norwegian electrical engineer/physicist who invented the linear accelerator in 1928. This device was the forerunner of all particle accelerators, and directly inspired Ernest Lawrence's invention of the cyclotron.

Wigner, Eugene (1902–1995) Hungarian-American theoretical physicist; Nobel Prize for Physics 1963 (shared with Hans Jensen and Maria Mayer) for contributions to the theory of nuclear physics and elementary particles. Wigner participated in the preparation of the Szilard-Einstein letter to President Roosevelt, and made seminal contributions to reactor engineering.

Chronology

1895

Rutherford arrives in Cambridge.
December: Wilhelm Röntgen discovers X-rays.

1896

February: Antoine Henri Becquerel discovers uranium radioactivity.
Rutherford coins "alpha" and "beta" terminology.

1897

Thomson discovers the electron. Marie Curie begins radioactivity research.

1898

July: Curies announce discovery of polonium.
Rutherford to McGill.
December: Curies announce discovery of radium.

1900

Rutherford and Soddy discover radioactive decay and half-life.

1903

Curie & Laborde, and Rutherford & Soddy quantify decay energy.

1907

Rutherford to Manchester.

1909

Francis Aston to Cambridge; begins mass spectroscopy work.

© Springer Nature Switzerland AG 2020
B. C. Reed, *Manhattan Project*,
https://doi.org/10.1007/978-3-030-45734-1

Rutherford and Royds determine alpha particles are ionized helium nuclei; Geiger and Marsden begin alpha-scattering experiments.

1911

Rutherford publishes nuclear model of atom.

1913

Bohr publishes atomic model; Soddy conceives of isotopy.

1919

Rutherford discovers artificial transmutation of nitrogen.

1928

Rolf Wideröe develops linear accelerator.

1929

Lawrence and Szilard conceive cyclotron.

1930

Bothe begins light-element bombardment experiments.

1932

February: Chadwick discovers neutron.

1933

September: Leo Szilard conceives of chain reaction.

1934

March: Fermi produces neutron-induced artificial radioactivity.

June: Fermi announces possible discovery of transuranic elements.

July 4: Marie Curie dies; Szilard files patent application for chain reaction.

September: Ida Noddack criticizes Fermi analysis.

October: Fermi discovers slow-neutron enhancement of induced radioactivity.

1935

Dempster discovers U-235.

1937

Rutherford dies October 19.

1938

July 13: Lise Meitner flees Berlin to Holland and thence to Sweden.

December 19: Hahn writes to Meitner re discovery of barium in neutron bombardment of uranium.

December 24: Meitner and Frisch conceive of fission. Enrico Fermi departs for America.

1939

January 3: Frisch informs Bohr about fission.

January 6: Hahn & Strassman fission paper published.

January 7: Bohr departs for America.

January 13: Frisch verifies fission; tests uranium and thorium with fast and slow neutrons.

January 15: Alfred Nier publishes U-234 abundance measurement.

January 16: Bohr and Rosenfeld arrive in America; Rosenfeld mentions discovery of fission.

January 20: Bohr paper to *Nature* establishing Meitner/Frisch priority and giving first discussion of energetics of fission.

January 25: Herbert Anderson detects fission at Columbia University.

January 26: Bohr announces discovery of fission at George Washington University.

January 28: Fission demonstrated at Johns Hopkins University and Carnegie Institution of Washington.

January 29: *New York Times* reports on discovery of fission.

January 31: Luis Alvarez detects fission at Berkeley.

February 7: Bohr paper on roles of isotope parity and neutron speed in fission; published February 15.

March 17: Fermi meets with Navy officials in Washington.

March: Columbia groups report detection of secondary neutrons; Fermi constructs first pile.

April 24: Paul Harteck and Wilhelm Groth inform German War Office of fission.

April 29: Reich Research Council conference on fission in Berlin. Niels Bohr discusses fission at American Physical Society meeting in Washington.

May 1: Francis Perrin publishes analysis of criticality in *Comptes Rendus*.

June 9: Siegfried Flügge publishes analysis of criticality in *Naturwissenschaften*.

June 28: Bohr and Wheeler fission analysis paper received by *Physical Review*.

July 16: Szilard and Wigner speak with Einstein regarding possibility of chain reaction.

August 2: Date of Einstein letter to Roosevelt.

September 1: Bohr and Wheeler fission analysis paper published. Germany invades Poland; World War II begins

September 16: German War Office conference on uranium as source of power or explosives. Werner Heisenberg attends a second conference 10 days later.
October 11: Alexander Sachs meets with Roosevelt; Uranium Committee established.
October 21: First meeting of Uranium Committee.
October: Peierls publishes analysis of criticality in *Proceedings of the Cambridge Philosophical Society.*

December 6: Heisenberg report to German War Office discusses power production and explosives.

1940
January 15: Paul Harteck advocates production of heavy water in a letter to Heisenberg.

February 28/29: Alfred Nier separates minute sample of U-235; Bohr theory of slow-neutron fissility of uranium verified. Werner Heisenberg overestimates critical mass.

March 4: Jacques Allier meets with Director of Norsk Hydro to secure heavy water.
March 15: Nier et al. verification of slow-neutron fissility of U-235 published.
March 19: Frisch-Peierls memorandum reaches Sir Henry Tizard.

April 9: Germany invades Norway.
April 10: First MAUD Committee meeting.
April 27: Second meeting of Uranium Committee.

May 5: William Laurence *New York Times* article on uranium research.
May 10: Germany invades France, Netherlands and Belgium; Winston Churchill appointed Prime Minister.
May 27: McMillan and Abelson report on elements 93 and 94 published.
May 29: Louis Turner manuscript speculating on fissility of element 94; withheld from publication until 1946.

June: John Williams *Harper's Magazine* article on uranium research.
June 14: Germans capture Paris.
June 27: National Defense Research Committee established.

July: Battle of Britain begins. Philip Abelson carries out first thermal diffusion experiments at Carnegie Institution. Carl Friedrich von Weizsäcker conceives of using element 93 as an explosive.

September: William Laurence *Saturday Evening Post* article on uranium research.

October: German "Virus House" facility ready.

1941

January: Walther Bothe concludes graphite to be a poor moderator.

January 28: Seaborg at al. report discovery of plutonium; withheld from publication until 1946.

March 28: First test of slow-neutron fissility of plutonium.

April 9: Peierls reports on feasibility of bomb to MAUD Committee.

May 17: First National Academy report to Frank Jewett.
May 29: Slow-neutron fissility of plutonium reported; withheld from publication until 1946.

June 22: Germany invades Russia.
June 28: Office of Scientific Research and Development established.

July 11: Second National Academy report to Frank Jewett.
July 15: MAUD report finalized.

August 27: Lindemann memo to Churchill re MAUD committee.

September: Heisenberg envisions "open road" to atomic bomb.
September 3: British Chiefs of Staff approve MAUD project.
September 15–21: Heisenberg visit to Copenhagen.

October 2: Battle of Moscow begins.
October 9: Bush meets with Roosevelt; Top Policy Group established.
October 12: Roosevelt letter to Churchill re cooperation on atomic programs.

November 17: Third National Academy report to Frank Jewett.
November 21: Meeting between OSRD representatives and British officials in London.
November 27: Third National Academy report to Roosevelt and Top Policy Group.

December 6: S-1 Section meeting; Arthur Compton advocates pile program.
December 7: Japanese attack Pearl Harbor.
December 16: Conference of directors of German program results in reduction of Army funding.
December 22: First Washington Conference begins.

1942

January 19: Roosevelt OKs conclusions of third National Academy report.

February 20: Conant report to Bush on enrichment methods.
February 26–28: German Army and Reich Research Council conferences.

March 9: Bush forwards Conant report to Top Policy Group; advocates Army control.

May 4–8: Battle of Coral Sea.

May 23: Program Chiefs meeting; General Styer attends; S-1 recommends proceeding with $85 million program.

June 6: Heisenberg meets with Albert Speer; "pineapple"-sized estimate of critical mass. Speer limits approvals to construction projects.

June 4–7: Battle of Midway.

June 17: Status report to Roosevelt; project to be handled by Army Engineers; Col. Marshall assigned to head program.

June 18: Metallurgical Laboratory Engineering Council begins to consider plutonium production-pile designs.

June 20: Roosevelt and Churchill discuss interchange in private meeting.

June 22: Philip Abelson achieves uranium enrichment with experimental diffusion column.

June 23: Speer briefs Hitler on German bomb project; Leipzig-IV pile fire.

June 25: Army/S-1 meeting to discuss plant requirements.

July: Berkeley meeting on bomb physics and design; fusion bombs discussed. Navy authorizes 48-column thermal diffusion pilot plant. Harold Urey advocates that Navy program be tied to S-1 Committee.

July 30: Sir John Anderson advocates merging British and American programs.

August 3: Kenneth Nichols meets with Treasury officials regarding use of silver for Y-12.

August 7: Battle of Guadalcanal begins; to February 1943.

August 13: General Order establishing Manhattan Engineer District issued.

September 13–14: Bohemian Grove meeting authorizes pile and electromagnetic plants.

September 17: Groves placed in command of Manhattan District.

September 23: Groves promoted to Brigadier General; Military Policy Committee established.

October 1: Heisenberg "Director at" Kaiser-Wilhelm Institute for Physics.

October 8: Groves meets Oppenheimer in Berkeley.

October 19: Groves approves idea of centralized research laboratory.

October 26: Stone and Webster produces first plan for Oak Ridge townsite.

October 30: First silver bullion bars withdrawn from West Point Depository for use at Y-12.

November 3: Seaborg alerts Oppenheimer to possible plutonium predetonation problem.

November 8: Allied invasion of Africa begins.

November 10: Groves, Nichols, and Compton visit DuPont headquarters.

November 16: Construction of CP-1 begins. Groves and Oppenheimer visit Los Alamos site.

November 18: Lewis Committee established to review program. Lawrence tests double-source ion beams in 184-in cyclotron.

November 19: Operation *Freshman* disaster.

November 27: Groves pitches pile project to DuPont Executive Committee.

December 2: CP-1 achieves chain reaction in Chicago.

December 3–17: Navy thermal diffusion pilot plant runs for two weeks without shutdown.

December 7: Lewis Committee report advocates diffusion, pile, and electromagnetic programs.

December 10: Military Policy Committee endorses Lewis Committee report.

December 11: James Conant informs Wallace Akers of American position re cooperation with British.

December 12: CP-1 operates briefly at 200 Watts.

December 14: Nichols, Compton, and Matthias discuss plutonium production pile requirements with DuPont.

December 16: Bush reports Military Policy Committee decisions to Roosevelt.

December 21: DuPont signs contract for pile work.

December 28: Roosevelt endorses Bush report of December 16.

December 31: Matthias reports to Groves on possible production-pile sites.

1943

January 14: Berkeley meeting to begin planning of Y-12 facility. Casablanca conference opens.

January 16: Groves inspects Hanford site.

January 23: S-1 group visits Navy thermal diffusion pilot plant.

February 9: Acquisition of Hanford site authorized by Undersecretary of War. DuPont decides on water-cooled piles appx. mid-month.

February 16: Operation *Gunnerside* commences.

February 18: Ground broken for first Alpha-enricher building at Y-12.

February 25: Oppenheimer appointed Director of Los Alamos Laboratory.

February 27: Operation *Gunnerside* sabotages Vemork plant.

March 10: First tract of land acquired at Hanford.

March 17: Groves authorizes first two Y-12 Beta enrichers.

March 18: Seaborg speculates on possibility of plutonium-240 impurity issues.

March 22: DuPont opens Hanford employment office in Pasco, Washington.
March 27: Tolman describes basis of implosion technique to Oppenheimer.

April 1: Oak Ridge site closed to public access. Los Alamos activated as military post.
April 5–14: Serber orientation lectures at Los Alamos.
April 6: Work on Hanford construction camp barracks begins.
April 15: Los Alamos planning meetings begin (Lewis Committee). Survey of B-pile area completed.
April 27: Excavation for X-10 pile building begins.

May: CP-2 goes critical.
May 22: SED created.

June 18: Oppenheimer outlines polonium requirements in letter to Groves. Met Lab Engineering Council considers pile designs.
June 22: Ground broken for 221-T Queen Mary plant at Hanford.

July 4: First implosion test shot at Los Alamos.
July 10: Groves requests Conant to review Navy thermal diffusion pilot plant.
July 20: Groves orders Oppenheimer security clearance; Roosevelt orders Bush to resume exchange with British.
July 28: Firebombing of Hamburg.

August 13: MPC meeting reviews diffusion barrier research.
August 14: First drop-test of gun-bomb "sewer-pipe" mockup.
August 17: First experimental Alpha unit operates at Y-12.
August 19: Quebec Agreement signed.

September 1: Graphite stacking for X-10 pile begins.
September 3: Allied invasion of Italy begins.
September 10: Construction of main K-25 process building begins.
September 17: First gun-bomb test shot at Los Alamos.

October 9: Layout of B-pile building begins.
October 28: Los Alamos Governing Board decides to strengthen implosion research.

November 4: X-10 pile goes critical.
November 13: First alpha-track startup at Y-12; shutdown soon thereafter.
November 16: Vemork hydroelectric plant bombed.
November 17: Navy authorizes 300-column thermal diffusion plant at Philadelphia Navy Yard.

December 1: Silverplate bomber-modification program initiated.
December 10: Milling of B-pile graphite bricks begins.

December 13: First British Mission scientists arrive at Los Alamos.

December 15: Groves visits Y-12 to review problems with magnets.

December 16: *Alsos* mission departs for Naples.

December 22: Groves meets with British scientists to review diffusion barrier research.

December: X-10 produces 1.5 milligrams of plutonium.

1944

January 1: Work on Hanford construction camp goes to three shifts per day. Walther Gerlach assumes control of German program. Construction of thermal diffusion pilot plant begins in Philadelphia.

January 16: Groves decides in favor of Johnson diffusion barrier.

February 15: Kaiser-Wilhelm Institute for Chemistry struck by bomb.

February 20: *Norsk Hydro* sinking. Prototype B-29 arrives Muroc Field, California.

March 3: First re-built alpha enricher enters service at Y12.

April 5: First X-10 plutonium tested for spontaneous fission.

April 17: First six-stage cell tested at K-25.

May: CP-3 goes critical. X-10 operates at 1,800 kW; later achieves 4,000 kW.

June 1: Laying of B-pile graphite completed.

June 6: D-Day invasion of Europe.

June 24: Groves decides to proceed with construction of S-50 thermal diffusion plant.

July: Hanford construction camp houses 45,000 workers.

July 4: Spontaneous fission crisis leads to reorganization of Los Alamos.

July 20: Pressure tests of B-pile begin. Assassination attempt against Adolf Hitler. Reorganization of Los Alamos in response to spontaneous fission crisis.

July 28: Farben heavy-water plant destroyed in bombing raid.

August: Installation of equipment at S-50 thermal diffusion plant begins.

August 25: Frédéric Joliot captured in Paris.

September 1: Paul Tibbets undergoes final security questioning.

September 4: Brussels liberated.

September 13: First fuel loaded into B-pile.

September 19: Bush/Conant memo to Stimson re postwar planning; second memo Sept. 30.

September 26: First official operation of B-pile; xenon-poisoning causes shutdown.

October 9: Test runs at 221-T Queen Mary plant begin at Hanford.

October 18: First process material introduced into S-50 thermal diffusion plant.
October 21: First 509th Composite Group test flight at Wendover Field.
October 30: First product drawn from S-50 thermal diffusion plant.

November 18: Jeffries report completed.
November 30: B-pile achieves 125-MW operation.

December 14: Implosion test shot at Los Alamos shows encouraging evidence of symmetry.
December 16: Battle of the Bulge; to January 25.
December 17: D-pile goes critical at Hanford. 509th Composite Group activated.
December 25: First irradiated fuel discharged from B-pile.
December 28: Tolman Committee report to Groves.

1945
January 20: First process gas introduced to K-25 plant. Dragon machine at Los Alamos produces first fast-neutron chain reaction.
January 30: Walther Gerlach orders German program to evacuate Berlin.
January: X-10 plutonium production (326 g) ceases.

February 4: B-pile achieves power of 250 MW.
February 5: First Hanford plutonium to Los Alamos.
February 17: Schedule for first test of bomb developed at Los Alamos.
February 19: Battle of Iwo Jima; to March 26.
February 25: F-pile goes critical at Hanford.
February 28: Groves and Oppenheimer decide on Christy-core design for implosion bomb.

March 3: First Cowpuncher Committee meeting, Los Alamos.
March 9–10: Firebombing of Tokyo.
March 10: First 102 diffusion stages in operation at K-25. Hanford briefly shut down by Japanese balloons.
March 15: All S-50 columns yielding enriched uranium. Uranium production factory in Oranienburg, Germany, bombed.
March 28: All three piles at Hanford operate at 250 MW for first time.
March 30: Sam Goudsmit locates Walther Bothe in Heidelberg.
March 31: Groves authorizes construction of K-27 diffusion plant.

April 4: Uranium hemispheres brought within 1% of criticality at Los Alamos.
April 12: Franklin Roosevelt dies; Harry Truman sworn in as President.
April 23: Haigerloch captured; German pile destroyed.

April 25: Stimson and Groves brief Truman on Manhattan Project.
April 27: First Target Committee meeting, Washington.
April 28: First slightly-enriched uranium from S-50 plant fed to K-25 plant.
April 30: Hitler commits suicide.

May 1: Walther Gerlach captured in Munich.
May 3: Heisenberg captured.
May 7: 100-ton TNT test at *Trinity* site. Germany surrenders.
May 9: Paul Tibbetts picks out *Enola Gay* at Martin Omaha plant. First meeting of Interim Committee.
May 10–11: Second Target Committee meeting, Los Alamos.
May 14: Second meeting of Interim Committee.
May 18: Third meeting of Interim Committee.
May 19: First 509th Composite Group personnel arrive at Tinian.
May 28: Third Target Committee meeting, Washington. Leo Szilard meets with James Byrnes.
May 30: Stimson deletes Kyoto from target list.
May 31: Meeting of Interim Committee with Scientific Panel recommends use of bomb against "vital war plant".

June 12: Franck report delivered to Compton.
June 18: Joint Chiefs of Staff meeting briefs President Truman on Japan invasion plans. First Los Alamos bomb-preparation personnel depart for Tinian.
June 21: First production Urchin initiator completed. Interim Committee considers Franck Report.
June 27: Ralph Bard memorandum advocates approaching Japanese re surrender.

July 1: Churchill assents to use of bomb.
July 2: Stimson memo to Truman on "Proposed Program for Japan" advocates retaining constitutional monarchy. *Trinity* plutonium hemispheres fabricated.
July 3: German scientists flown to Farm Hall.
July 4: Combined Policy Committee informed of pending use of bomb.
July 6: *Trinity* uranium tamper configured.
July 11: *Trinity* plutonium hemispheres delivered to test site.
July 13: Final assembly of *Trinity* device begins. Met Lab poll on use of bomb.
July 16: *Trinity* test. Gun Bomb components depart San Francisco on *USS Indianapolis.*
July 17: Potsdam Conference begins; to August 2.
July 18: Truman discusses *Trinity* test with Churchill.
July 21: Stimson receives Groves' report on *Trinity* test.
July 23: Little Boy L1 test bomb dropped.
July 24: Truman informs Stalin of bomb.

July 25: Bombing orders authorized.
July 26: Potsdam Declaration.
July 28: Little Boy and Fat Man components begin arriving at Tinian; Japan rejects Potsdam Declaration. Smyth Report completed.
July 29: *Indianapolis* torpedoed.
July 31: Little Boy assembly complete.

August 1: First Fat Man test bomb (F13) dropped.
August 6: Hiroshima bombed (Japan time; Aug. 5 in Washington). Truman statement on bomb released.
August 8: Russia declares war on Japan.
August 9: Nagasaki bombed (Japan time; Aug. 8 in Washington). Japan offers to surrender if Emperor Hirohito can remain Sovereign Ruler. Truman authorizes release of Smyth Report.
August 10: Truman orders halt to any more atomic bombings.
August 14: Japan accepts surrender terms. Heisenberg calculates critical mass at Farm Hall.
August 21: Harry Daghlian accident at Los Alamos; dies September 15.

September 2: Surrender documents signed in Tokyo Bay.
September 5: ZEEP reactor goes critical.
September 8: Manhattan Project Atomic Bomb Investigating Group visits Hiroshima; visits Nagasaki Sept. 13.

October 16: Oppenheimer resigns as Director of Los Alamos; succeeded by Norris Bradbury.

November 1: Invasion of Japan (Operation Olympic) scheduled to begin.

1946
March 1: Operation Coronet scheduled to begin.

May 21: Louis Slotin accident at Los Alamos; dies May 30.

Sources

Abbreviations

H&A   Hewlett & Anderson
HAER   Historic American Engineering Record: B Reactor (105-B) Build-
        ing, HAER No. WA-164. DOE/RL-2001-16; http://wcpeace.
        org/history/Hanford/HAER_WA-164_B-Reactor.pdf   (accessed
        July 27, 2018)
HHMW   Hoddeson, Henriksen, Meade & Westfall

Phrases from the text appearing in these citations are to be taken as indicating
general textual areas to which citations apply. Some citations are to sources
which adopted material from other sources; these are included to cover a
wide variety of sources that readers might have available. Microfilm images
are cited in the format microfilm set (reel), image number(s) on the DVDs
supplied to this author by the NARA. For example, M1392(1), 0296–0299
indicates NARA microfilm set M1392, reel 1, images 0296 through 0299.
Publication years for books are not normally cited except for sake of clarity
regarding the specific edition used or if a number of books have been published by the same author.

© Springer Nature Switzerland AG 2020   **461**
B. C. Reed, *Manhattan Project*,
https://doi.org/10.1007/978-3-030-45734-1

The Manhattan District History is available online at https://www.osti.gov/opennet/manhattan_district.jsp (July 27, 2018).

Dates following web links are that of last access.

Prologue

vii   Casualty statistics: Jones 547; Rhodes 734

viii   "One huge factory": Rhodes 294

viii   Labor and cost statistics; dozen individuals: U. S. News & World Report **119**(5), 44–59, Jul. 31, 1995; H&A 723; Kelly (2007) 93

Chapter 2—From Atoms to Nuclei: An Inward Journey

16   Discovery of radioactivity: Becquerel (1896; two papers); Badash (1965, 1966, 1996)

18   Discovery of Polonium: Curie & Curie (1898)

18   Discovery of radium: Curie, Curie, & Bémont (1898)

19   Discovery of electrons: Thomson (1897)

20   Alpha and beta rays: Rutherford (1899)

20   Beta-rays as electrons: Becquerel (1900; two papers in *Comptes Rendus*)

20   Thorium emanation: Rutherford (1900)

20   Half-life: Rutherford (1902); Kragh http://arxiv.org/abs/1202.0954 (July 27, 2018)

23   Isotopes: Soddy (1913)

23   Mass spectroscopy: Squires (1998); Weinberg Chap. 3

24   Thomson mass spectrometer: Thomson (1907)

24   Aston mass spectrograph (1919 and 1920; 4 papers in *Nature* and *Phil. Mag.*)

24   Uranium 238: Aston (1931)

27   Alpha particles as helium nuclei: Rutherford & Royds (1909)

28   "15-in shell": Preston, p. 36, attributed to Chadwick

28   Geiger and Marsden (1909)

28   Rutherford nuclear model: Rutherford (1911)

28   Nucleus terminology: Nicholson (1911)

32   Hydrogen scintillations: Rutherford (1919)

35   Linear accelerator: WidERöe (1928)

37   Cyclotron: Lawrence & Edlefsen (1930); Lawrence & Sloan (1931); Lawrence & Livingston (1931); Cassidy (2011) Chap. 3

38   Szilard patents on linear accelerator and cyclotron: Lanouette 105; Telegdi (2000); Dannen (2001)

39   Neutron discovery: Bothe & Becker (1930); Curie & Joliot (1932); Chadwick (1932; two papers); Kuhn (1932); Frisch (1967); Brown (1997); Reed (2007a)
44   Moonshine: Rhodes 27; Lanouette 133; Jenkin (2011) 128
46   "Silly experiment": Frisch (1967)
46   Italian historians: Acocella et al. (2004); Guerra et al. (2006, 2009)
46   Fermi neutron-induced radioactivity and transuranic elements discovery: Fermi (1934; three papers)
48   13-min activation: Fermi et al. (1934; *Proc. Roy. Soc.*)
48   Ida Noddack: Noddack (1934). A translation by Graetzer is available at www.chemteam.info/Chem-History/Noddack-1934.html (July 27, 2018).
50   "Carping criticism": Rhodes 231
50   Strong-medium-weak: Segrè (1970) 79–85
52   Extensive results: Amaldi et al. (1935)
52   Fermi arrival in America: Segrè (1970) 98–100; Badash et al. (1980) 90
52   Discovery of U-235: Dempster (1935).

Chapter 3—Fission

53   Hahn & Meitner collaboration renewal: Sime (1989) 373
55   Nine distinct half-lives: Sime (2000); Frisch (1967)
55   3.5-h beta-decay: Curie & Savitch (1937, 1938)
56   Strassmann search for Thorium: Sime (1996) Chap. 7, esp. 179–181
56   Amaldi search for alpha-reactions: Rhodes 221
56   3.5-h beta-decay behaves as lanthanum: Sime (1996) 183, 221; Hermann (1990) 501
57   Intellectual leader: Sime (1989) 374; Sime (1996) 367
57   Meitnerium: Crawford et al. (1997)
57   Hahn-Meitner letters: Sime (1996) 222–224; Hahn to Meitner, Dec. 19, 1938: In Hermann (1990) 494
58   Discovery of fission: Hahn & Strassmann (1939); Meitner & Frisch (1939); Frisch (1939); Bohr (1939, *Nature* and *Phys. Rev.*); Graetzer (1964); Frisch (1979); Sime (1989); Sime (1996) Chap. 10; Stuewer (1985); Hermann (1990)
59   Liquid-drop model: Stuewer (1994, 1997); Sime (1996) 237–238; Hermann (1990) 498
61   Fermi modifies Nobel lecture: Segrè (1970) 99–100
62   Possibility of chain reaction: Frisch (1979) 118
62   Bohr hits self on head: Frisch (1979) 116

62 Frisch to Hahn, January 4, 1939: Hermann (1990) 498

62 Curie detects light elements in U-bombardment: Sime (1996) 245

64 Bohr and Anderson: Rhodes 269–271; Stuewer (1985); Anderson (1974) 57

66 George Washington University: Stuewer (1985); Halpern (2010) 452–457

67 Reports of fission verification: Roberts et al. (1939); Green & Alvarez (1939); Fowler & Dodson (1939); Abelson (1939), Anderson, Booth, et al. (1939), New York Times (1939)

67 "Neutron emission had to be observed experimentally": Anderson (1974) 60

67 "Little doubt in my mind": Rhodes 292

67 Neutron emission from fission: von Halban et al. (1939); Anderson, Fermi, & Hanstein (1939); Szilard & Zinn (1939)

68 Fermi 47 papers: Schwartz (2017) 180

68 Bohr realization of fission due to 235: Bohr (1939, *Phys. Rev.*); Bohr Collected Works, Vol. 9, pp. 64–65

70 "Drawing of a bomb": Rhodes 274–275

70 Wheeler in Copenhagen: Ford (2009)

71 Theory of fission: Bohr & Wheeler (1939); Reed (2003, 2009, 2011)

72 Compound nucleus model: Bohr (1936)

73 Slowing of inelastically scattered neutrons on U-238: Fetisov (1957)

75 Criticality: Perrin (1939); Flügge (1939); Peierls (1939)

76 Frisch arrival in Birmingham: Frisch (1979) 120–121

77 Olpihant role in American program: Frisch (1979) 123

77 Peierls refines Perrin calculation: Peierls (1985) 128, 146, 153; Reed (2008)

78 Nier U-234 abundance measurement: Nier (1939)

78 Fermi encourages Nier: Nier (1989)

78 Airmail special delivery: Rhodes 332

79 Uranium fissility: Nier et al. (1940, two papers in *Phys. Rev.*)

79 Few hundred dollars: Nier (1989)

80 Laurence *Saturday Evening Post* article: Williams 94

81 Diffusion: Clusius & Dickel (1938, 1939)

81 Royal Society for Chemistry: Frisch (1940); Frisch (1979) 124–125

81 Frisch thought atomic bomb impossible: Frisch (1979) 126

81 "We stared at each other": Frisch (1979) 126; Peierls (1985) 154–155

82 Frisch-Peierls memorandum: Serber 79–88; Bernstein (2011)

84 Tizard receives Frisch-Peierls memorandum: Clark (1965) 214–217

84 Chadwick reconsiders bomb: Clark (1961) 46–47

84   Chadwick and Appleton: Farmelo 123–126
84   Thomson group unsuccessful chain reaction: Gowing (1964) 37–39
84   Policy and Technical Committees: Gowing (1964) 48; Farmelo 162
85   Naming of MAUD Committee: Clark (1961) 77; Clark (1965) 221; Farmelo 161
85   MAUD April 10 meeting: HHMW 18
85   Thomson to Chadwck April 16: Farmelo 144
85   "Many sleepless nights": Chadwick oral history interview (Session IV, April 20, 1969), http://www.aip.org/history/ohilist/3974_4.html. (July 27, 2018). Also Farmelo 179
86   Size of U sphere manageable: Gowing (1964) 68
86   Louis Turner: Turner (1940, 1946)
86   "Inertia of man": Feld et al. (1972) 188
86   McMillan experiment: McMillan (1939)
87   Segrè and fission products: Segrè (1939)
88   Discovery of Np: McMillan & Abelson (1940)
88   Chadwick protest: Brown (1997) 206; Gowing (1964) 60n
88   Search for element 94: Kathren et al. (1994) 12–14
89   2.1-day decay of Np: Kathren et al. (1994) 15
89   Discovery of Pu: Seaborg et al. (1946a, b)
89   1.2-kg sample of UNH: Kathren et al. (1994) 26–30, HHMW 22
89   0.3-μg sample: Kathren et al. (1994) 32
89   Poor sample geometry: Kathren et al. (1994) 34
89   Slow-neutron fissility of Pu: Kathren et al. (1994) 40–41; Kennedy et al. (1946)
90   Properties of Pu: HHMW 224; Bernstein (2007) 105.

Chapter 4—Organizing: Coordinating Government and Army Support 1939–1943

91   $1500 to Fermi: Rhodes 293–295; HHMW 13–15; Segrè (1970) 111; Ahern (2003)
91   Wigner as nuclear engineer: Weinberg (2002)
92   Enlisting help of Einstein: Isaacson Chap. 21; Rhodes 304–307; Lanouette Chap. 14
94   "Unenvisaged potency and scope": Sachs Exhibit 3
95   "To see that the Nazis don't blow us up": H&A 17
97   October 21, 1939 Uranium Committee meeting: H&A 19–20; Sachs 6–7
97   $6000 to Fermi and Science Advisory Subcommittee: H&A 21; Briggs to Bush July 1, 1940: M1392(1) 0272; Weinberg (2002) 42

97   Harold Urey: Cohen et al. (1983) 635
97   Szilard to Briggs, October 26, 1939: Feld 204–206
98   "Destructiveness vastly greater": H&A 20; Sachs Exhibit 5
98   Briggs to Watson, Feb. 20, 1940: Smyth Sect. 3.6
98   Sachs to Watson, Feb. 15, 1940: Sachs Exhibit 6c
98   Einstein to Sachs, Mar. 7, 1940: Sachs Exhibit 7a
98   Sachs to Roosevelt, Mar. 15, 1940: Sachs Exhibit 7b
99   Sachs meets with FDR early April: Sachs 12
99   Importance of Belgian ores and long-term planning: H&A 23
99   Roosevelt and Watson to Sachs, Apr. 5, 1940: Sachs Exhibits 8a, 8b
99   Einstein to Briggs, Apr. 25, 1940: Sachs Exhibit 12a
99   April 27 Uranium Committee meeting: H&A 23; Sachs 20–22
99   Uranium lattice: HHMW 20
99   Sachs to Roosevelt, May 11, 1940: Sachs Exhibit 14a; Smyth Sects. 3.6, 3.7; cost estimates: Sachs 20–21
99   Sachs to Watson, May 15, 1940: Sachs Exhibit 15a
99   Sachs to Watson May 23, 1940: Sachs Exhibit 16a
100 Briggs to Sachs, June 5, 1940: Sachs Exhibit 18
100 Censorship of publications: Smyth Sect. 3.3; Weart (1976)
100 "Import of War Developments": Sachs Exhibit 17
101 NDRC: Norris 165; Cassidy (2011) Chap. 4; Thomas (2017) Chap. 4
102 Roosevelt to Briggs, Jun 15, 1940: Sachs Exhibit 19; Briggs to Bush July 1, 1940: M1392(1), 0271–0272
102 Briggs to Bush, July 1, 1940: M1392(1), 0271–0272; Smyth Sect. 3.4
103 NDRC/OSRD contracts: Smyth Sect. 3.12
103 Compton to Bush, Mar. 17, 1941: M1392(1), 0296–0299
103 "swimming in syrup": Weinberg (2002) 42
103 Bush to Compton, Mar. 21, 1941: M1392(1), 0295
103 Bush request to Jewett: Norris 166; H&A 36
104 Compton committee: Goldberg (1992) 438 ff
104 Compton May 17 report: M1392(1), 0370–0377
105 Jewett to Millikan, May 28, 1941: M1392(1), 0391–0392
105 Millikan to Jewett, May 31, 1941: M1392(1), 0389–0390
105 Jewett to Bush, June 6, 1941: M1392(1), 0384–0385
106 Bush to Jewett, June 7, 1941: M1392(1), 0380–0383
106 Briggs to Bush, June 11, 1941: M1392(1), 0874–0877
107 NDRC minutes, June 12, 1941: M1392(1), 0393–0394
107 Briggs to Conant, July 8, 1941: M1392(1), 0396–0401
107 NDRC minutes, July 18, 1941: M1392(1), 0419–0422; Smyth Sect. 3.16

107 July 11 NAS report: M1392(1), 0403–0407

108 Fermi report on production of energy by a chain reaction: Collected Papers, Vol. 2, 86–90. Also in M1392(1), 0595–0602

108 Bush to Conant, July 21, 1941: M1392(1), 0591.

108 OSRD: H&A 41; Norris 259

109 British contribution: Groves 408; Gowing (1964); Clark (1961); Lee (2006)

109 Tizard mission: Farmelo 164

109 British Commonwealth Scientific Office: Clark (1961) 163–165; Gowing (1964) 66

109 Conant visit to Britain: Farmelo 173–175; briefed by Lindemann on Frisch-Peierls memorandum Farmelo 182–183

109 July 1 MAUD meeting: Clark (1961) 131, 135; Farmelo 183 gives the date of this meeting as July 2.

109 Cahdwick works 20 h per day: Farmelo 184

109 Lauritsen briefs Bush: H&A 42

110 MAUD report recommendations: Farmelo 184–187

110 Thomson, Bush & Conant: Clark (1961) 166; H&A 44

110 Thomson to Conant, Oct. 3, 1941: M1392(1), 0446

111 Reorganization of uranium committee: Briggs to Conant, July 30, 1941: M1392(1), 0424–0425; Conant to Briggs, July 30, 1941: M1392(1), 0426–0427

111 Oliphant visit to United States; meeting with Briggs: Farmelo 196–199; Oliphant (1982)

111 Influence of Oliphant: Rhodes 372; Coolidge to Jewett, Sept. 11, 1941: M1392(1), 0443. See also Jewett to Bush, Sept. 12, 1941: M1392(1), 0442; Jewett to Coolidge, Sept. 12, 1942: M1392(1), 0445; Conant to Bush, undated, M1392(1), 0444

112 "If you tell me this is my job": H&A 43–44; Compton (1956) 6–9

112 Imperial Chemical Industries; Wallace Akers: Farmelo 199–202, 211

112 October 9, 1941: Bush to Conant, Oct. 9, 1941: M1392(1), 0605–0606; H&A 44–49; Rhodes 377–379; Farmelo 198–199

113 Bush to A. Compton, Oct. 9, 1941: M1392(1), 0477–0478

114 Bush to Briggs, Oct. 9, 1941, M1392(1), 0479

114 Schenectady meeting, Oct. 21, 1941: H&A 46

114 Third NAS report: M1392(1), 0491–0551. Also Jewett to Bush, Nov. 3, 1941: M1392(1), 0480–048, and Bush to Jewett, Nov. 4, 1941: M1392(1), 0482–0484. Compton to Jewett, Nov. 17, 1941: M1392(1), 0485; Compton to Bush, Nov. 17, 1941: M1392(1), 0489–0490

115   Conant draft history: M1392(1), 0302–0331

116   Bush to Roosevelt, Nov. 27, 1941: M1392(1), 0552–0553; FDR "OK" note: M1392(1), 0945

117   Murphree to Bush, Nov. 27, 1941: M1392(1), 0610–0611

117   Bush to Murphree, Nov. 29, 1941: M1392(1), 0612–0616

118   Tubealloy organizational chart: M1392(1), 1021; Smyth Sects. 5.2–5.4

118   Urey to Bush, Dec. 1, 1941, M1392(1), 0557–0562; Smyth to Bush Dec. 1, 1941, M1392(1), 0563

118   Bush to Briggs, Dec. 2, 1941, M1392(1), 0697–0699; see also Bush to Smyth, Dec. 2, 1941, M1392(1), 0566

118   Murphree to Bush, Dec. 3, 1941, M1392(1), 0710–0713

119   December 6, 1941 meeting: Compton 70; Bush to Murphree, Dec. 10, 1941, M1392(1), 0629–0631

120   "An afterthought": Smyth Sect. 5.7

120   Columbia pile experiments: Smyth Sects. 4.6, 4.13, 4.17–4.21, 6.11; HHMW 19–20; $k = 0.87$ pile described in Schwartz 184–185; last Columbia pile with cylindrical slugs described in Schwartz 190–192; $k = 0.98$ value in May 1942: Smyth Sect. 6.11. The $k = 0.87$ and 0.92 piles are described in Fermi's Collected papers, Vol. 2, 128–143.

120   Murphree to Bush, Dec. 10, 1941, M1392(1), 0648–0652

120   Bush to Murphree, Dec. 13, 1941, M1392(1), 0645–0647

120   Bush to Compton, Lawrence, and Urey, Dec. 13, 9141, M1392(1), 0570–0581

121   Army involvement and funding: Bush to Conant, Dec. 16, 1941, M1392(1), 0653–0654; Smyth Sect. 5.9; H&A 289–290; Jones 56, 272–274

121   Compton to Bush, Conant, and Briggs, Dec. 20, 1941, M1392(1), 0862–0867

121   Conant to Lawrence, Dec. 20, 1941, M1392(1), 0734–0735

122   Compton to Conant, Jan. 22, 1942, M1392(1), 0868–0871

122   Lawrence to Conant, Jan. 24, 1942, M1392(1), 0872

122   Conant report, Feb. 20, 1942, M1392(1), 0775–0781; Smyth Sect. 5.13

123   75-μg samples: HHMW 28

123   Lawrence to Bush, Mar. 7, 1942, M1392(1), 0782; Lawrence to Conant, Mar. 13, 1942, M1392(1), 0783–0784; Lawrence to Conant, Mar. 26, 1942, M1392(1), 0790

123   Bush to Roosevelt, Mar. 9, 1942, M1392(1), 1007–1023

124   Reich Research Council: Rhodes 403

124   Roosevelt to Bush, Mar. 11, 1942, M1392(1), 0785

125 General Styer: M1392(1), 0788; H&A 72. Bundy's memo refers to Styer as a Colonel, whereas Hewlett & Anderson and Jones have him as a Brigadier General

125 Conant to Bush, Apr. 1, 1942, M1392(1), 0791–0800

125 "Court of public opinion": Conant to Bush, May 14, 1942, M1392(1), 0812–0814; Smyth Sect. 5.15

126 Bush to Conant, May 21, 1942, M1392(1), 0886

126 Breit resignation: Breit to Briggs, May 18, 1942, M1392(1), 0817–0819; HHMW 27, 41

126 Conant to Bush, May 25, 1942, M1392(1), 0821–0825

127 Bush to Styer, June 11, 1942, M1392(1), 0844; Bush to Conant, June 11, 1942, M1392(1), 0845

127 Bush and Conant to Wallace, Stimson, and Marshall, June 13, 1942, M1392(1), 1024–1029

128 Bush to Roosevelt, June 17, 1942, M1392(1), 0944; Smyth Sects. 5.21, 5.22

128 Fate of Col. Marshall: Norris 189, 615n8; Nichols 101, 114; Fine & Remington 681

128 Army organization: Norris 158–161

129 Marshall diary: Norris 609n19; I am most grateful to Mr. Norris for providing me with a copy of Marshall's diary

129 Marshall having authority of Division Engineer: Jones 41

129 Manhattan District versus Manhattan Project: Norris 189

130 Program Chiefs meeting of June 25: Fine & Remington 654

130 Project reorganization: HHMW 30; Bush to Conant, June 19, 1942, M1392(1), 0848 and 0850–0854; Bush to Briggs, June 19, 1942, M1392(1), 0849; Smyth Sect. 5.17; H&A 75

130 Two or three experienced men: HHMW 42

130 Berkeley conference: HHMW 43; Hawkins 2 gives the time of the Berkeley conference as late June. Other participants included John van Vleck, Emil Konopinski, Stanley Frankel, Eldred Nelson, and Felix Bloch

131 Impurity issue: Smyth Sect. 12.2; HHMW 43–47

131 Groves suggests Stone & Webster: Norris 170; Jones 126

132 270 Broadway: Kelly (2007) 220; Kelly & Norris 46–47

133 Senator McKellar: Quoted in Kelly (2004) 6–7

135 Establishment of MED: H&A 74; Smyth Sect. 5.23; Norris 189; Fine & Remington 659 give the date as Aug. 16

135 Bohemian Grove meeting: HHMW 57; Compton 150–154. Compton gives the date as 15–16 September, different from all other sources. The

minutes of the meeting can be found in M1392(9), 0077–0082, and confirm the September 13–14 dates

136 "Oh, that thing": Groves 4

136 Groves promotion: Norris 180

137 Groves responsibilities: Groves 3–4; Norris 161–162, 606n95, 611n42

138 Groves' offices: Kelly 222; Norris 2–3, 560n3; http://history.state.gov/departmenthistory/buildings/section28 (July 27, 2018)

138 "Better than our intrinsic abilities": Kelly (2007) 122; Norris 235–6

138 Nichols on Groves: Nichols 108, cited in Norris 210

139 "In the soup": Groves 20; Norris 178

139 Groves and Nelson: Groves 22

139 Groves and Navy: Groves 23

139 Military Policy Committee: Smyth Sects. 5.25–5.27; H&A 82–83; Norris 180–181; Nichols 58–59, 115; HHMW 31

140 65% of production facilities: Thayer 34; Norris 151; Nichols 61

140 DuPont October 3 contract: H&A 186

140 Ordinary and heavy water still in running: Marshall diary 183 (Oct. 2, 1942)

140 Groves first visit to Met Lab: Groves 39–41

141 Plutonium program to be overseen by single firm: Jones 97; Groves 42

141 October 31 meeting with Harrington & Stine: Jones 98; Groves 46; M1109(1) 119–121

141 November 10 meeting: Jones 98–101; Groves 48

141 November 27 meeting: M1109(1) 119–121

141 DuPont indemnification: Jones 98–106; Norris 213; H&A 187; Groves 48–50. Marshall diary 225–7 (Nov. 10, 1942); Nichols 63

141 $20 million trust fund: Groves 59; Nichols 82

141 DuPont $1 fee: Groves 59; Compton 165; Cannon 2-1.23; Carlisle & Zenzen 19–20; Thayer 85; Jones 106

142 DuPont subdivisions: Jones 198–199; Thayer 38

142 Pu impurities: HHMW 35; Kathren et al. 201–202. Seaborg also anticipated possible spontaneous fission issues with Pu: HHMW 230–231

142 November 18 DuPont review committee: Jones 101; Groves 52

142 Greenewalt: Groves 52, 79; H&A 188; Compton 164; HHMW 36–38; Nichols 65

143 49 Project report: M1392, Roll 3, Target 4, Folder 17, "S-1 Technical Reports [1942–1944]". Enrico Fermi's contribution to this report, "Feasibility of a Chain Reaction," can be found in his Collected papers, Vol. 2, 263–267

143 Lewis report to Groves: Nichols 68

144 MPC meeting, Dec. 10, 1942: Jones 105–109; Norris 213; H&A 120; MPC minutes Nov. 12, 1942

144 January 4, 1943 contract: Jones 112; H&A 191; Compere 7

145 Pilot plant operating contract: Jones 111–115; Compton 172–174, 197; H&A 190–193; Carlisle & Zenzen 30

145 29-page report to FDR: H&A 114–115; Nichols 69

145 "OK-FDR": Bush to Roosevelt, Dec. 16, 1942 (amended Dec. 23), M1392(1), 0949–0951; Bush to Wallace, Stimson, and Marshall, Dec. 15, 1942, M1392(1), 1035–1063

147 British postwar commercial development: H&A 271

147 Roosevelt to Bush, Dec. 28, 1942, M1392(1), 0946

147 Conant and Tolman: Smyth Sects. 5.28–5.31; Groves 44

147 "Magnitude to be achieved": Conant draft history, M1392(1), 0302.

Chapter 5—Piles and Secret Cities

149 Metallurgical Lab: Compton 82

150 Fermi Columbia pile research: Segrè 121; Wattenberg (1993) 46–47; five months to move Columbia material to Chicago described in Schwartz 193

150 Uranium production: Compton 90–96; H&A 66, 87

150 Sixteen piles between September 15 and November 15: HHMW 32; 29 piles before CP-1 inferred from Fermi (1952); see also Schwartz 197

150 Lattice spacing and reactor self-start-up: HHMW 32; Department of Energy website (1982) 9; Anderson (1974) 43; Wattenberg (1982) 25

150 Begin planning for a critical pile: Segrè 127

150 Performing the experiment at the university: Compton 136–138

150 Squash court: Compton 138, Department of Energy (1982) 1–2

150 Pumpkin patch: Argonne National Laboratory website

150 Ellipsoid dimensions: Fermi (1952)

151 "Awesome number": Anderson (1974) 43

151 Balloon arrives November 16: Schwartz 206

151 Twelve-hour shifts: Segrè 127; Anderson (1974) 43

151 Hydraulic press: Anderson (1974) 43

151 Two layers per shift: Wattenberg (1982) 28

152 Drill-bit resharpening: Wattenberg (1982) 27

153 Ten horizontal slots: Carlisle & Zenzen 21

153 Vertical zip rods: Libby 119

154 49 people present: Argonne National Laboratory website

154 "It will not level off": Wattenberg (1993) 50

155   "No one present": Anderson (1974) 44

155   "as cool as a cucumber": Kelly (2007) 87

155   "Birth certificate of the nuclear age": Wattenberg (1993) 50

156   Fermi laconic report: Collected Papers, Vol. 2, 270

157   "Far-reaching consequences": Wigner (1979) 240

157    $k = 1.0006$: Fermi (1952); Collected Papers, Vol. 2, 272–307

157   Exponential rise time of 2.6 minutes: Wattenberg (1993) 50

157   Opening a window: Anderson (1974) 45

157   Minute steering adjustments: Libby 123

158   X-10 missions: https://www.ornl.gov/content/graphite-reactor (July 27, 2018)

159   Clinton Laboratories site: Jones 204

159   X-10 dimensions: H&A 208; A1218(6, Vol. 2, part 2) p. 4.3

159   Graphite brick dimensions: A1218(6, Vol. 2, part 2) p. 4.2

159   120-ton fuel load: H&A 195

159   Graphite thermal column: A1218(6, Vol. 2, part 2) p. 4.4

160   Loading elevator: H&A 194, 208

160   183 DuPont employees: Carlisle & Zenzen 30

161   Weight-driven shim rods and experimental channels: H&A 196

161   Channel plugs: A1218(6, Vol. 2, part 2) p. 4.2

161   200-foot exhaust stack: A1218(6, Vol. 2, part 2) p. 4.5

162   DuPont begins pile building excavation: H&A 208

162   Fermi inserts first X-10 slug: H&A 211

162   X-10 goes critical: Kathren et al. 334

162   90% chemical separation efficiency: Jones 209

162   X-10 power level to 1.8 MW: H&A 211

162   4 MW power level: Jones 209; A1218(6, Vol. 2, part 2) pp. 4.8–4.9

162   299 batches of slugs: Jones 209

163   Growth of Oak Ridge: Wilcox (2002) 8; Jones 432–449

164   Skidmore, Owings, and Merrill: Jones 435

164   Development of Oak Ridge: Jones 435–440; H&A 116–117; Oak Ridge Operations Chap. I; Nichols 124

164   100 miles paved streets plus 200 miles for production sites: Oak Ridge Operations 23

165   Houses and dormitories: Jones 438–439; H&A 119; Oak Ridge Operations 7–9

165   163 miles of boardwalk: Wilcox (2002) 9

165   Oak Ridge construction cost: Jones 439–440; Robinson 45; Nichols 125

165 Cemesto panels; sturdiness of housing: Fine & Remington 671; H&A 118; Robinson 49
166 Roane-Anderson: Jones 445–446; Groves 425
166 Bus system: Wilcox (2009) 18; Oak Ridge Operations 16; Nichols 121
166 Richland statistics: Jones 457–463; Groves 89; H&A 303; Cannon 2-11.8
168 Police kept a copy of a key to every house: Toomey 81
168 Construction camp conditions: HAER 17–18; Cannon 1.18
168 Hanford Site patrol: Cannon 2-8.10
168 "DuPont didn't have that job": Thayer 17, 35, 89
168 Over 260,000 applicants: HAER 15; H&A 216, 303; Thayer 93; Toomey 60
168 145,000 ID photos: Toomey 51
169 Wages and Espionage Act: Jones 454–462; HAER 57–58; Harvey 7, 15; Cannon 2.8–16.

Chapter 6—U, Pu, CEW and HEW: Securing Fissile Material

171 500 tanks: H&A 147
172 Y-12 site: Jones 130
172 Y-12 employment and building statistics; perimeter fence: Jones 130–132; A1218(10), 0691
174 2000 sources to separate 100 g/day: H&A 143
174 Vacuum requirements: Yergey & Yergey (1997) 947; Jones 120; H&A 143
174 Chemical separation operations: Wilcox (2009) 13; Groueff 244
174 Two sources per tank: H&A 144
175 Cubicle operators: H&A 150; Jones 134; Groves 110; Nichols 87–88; Fine & Remington 672; Kiernan (2013)
175 Groves authorizes first two beta units: H&A 151, 157, 295; Wilcox (2009) 11; Jones 128
175 Beta unit geometry: Yergey & Yergey (1997) 952
176 Four-bean sources in fifth Alpha unit: H&A 160. For further statistics on Alpha and Beta operating parameters, see Quist (1999)
176 Electromagnetic program research cost: Jones 122–123
176 Y-12 Ground broken Feb. 18, 1943: H&A 152; Wilcox (2009) 9; Nichols 88
177 Four new Alpha II units: Nichols 89–90
177 Two more Beta tracks authorized at the same time: H&A 161; Jones 129; A1218(10), 0123

177   Westinghouse, GE, Allis-Chalmers: Nichols 86

177   Treasury Department silver: Reed (2009b)

179   67 million man-hours of labor: Groves 98–100

179   Tennessee Eastman: H&A 148; Jones 140

179   4,800 operators: H&A 161–162

179   "We must do it": H&A 154–155

180   Eight fatal accidents: Groves 110

180   Securing tanks to floor with straps: H&A 162; Groves 106

180   Coil refurbishing: H&A 163

180   200 g of material to 12% U-235: H&A 164; Jones 143; Compere 10

181   Lawrence proposes four additional Alpha tracks: H&A 165

181   Copper winding in third Beta building: H&A 299; A1218(10), 0123

181   Feed material losses; abandoned material: Yergey (1997) 948; Groueff 236; H&A 295–6; Jones 144

181   Electrocuted bird: A1218(10), 0754; Groves 106

181   Improvements accumulate through 1944: Nichols 129

181   By December 15: H&A 299

182   Optimization of enrichment methods: Wilcox (2009) 16; H&A 301; Nichols 159; Groves 123

183   Enriched uranium to New Mexico: Wilcox (2009) 17; H&A 300

183   200 g/day to 80%: Yergey (1997) 943

183   Electricity consumption: Yergey (1997) 948; Jones 391

185   Half of material returned to next-lower diffusion stage: Keith (1964) 115

185   Feed point one-third of way along cascade: Smyth Sects. 10.7–10.13; Nichols 90

185   Fused glass: Jones 10; Rhodes 380; Cohen (1983) 636

185   Zinc/brass etching: Rhodes 381; Cohen (1983) 683

185   Franz Simon diffusion experiments: H&A 37; MAUD report

185   1 kg/day plant recommended May 25, 1942: Jones 38

186   Booth 12-stage demonstration system: H&A 101

186   Cabinet eight feet square: H&A 128; Groves 113

186   SAM laboratory: Jones 150; Cohen (1983) 641; H&A 136; Groves 111

186   Diffusion barrier as most difficult aspect of Project: Jones 149; Norris 207

186   Foster Nix: H&A 101

186   Houdaille-Hershey, Norris-Adler, & Chrysler: H&A 127; Rhodes, 493–4; Cohen (1983) 642; Jones 154; Groueff 170, 176, 180, 280

187   Express purpose of carrying out one job: Smyth Sect. 10.24-a

187 Jersey City pilot plant: H&A 101–102, 120, 128
187 British cascade design: Jones 153; Cohen (1983) 638
187 Teflon: Rhodes 494
187 Pump cooling: H&A 124; Smyth Sects. 10.15–10.16; Groueff 230
188 Plagued with pinholes: Jones 155; H&A 132
188 July 1, 1945 target date: Rhodes 495; Jones 156; H&A 135–137; Groves 118; Groueff 269; Nichols 144
188 Groves announced his decision: Jones 156; Norris 209; Groueff 271–272
188 Carbide & Carbon contract: Norris 207; Jones 165; Groves 113; Groueff 157
189 A city the size of Boston: Groueff 215
189 Power plant online March 1, 1944: Jones 160–161; Norris 206; Groves 112; McBride 82; Groueff 119, 121–122, 215; Nichols 94
189 5, 15, 36.6 and 90% plants: Jones 157; H&A 129
189 5,000-acre tract: Norris 206; Jones 159; H&A 130
189 Happy Valley: H&A 130–131; Nichols 127
189 K-25 process building statistics: Norris 207; H&A 131, 140–1; Nichols 146; Keith (1964) 120; Fine & Remington 680
189 2,892 stages: Jones 158
190 130,000 monitoring instruments: Jones 158–159, 168
190 K-25 Cases: Jones 162, 164
190 Cleaning requirements: Groves 116; Groueff 216
191 1,200 welding machines: Keith (1964) 119
191 Nier mass spectrometers: Jones 164; Nier (1989); Groueff 194
191 Pipe electroplating: Groves 114–115; Norris 123
191 Six-stage cell: Nichols report, May 9, 1944
191 Ready to be turned over: Jones 164–167; H&A 141, 298–299; Groves 113; Nichols 147
191 First process gas Jan. 20, 1945: Jones 168; H&A 300
192 2,892 stages in operation Aug. 15, 1945: Jones 169
192 Enrichment increased to 23%: Jones 169; Groves 120
192 Full operation Feb. 1946: Jones 159; H&A 302
192 S-50 facility: Reed, *Liquid Thermal Diffusion* (2011)
202 S-50 output to K-25 plant, Apr. 1945: Nichols 160
203 Shortening the war by about nines days: Nichols 150
203 "Truly pioneering": Groves 38
204 Franklin Matthias: Fine & Remington 667; Thayer 2
204 Hanford site requirements; village 30 miles from plant: Marshall diary 252 (Dec. 14, 1942); Cannon 2-1.2

204   44 by 48 mile buffer area: Harvey, 3, 8; Cannon 2-1.3

205   400,000-acre site: Jones 331–333; Groves 75

205   Last site selected for Manhattan Project: Cannon 1.12; Thayer 26

205   Hanford site characteristics: Jones 110–111; Groves 73–75; H&A 189; 212–214; Libby 167; Toomey 63

205   Low property valuations: Cannon 1.14

206   Reappraisal of land tracts: Jones 334–339; H&A 213

206    $5 million land acquisition: Jones 340–342; Groves 77

206   Met Lab: Compton 161

206   Engineering Council and Thomas Moore: H&A 174–175; Compton 162

206   Pile configurations: H&A 175–176

207   Drawbacks of liquid cooling: H&A 176–177; Jones 190–191

207   Wheeler & Wigner water cooling: H&A 179

207   Bismuth-phosphate process: Jones 193–194; Thayer 76

207   Wigner 500-MW pile: Jones 192; Weinberg (2002) 43; H&A 193–194; Compton 167; HHMW 32

207   Choice of water-cooling: Jones 193; Groves 80–81; Weinberg (2002) 43; H&A 198; Compton 169–170; Cannon 2–3.3; Carlisle & Zenzen 29

208   Wigner reviews blueprints; patents: Weinberg (2002) 44; Snell (1982)

208   Irradiated slugs to Queen Marys: Jones 214–218; Warriner (2013)

208   100, 200, and 300-areas: Jones 211–214

208   Support facilities: Cannon 1.15, 1.18

209   Eight piles down to three: HAER 19; H&A 215

210   2.5 million cubic feet of Masonite, etc.: HAER 21–25; H&A 216–217

210   Land excavation: Thayer 16; Warriner (2013)

210   Pile structure: Thayer 56

210   Hanford a year ahead of schedule: Thayer 55, 63, 82

211   Weld quality: HAER 57–58; Groves 84; H&A 216; Thayer 72

212   Graphite bricks: HHMW 34

213   Graphite milling and stacking: HAER 26–30; H&A 217; HHMW 33–34; Thayer 216

213   Eight to ten months to revise design: Bankoff (2004); Thayer 53

213   Normal operation of 32 slugs per channel: HAER 33

213   14 gallons of water: Thayer 5

214   Radioactive decay of discharged slugs: HAER, 38–40

214   1.3 million inhabitants: HAER 20; H&A 216; Thayer 16

215   Cooling system pumps and 300,00-gallon tank backup: HAER 32–35; 41–48; Groves 82–83

215 Pressure tests begin July 20, 1944: HAER 36–37

216 Japanese balloons: HAER 49–51; 94; Cannon 2–12.12; https://en.wikipedia.org/wiki/Fire_balloon#cite_note-24 (July 27, 2018)
B-Reactor Museum Association newsletter, Spring 2014, pp. 5–6; http://b-reactor.org/wp-content/uploads/2016/06/mod2014-sprg.pdf (July 27, 2018)

216 Power level determined by water temperature differences: HAER 51–55

217 No serious cases of radiation exposure: HAER 89–94; Groves 87–88; Compton 180–181; Libby 174

217 Dry criticality with 400 tubes and wet criticality with 838 tubes: A1218(9), 106

217 9 MW achieved 1:40 a.m. Sept. 27: HAER 65–68; H&A 305; A1218(9), 107

219 Groves informed of xenon poisoning: HAER 68–73; Jones 222; Weinberg (2002) 44; Snell (1982); H&A 307

219 Compton to Hanford: Jones 222

220 250 MW achieved Feb. 4: Bankoff (2004). Note that A1218(9), 107 gives different figures for number of tubes/power level. Matthias diary confirms 150 MW on Dec. 29.

220 D and F piles go critical: HAER 72–73; Matthias diary Dec. 17, 1944

220 F pile at 190 MW by March 1: Matthias diary March 1, 1945

220 Groves orders five kilograms every 10 days: Matthias diary; May 3 and June 1, 1945

222 Operators take over 221-T: Cannon 1.26; Gerber (1996) 3-1. Details on 221-T in particular can be found in Gerber (1994)

222 First Pu to Los Alamos: HAER 76–77. H&A 310 give the date of the first plutonium reaching Los Alamos as February 2. See also Warriner (2013)

222 Processing yield 90%: H&A 220, 309; Rhodes 603–604; DSM July 7, 1945; see also http://www.atomicarchive.com/History/mp/p4s24.shtml (July 27, 2018)

222 Feed materials: Reed (2014).

Chapter 7—Los Alamos, *Trinity*, and Tinian

228 Anticipated production schedules in Tennessee and Washington: Hawkins 16–23

228 OSRD contracts with nine universities: Seidel, 8; Badash et al. 25

228 Fermi and others meet Sept. 19–23, 1942: HHMW 57

228   Groves familiarization tour: HHMW 57; H&A 228; Jones 83; Groves 60–61; Seidel 17

229   Possible Los Alamos sites: Badash et al. 4

229   "The poplars": Los Alamos Historical Society 5; Seidel 18–20; Szasz (1984) 17

229   Los Alamos Ranch School; Oppenheimer ranch: Conant, 73; H&A 229; Jones 84; Truslow 2. Jones gives Dudley's rank as Major

229   Cost of Los Alamos land: Jones 328, 331; Truslow 9, 24; Seidel, 31

229   Groves obtains right of entry: Badash et al. 6; Seidel 46; H&A 230; Jones 84–85

229   Combination Army camp and mountain resort: Kelly (2006) 66

229   Fenced Technical Area: Jones 329, 465–467; Hawkins 4

229   $26 million construction costs: Seidel 45.

231   Lawrence reaction to Oppenheimer appointment: Hawkins 1; Seidel 14–15; Conant 38

231   Reactions to Oppenheimer appointment: Kelly (2006) 106; Kelly (2007) 135; Pais & Crease 139; Alvarez (1987) 78

231   Groves orders Oppenheimer clearance: Groves 61–63

231   Weisskopf on Oppenheimer and Los Alamos: Weisskopf (1967) 40

232   Oppenheimer recruiting efforts: Seidel 35

232   Robert Bacher and Isidor Rabi: HHMW 59; H&A 230–232; Jones 86; M1392(9), 0777–0779; M1392(9), 0750–0751

233   Los Alamos to initially be civilian, then military: H&A 231; Hawkins 5; Conant and Groves to Oppenheimer, February 25, 1943: https://www.atomicheritage.org/key-documents/groves-conant-letter-oppenheimer

233   Division of responsibility between Commanding Officer and Director: Hawkins 35; Groves 154

234   Oppenheimer, Wilson, McMillan, Manley, Serber and Condon plan laboratory: HHMW 68

234   Oppenheimer considers self to head Theoretical Division: HHMW 92, 209, 247; Seidel 43

234   Governing Board deals with housing, etc.: Groves 164

235   Ordnance program machine shop: Badash et al. 54

235   Health Group and decontamination statistics: HHMW, 104–105; Hawkins 58–70, 185

235   Los Alamos Primer: HHMW 69; Serber (1992); https://fas.org/sgp/othergov/doe/lanl/docs1/00349710.pdf (July 27, 2018)

236   Serber lectures attended by about 30 people: Segrè 135

236   Condon resignation: Seidel 65

236 Lewis review committee: HHMW 69, 82; Hawkins 9, 23; Jones 490; H&A 236; Groves 162-163 attributes the idea of a review committee to Conant

237 Parsons suggested for Director of Ordnance: Groves 159–160. A proximity fuse is a radar unit small enough and robust enough to fit into a ground-fired shell or missile, and was one of the most important electronic developments of World War II

237 Unsung hero: Badash et al. 82

237 "What we were trying to do": Badash et al. 28

238 Pole piece of Harvard cyclotron laid April 14: H&A 233; Jones 86; Hawkins 7, 71, 102

238 Initially only one gram of U-235: HHMW 77

238 165-μg sample of Pu: HHMW 79; Seidel 78, 81; H&A 240; Hawkins 104; Groves 159; Williams (1943); Taschek (1943)

238 Roosevelt to Oppenheimer, June 29, 1943: http://lcweb2.loc.gov/cgi-bin/query/r?ammem/mcc:@field(DOCID+@lit(mcc/083)) (July 27, 2018). Also Hawkins 33, 36; Groves 167

238 Klaus Fuchs in Theoretical and Explosives divisions: Badash et al. 37, 62

238 Fuchs not tailed by American security: Farmelo 280

239 Groves' compartments hardly sealed: HHMW 94–95; H&A 238–239

239 Oppenheimer plan to run laboratory with staff of couple hundred: Hawkins 7; Groves 151

239 Staff at 1,100 by end of 1943: Jones 487, 498; Seidel 65; Conant 215, 262; Kelly (2006) 76. Birth figures from Szasz (1984) 18; poem from Kelly (2007) 170

240 Quality of housing and conditions at Los Alamos: Conant 122, 138, 140, 143, 214; Goodchild 125; Badash et al. 138; Los Alamos Historical Society 12; Kelly (2007) 167

240 Dorothy McKibben: Conant 56

241 Women as 30% of Laboratory staff: HHMW 99

241 SEDs: HHMW 97; Seidel 91–92; Badash et al. 57; Jones 141, 181, 208, 349, 359–360, 469, 497; Hawkins 43–46; H&A 310; For personal views of life as a SED, see Bederson (2001) and Hull (2005)

241 Security regulations restrict travel: HHMW 109; Seidel 46; Conant 111–118

241 All residents over age six issued security passes: Badash 140; HHMW 103, 107; Hawkins 37-41, 48; Groves 153, 166, 168; Truslow 24, 85; Hunner 29

242 Groves restricts liaison with other parts of Project: HHMW 209; Hawkins 33–35

242 Lindemann as politically influential, but poor grasp of modern physics; appointed advisor to Churchill, who does not appreciate strategic implications of bomb: Farmelo 5–6, 46, 114–116, 156, 163; Ruane 23–29

242 Scientific Advisory Committee: Farmelo 170–171, 185–187

243 Lindemann as Lord Cherwell: Farmelo 176

243 Hankey dismissal: Farmelo 208

243 Lindemann memo of August 27 to Churchill: Irving 110; Farmelo 188–189

243 Churchill and Chiefs of Staff: Farmelo 190–192; Ruane 29. The September 3 date is from Irving 112

243 Sir John Anderson: Ruane 30; http://en.wikipedia.org/wiki/John_Anderson,_1st_Viscount_Waverley (July 27, 2018)

244 Scientific Advisory Committee report: Farmelo 193–194

244 Roosevelt to Churchill October 12, 1941: Farmelo 194–195

244 British reluctance to tie program to America: Bernstein (1976) 206

244 Anderson and Lindemann meeting with OSRD representative: Farmelo 203–204

244 Hankey's Private Secretary as spy: Farmelo 302

245 Akers visit to United States early 1942: Farmelo 212–215

245 Churchill and Roosevelt discuss project June 20, 1942 at Hyde Park: H&A 260–284; Farmelo 208-210. Ruane 43–45 gives a critical analysis of Churchill's later recollection of the meeting

245 Anderson to Churchill, July 30, 1942: Farmelo 214; Bernstein (1976) 208

246 Bush October 1, 1942 noncommittal reply to Anderson: H&A 263–264; Ruane 46–47

246 Akers-Conant meetings, November/December 1942: Jones 99; H&A 265; Conant (2017) 282–284

246 Roosevelt initials December 15, 1942 MPC report: Fakley (1983); Szasz, *British Scientists*

246 January 13, 1943 meeting: Farmelo 222

247 Bush-Roosevelt meeting of June 24, 1943: Farmelo 229

247 FDR directive to Bush of July 20, 1943: Ruane 60–61

247 Bush-Churchill meeting of July 15, 1943: Farmelo 230

247 Combined Policy Committee: Farmelo 237; Ruane 64

248 Quebec Agreement: Stoff et al. 46–47; Farmelo 240–241; Ruane 62–70

248 Lindemann skeptical of bomb: Farmelo 285–286

248 Akers and British Mission: Farmelo 249

248 Bethe on British Mission: Fakley (1983); http://permalink.lanl.gov/object/tr?what=info:lanl-repo/lareport/LA-UR-83-5078 (July 27, 2018)

248 60 British scientists: Ruane 71

248 C. D. Howe: Williams (2000) 261

248 Staff of 300; heavy-water reactor: Atomic Heritage Foundation web page

249 von Halban replaced by Cockcroft: Farmelo 277; Williams (2000) 261

249 ZEEP reactor: https://en.wikipedia.org/wiki/ZEEP (July 27, 2018)

249 Diffusion theory and critical mass: Reed, *Physics of the Manhattan Project*, Chap. 2 and App. 6.7

249 Calculation of critical mass with tamper: Reed, *Physics of the Manhattan Project*, Sect. 2.3; Reed (2009); Hawkins 85–87

251 Muzzle velocity 1,000 m/s: Serber 56

252 Implosion; Tolman to Oppenheimer March 27, 1943: HHMW 55, 87; Serber 59; Rhodes 466-467; Hawkins 23, 138

253 Implosion discussed at March 30 and April 2 meetings: HHMW 86

253 Oppenheimer assigns Neddermeyer to research implosion: HHMW 67, 86–87

254 Range of possible weapon efficiencies: Hawkins 13

256 Initiators used 50 Curies Po: http://nuclearweaponarchive.org/Nwfaq/Nfaq8.html (July 27, 2018)

256 Production of polonium by bismuth bombardment: Hawkins 80; Hull 49. For information on Monsanto's Dayton operations, see http://moundmuseum.com/ (July 27, 2018)

257 Charles Allan Thomas: Thomas (2017)

257 Properties of Pu and bomb cores: Hawkins 154, 256–257; HHMW 285, 330; Chemistry and Metallurgy Division tasks: Hawkins 162–164; Smith (1981)

259 6 kg Pu in *Trinity* and Nagasaki bombs; predetonation probability calculations: Coster-Mullen (2010) 47; Reed (2010)

259 Seaborg speculation on Pu-240 spontaneous fission: HHMW 231; Kathren et al. 257

259 Discovery of spontaneous fission in uranium: Flerov & Petrzhak (1940); HHMW 229

259 Plutonium spontaneous fission 18 per gram per hour: Seidel 74

259 Oppenheimer invites Segrè to Los Alamos: HHMW 231-232

259 Pajarito Canyon laboratory: HHMW 3, 232, Segrè 137

260   Six counts over five months: HHMW 236

260   Relaxation of gun-bomb assembly speed: HHMW 235–236; H&A 241

261   "As white as a sheet of paper": HHMW 3, 227–228, 238–240

261   Gun-assembly method for Pu would have to be abandoned: HHMW 240; Hawkins 106–107

261   "At the present time": M1392(1), 0911–0917; Kelly (2007) 143

261   "The choice was to junk": HHMW 242

261   Reorganization plan: HHMW 129–130; 243–248

261   X, G, and R Divisions: HHMW 4, 228–229; Hawkins 75, 82, 84, 144, 173, 205–212; H&A 310–312

262   Teller's group within Theoretical Division; numerical calculations: HHMW 157

262   Rudolf Peierls replaces Edward Teller: HHMW 157–162, 179

263   Gun method engineering program: Hawkins 127–128; Jones 505

263   Albert Francis Birch: HHMW 250

264   *Little Boy* 6.5-in bore: HHMW 115; Coster-Mullen (2010) 123

264   Neutron-reflecting properties of gun barrel: HHMW 83

263   Tamper to stop projectile: HHMW 83, 117

263   Sept. 17, 1943 shot: Badash et al. 18

263   Washington Navy Yard: HHMW 116; Hawkins 128–130

264   Three new *Little Boy* guns ordered: Hawkins 130–131, 144; Jones 506

264   Dummy guns not intended for test-firing: Hawkins 223

264   Beryllium as tamper: Hawkins 162–164

264   Tungsten-carbide as "Watercress": Hawkins 224; Coster-Mullen (2010), 22–23

266   Concern with detonation altitude: HHMW 261, 344

266   Firing-process sequence: Coster-Mullen (2010) 19–22

266   Gun barrel 6 feet long, 1,100 pounds: Coster-Mullen (2010) 18; Serber 57

266   Dimensions of *Little Boy* core: Norris 409; Coster-Mullen (2010) 27

267   *Little Boy* nose nut: Coster-Mullen (2010) 27–28

267   Groves orders gun bomb to be ready July 1, 1945: HHMW 255

267   Shooting concept as first combat nuclear weapon: HHMW 262–264

267   Implosion method enjoys increasing resources fall 1943 onward: Hawkins 77, 139; HHMW 88, 135; H&A 247

267   Von Neumann suggests higher implosion velocity: HHMW 134; Hawkins 139; H&A 246

268   Kistiakowsky as buffer between Parsons and Neddermeyer: HHMW 137; H&A 247, 311–312; Hawkins 125–126, 140

268 "There was not a single experimental result": HHMW 130, 177; Hawkins 143

268 "Kistiakowsky goes nuts": HHMW 140

268 James Tuck suggests implosion lenses: HHMW 163; http://bayesrules. net/JamesTuckVitaeAndBiography.pdf (July 27, 2018)

268 Lenses about a foot across: Hull 31

268 Comp B: Hull 50

271 High explosive 5,300 pounds, bomb casing 1,100 pounds: Coster-Mullen (2010) 47, 52

271 U-238 contributes 30% of *Fat Man* yield: HHMW 161; Libby 211; Coster-Mullen (2010) 45

271 Christy core: Hawkins 91, 95, 202; Lippincott (2006, Part II) 414; Coster-Mullen (2010) 48

272 X-Division staff of some 600: Goodchild 119

272 Responsibilities of X-Division: Hawkins 241–242; HHMW 139

272 "figure out how to cast the lenses": Hull 30

274 Men work three shifts to produce lenses: Hull 50–57

274 "I used a stirrer": Coster-Mullen (2010) 43; Hull 52

274 "it just goes to show the incompressibility of water": Hull 56–57

274 One gram could finish off a hand: HHMW 320; Badash et al. 51

274 Diagnostic methods: Terminal observations: HHMW 280, 296; Magnetic method: HHMW 272; Electric method: HHMW 143–156, 271–272; Hawkins 140–142, 151, 231–235, 259; gamma-ray (radio-lanthanum) method and test shots: HHMW 268–269; external gamma-ray (betatron) method: HHMW 274–277

275 Detonator simultaneity requirements: Hawkins 237

275 Donald Hornig trial-and-error approach to detonators: Coster-Mullen (2010) 63

275 Spark-gap switches: Coster-Mullen (2010) 65

275 Timing spread down to several hundredths of a microsecond: HHMW 302; Coster-Mullen (2010), 64

275 Detonator refinements and production: HHMW 171–173, 321, 324

276 Godiva experiments: Hawkins 229–230; HHMW 341

276 Estimating critical mass by extrapolating neutron numbers with uranium spheres: HHMW 337–339; Serber (1992) 33

276 Plutonium-water solution brought to criticality: HHMW 340

276 Daghlian and Slotin accidents: Libby 202; HHMW 341–342; Hull, 106; http://en.wikipedia.org/wiki/Harry_K._Daghlian,_Jr; http://en. wikipedia.org/wiki/Louis_Slotin (July 27, 2018)

277   Dragon experiments: HHMW 346–348; R. E. Malenfant, "Experiments with the Dragon Machine," Los Alamos publication LA-14241-H (August 2005) http://www.osti.gov/energycitations/purl.cover.jsp?purl=/876514-I1Txj9/ (July 27, 2018)

278   50:50 chance of working for May 1, 1945: HHMW 293

278   "My own bets are very much against it": Hawkins 193; H&A 313; M1392(1), 0935–0936

278   July 20 target date for test: Hawkins 193–194; HHMW 312

278   Feb. 28 decision on Christy-core and Comp B design: HHMW 294, 300, 312

279   Norman Ramsey: HHMW 378

280   B-29 bomb load and range: Campbell 6; Polmar 5, 6, 73

280   Silverplate aircraft: Campbell, 6–8, 21–23, 107; HHMW 380; Coster-Mullen (2010) 13; Jones 521

280   "an ominous and spectacular failure": Campbell 42; HHMW 380; Ramsey *History of Project Alberta* 4; Russ 13

281   *Thin Man, Fat Man,* and *Little Boy*: Ramsey 4, 6; Coster-Mullen (2010) 378n23

281   1,500 bolts cut to 90: Coster-Mullen (2010) 52; HHMW 382

282   March 16 drop-test accident: HHMW 381–382; Ramsey 5; Hawkins 145–146; Campbell 9, 43, 76

282   Prototype bomber reconfigured back to original configuration: Campbell 10

282   400-lb *Fat Man* tail-end: Coster-Mullen (2010) 67

282   In-fight arming of *Little Boy*: Hawkins 225; HHMW 263

282   Army Air Force recommends freezing design, Aug. 11, 1944: Campbell 43

283   Groves & Arnold decide to organize self-sustaining Air Force unit: Groves 258

283   First B-17 across English Channel: Thomas & Morgan-Witts 20

283   Twenty-five combat missions: Groves 258; Norris 318; Thomas & Morgan-Witts 20

283   Ramsey briefs Tibbets: Thomas & Morgan-Witts 20–21

284   B-29 test flights begin October, 1944: Campbell 11–12

284   Second group of B-29's: Campbell 1, 15, 26, 159

284   *Enola Gay* and *Bockscar*: Campbell 26, 159–160, 172, 191; Thomas & Morgan-Witts 95; Norris 318, 323

284   155 test-drops at Wendover: HHMW 383–384; H&A 319; Ramsey 9–11

284   X-units with 64 cables: Coster-Mullen (2010) 66

285   *Fat Man* contact fuses: Coster-Mullen (2010) 27

285   Hardened armor plate: Russ 22

285   Tinian Island and air base: Russ 48; Jones 524

286   Assembly huts and pits on Tinian: http://www.globalsecurity.org/military/facility/tinian.htm (July 27, 2018); HHMW 386–387

286   "Silverplate" code word: HHMW 387–388

287   "Implosion gadget must be tested in a range": HHMW 174; Norris 395

287   No Indians to be displaced: Norris 396; Groves 289

287   Possible *Trinity* test locations: Hawkins 142, 267

287   "the most disagreeable man I ever met": HHMW 310; Szasz (1984) 28

289   Alamogordo Army Air Field: Jones 478; H&A 318; Norris 396

289   McDonald Ranch House: Bainbridge 3; http://en.wikipedia.org/wiki/McDonald_Ranch_House (July 27, 2018)

290   Naming of *Trinity*: Norris 397

290   Trinity in Hindu culture: Szasz (1984) 41

290   Base Camp: HHMW 310

290   Shot tower: Coster-Mullen (2010) 7; Szasz (1984) 34

291   Instrument stations: HHMW 311

291   Scientists at shelters: Los Alamos Historical Society 46; Bainbridge, 30

291   Campañia Hill observers: Rhodes 653, 668, 672

291   Compton declines to attend test: Hunner 65; Compton 213–214

292   Cowpuncher Committee: Hawkins 175–176, 194; HHMW 316

292   108-ton test shot: Bainbridge 8; HHMW 361

292   Instruments at scaled distances; *Trinity* test overwhelms many instruments: Hawkins 202; HHMW 332; Bainbridge 7, 9; Broyles (1982)

292   1,000 Curies of beta activity: Hawkins 270–271; H&A 376; Bainbridge 11

292   60,000 psi containment pressure: HHMW 366

293   Bainbridge on Jumbo: Bainbridge 5

293   "It was a very weighty albatross": Bainbridge in Wilson 222

293   Fate of Jumbo: Norris 398–399; Hawkins 247–248, 270; Groves 288; Neuenschwander (2004); Los Alamos Historical Society 31

293   Vetting of experiments: HHMW 351–352, 362; Bainbridge 25–26; Hawkins 273

294   Six chief groups of experiments: Bainbridge 60–69; Hawkins 277–280; HHMW 351–354

294   Gold foils to measure neutron flux: HHMW 357

294   Fission fragments in soil: HHMW 358

294   Pressure gauges to measure energy release: HHMW 359

294   500 miles of wires and cables: Libby 219

294   Films mailed to dummy addresses: Bainbridge 36

294   Security contingent of 160 men: HHMW 351–352, 362; Bainbridge 25–26; Hawkins 272

294   Eighteen-hour workdays: Jones 478–480; Hawkins 271–272

295   *Trinity* Pu hemispheres complete July 2: Norris 400; HHMW 330

295   *Trinity* tamper machined July 6, best lenses selected: HHMW 333, 365; Bainbridge 39

295   Meteorological service for 108-ton test excellent: HHMW 363; Szasz (1984) 68–69; Bainbridge 12

295   Hubbard's possible test dates: Hawkins 273; HHMW 364; Bainbridge 28

295   Churchill assent to use of bomb: Farmelo 304

295   British and Canadians informed of pending bomb use: Szasz (1984) 69; Nichols 183; H&A 372

296   Thunderstorms two hours before scheduled test: Szasz (1984) 72–75

296   Colonel Holzman: Norris 402

296   Groves dismisses forecasters: Groves 291–292; Norris 404

296   Rehearsal tests: These dates Bainbridge 28; other sources differ

296   Hemispheres conveyed July 11, initiators next day: HHMW 333

296   Bethe's analysis of magnetic-method test: HHMW 327

296   "From this crude lab": Los Alamos Historical Society 44

297   Kistiakowsky chooses final assembly time: HHMW 367

297   Final assembly of high-explosive components: Hawkins 196, 274

297   Core/explosive thermal equilibrium: HHMW 368–369

297   Bomb raised to tower: Hawkins 274; HHMW 370

298   Arming party; "Oppenheimer was really terribly worried": Hawkins 275; HHMW 371; Los Alamos Historical Society 49; Kelly (2007) 298

298   "or I will hang you": Szasz (1984) 76–78

298   Groves not amused by Fermi: Groves 297; Szasz (1984) 59

299   Rabi wins $102: Los Alamos Historical Society 44; Conant (2005) 299, 316

299   "My personal nightmare": Bainbridge in Wilson 226

299   Oppenheimer down to 115 pounds: McCullough 396

299   "The scene inside the shelter": Groves 435–436

299   Final *Trinity* countdown: Szasz (1984) 82

299   Time of *Trinity* test: The 5:29:15 time is taken from Bainbridge's official report of the test, p. 31. Error estimate was +20 s or −5 s. Los Alamos Historical Society 51–53; Seidel 33, 79

299 "As the time interval grew smaller": Groves 435–437

300 "The effects could well be called unprecedented": Hawkins 275; Groves 437–438; Norris 408

300 'The war is over": Seidel 100

301 Fermi's estimate of *Trinity* yield: http://www.atomicarchive.com/Docs/ Trinity/Fermi.shtml (July 27, 2018); HHMW 372

301 "A foul and awesome display": Bainbridge in Wilson 230

301 Bethe, Kistiakowsky and Bradbury descriptions of test in Los Alamos Historical Society 53, 54

301 Conant's description of test: Conant (2005) 309

302 Rabi's description of test: Serber xvii, quoted from Rabi 138

302 Emilio Segrè's description of test: Segrè 147

302 Robert Christy's description of test: Lippincott (2006, Part II) 416

302 Charles Thomas' description of test: Thomas 135–139

302 Farrell "long-hairs" comment: Lamont 237

302 "Oppie, you owe me $10": Badash et al. 60

304 "July 1945 at Alamogordo": Kelly (2007) 146

304 Fireball physics discussed in Barasch (1979)

306 *Trinity* prompt radioactivity release: Hawkins 276 estimates one trillion Curies; a more detailed calculation by the author gives ~14 trillion Curies: Reed, "Counting the Curies" (2016)

307 70 acres of Trinitite: HHMW 374; Szasz (1984) 137

308 "Operated on this morning": Norris 406

308 Stimson takes report to Marshall and Truman: Groves 304, 433–440; McCullough 430–431; Farmelo 298

308 "Discussed Manhattan": Ferrell 30

308 "Second Coming in Wrath": Farmelo 300

308 "It resulted from the atomic fission": Norris 408, 663n44

309 Truman informs Stalin of a new weapon: H&A 394; Szasz (1984) 147; McCullough 437, 442–443

309 Soviets knew of features of implosion bomb five months before test: Albright & Kunstel 121

309 Newspaper accounts of *Trinity*: Groves 301

309 5 tons per square inch: HHMW 376

309 Soil samples indicate 18.6 kt: HHMW 374–376; Szasz (1984) 117

309 21.4 ± 2.0 kt estimate from Semkow et al. "Modeling the Effects of the Trinity Test" (2006); 22.1 ± 2.7 kt figure from Hanson et al. "Measurements of extinct fission products in nuclear bomb debris: Determination of the yield of the Trinity test 70 y later" (2016); also http://www.fas.org/sgp/othergov/doe/lanl/la-1398.pdf (July 27,

2018). Officially, the *Trinity* yield is listed at 21 kt: Coster-Mullen (2010) 41. See United States Nuclear Tests July 1945 through September 1992. (U. S. Department of Energy, Nevada Operations Office, report DOE/NV-209 REV16, available at https://www.nnss.gov/docs/docs_LibraryPublications/DOE_NV-209_Rev16.pdf (July 27, 2018)

310 Fallout from *Trinity:* Szasz (1984) 115–117, 121

310 10 rems/h recorded at North shelter: Szasz (1984) 124–128; Bainbridge 31

310 Exposed cattle: Szasz (1984) 132–134; $1350 compensation: Hawkins 298

310 X-ray films contaminated by *Trinity:* Szasz (1984) 134–135; Badash et al. 75

311 Sept. 9, 1945 "media day": Szasz (1984) 160–163

311 Trinitite not water-soluble: Szasz (1984) 137–139

312 "Would have to eat some 100,000 kg": Szasz (1984) 166–170; Fey (1967)

312 "Exposure ... less than 0.2%": Hansen & Rodgers (1985)

313 *Little Boy* sent to Tinian in two shipments: Weintraub 74

313 *Indianapolis* arrives Tinian July 28: Norris 410–411; the date is indicated in M1109(1), 570

313 *Little Boy* target rings transported by C-54 aircraft: HHMW 389–390; M1109(1), 570; Ramsey (1982)

313 Theoretical Division estimate of 13.4 kt for *Little Boy:* HHMW 265–266

313 Japanese government debates Potsdam Declaration; "silent contempt": H&A 396; Rhodes 692–693

314 Three B-29's depart Kirtland with high-explosive implosion preassemblies: M1109(1), 570

314 Unit L6 test-dropped July 31: Campbell 46

314 *Fat Man* F13 and F33 units: Campbell 46; Coster-Mullen (2010) 16, 68

315 Training for 509th crews: Campbell 19–20, 26–27

315 *Indianapolis* disaster: Weintraub 296–297; Rhodes 694–695; http://www.ussindianapolis.org (July 27, 2018); *New York Times,* July 14, 2001, p. A9; *New York Times,* Aug. 15, 1945, p. 1

316 Handy orders to Spaatz: Norris 412–413; Groves 308–309

317 Truman diary entry for July 25: Ferrell 31. A copy of the diary entry appears in Coster-Mullen (2010) 298

318 Groves to Marshall, July 30, 1945: https://nsarchive2.gwu.edu//NSAEBB/NSAEBB162/45.pdf.

Chapter 8—The German Nuclear Program: The Third Reich and Atomic Energy

336   Bagge uranium sluice: Irving 98, 129, 218, 277

336   Houtermans: Irving 101–102; Khriplovich (1992)

337   Jones, Walsh, and Tronstad: Irving 107, 112

338   360 kg heavy-water by end of 1941: Irving 113

338   L-II pile with 142 kg uranium oxide: Irving 114

338   Heisenberg "open road": Irving 114

338   Heisenberg Copenhagen visit: Irving 115; Cassidy (2000, 2009, 2017)

340   heavy-water production 140 kg/month: Irving 124

340   Groth centrifuge explodes: Irving 129

340   Schumann December 5, 1941 letter: Irving 117

340   December 16, 1941 conference and Army decision to reduce funding: Irving 117

341   February 1942 conferences: Irving 118–122; Cassidy (2017) 54–55

341   131-page Army report: Irving 123

341   3.5 tons powdered uranium: Irving 130

341   L-III pile fire; L-IV shows 13% neutron increase: Irving 130–131

342   13,700 kg metallic uranium: Irving 151

342   June 6, 1942 meeting with Speer: Irving 132–134

342   Heisenberg "pineapple" comment: Irving 103, 134

342   Speer approves construction of Berlin shelter: Irving 135

343   Speer briefs Hitler June 23: Irving 136

343   Rudolf Mentzel: Irving 144

343   L-IV pile explosion: Irving 137–138

344   *Freshman* and *Gunnerside*: Irving 155–166; 178–196; Bascomb Chaps. 7–19

347   Heisenberg "director at" KWIP: Irving 166

347   Diebner cubes concept: Irving 167

347   6,802 cubical voids: Irving 168

347   Esau as Plenipotentiary: Irving 144, 175

347   Vögler and Esau, February 4, 1943: Irving 177

348   Esau budget: Irving 199

348   Diebner prefers 6.5-cm cubes: Irving 201

348   frozen heavy-water pile with 232 kg U: Irving 201

348   cubes on wires concept: Irving 202

349   evacuation of KWIP to Hechingen: Irving 219

349   500-kg lattice pile: Irving 223

349   Vemork bombing raid: Irving 225–226; Bascomb Chap. 21

349   Farben plant budgeted for 800,000 Reichsmarks: Irving 227

349   Gerlach approached to lead effort: Irving 232–234; 254–257

351   *D/F Hydro* sinking: Irving 236–248; Bascomb Chaps. 25–28

351 *Alsos* mission: Irving 257–258; Goudsmit (1947); Cassidy (2017) 59–75

352 *Alsos* mission in Italy: Irving 258–259

352 Maj. Calvert: Irving 260–263

352 KWIC bombing, Feb. 15, 1944: Irving 264

353 Farben works bombed July 28, 1944: Irving 281

353 sluice produces 2.5 g enriched uranium: Irving 277

353 Diebner relocates to Stadtilm: Irving 302

353 Berlin bunker pile with 124-cm cylinder: Irving 274–276

353 Gerlach's house bombed: Irving 278

353 Haigerloch chosen as pile site: Irving 279

354 Goudsmit: Irving 286; Goudsmit *Alsos*; Walker *German National Socialism*

354 Joliot captured: Irving 289–290

354 Goudsmit & Calvert in Brussels: Irving 292–294

355 Strasbourg liberated: Irving 305–308

355 B-VII pile in Berlin: Irving 312–313

355 B-VIII pile Irving 314

355 Gerlach orders B-VIII evacuation to Haigerloch: Irving 315; 318–322

356 B-VIII needs to be 50% larger: Irving 322

356 Bothe captured: Irving 327

357 Calvert traces ore to Stassfurt: Irving 332–333

357 Haigerloch pile dismantled: Irving 334–335

357 Heisenberg to Urfeld: Irving 335

357 Uranamium cubes: Koeth & Hiebert (2019)

357 von Weizsäcker's cesspit: Irving 338

358 Pash captures Heisenberg: Irving 342.

Chapter 9—Hiroshima and Nagasaki

361 63 and 58 missions: Thomas & Morgan-Witts 35

361 Units and staffing of 509th Composite Group: Campbell 25–26; Norris 319; Groves 259; Jones 521–522

362 34,000-foot ceiling: Thomas & Morgan-Witts 43

362 509th activated Dec. 17, 1944: Thomas & Morgan-Witts 44, 58; Coster-Mullen (2010) 12

362 LeMay headquarters on Guam: Thomas & Morgan-Witts 68; Rhodes 591

362 March 9–10, 1945 raid on Tokyo: Thomas & Morgan-Witts 75

363 5,000 American dying each week: Thomas & Morgan-Witts 56

363   Battle and casualty statistics: Giangreco (1997); Giangreco (2017). For a more detailed examination of revisionist analyses of the bombings, see Maddox (2007); for a collection papers examining President Truman's "decision" to use the bomb, see Walker (2016)

363   No Japanese unit had ever surrendered: McCullough 438

363   10-year guerilla war: Thomas & Morgan-Witts 120; Polmar 12

363   560,000 Japanese troops in Kyushu: Weintraub 399

366   Groves instructed to see to targeting himself: Groves 266–267

367   Good enough for visual bombing: Groves 268

367   Civilians assume military ranks: Jones 527; Groves 282; Serber & Crease 97

370   Stimson deletes Kyoto from target list: Jones 529

371   Groves tries to get Kyoto back on target list: Norris 386–388; Jones 530; H&A 365; Groves 274–276, 309

371   "All your local military advisors": M1109(1), 642 and 653

372   Jeffries report can be found in e.g., Sherwin 315–322

373   Tolman Committee on Postwar Policy: Compton 233; H&A 325

373   Bush-Conant report: Jones 567

374   Truman meets with Roosevelt only eight times: Thomas & Morgan-Witts 85–86

375   "An immense project": Norris 375

375   "Perfecting an explosive": Norris 375

375   Results fell squarely in Truman's lap: McCullough 379

375   Groves report to Stimson and Truman: http://www.gwu.edu/%7Ensarchiv/NSAEBB/NSAEBB162/3a.pdf (July 27, 2018)

377   Membership of Interim Committee: H&A 344–345; McCullough 390

379   Idea of a demonstration shot: Norris attributes to Compton 391; H&A attribute to Lawrence 358; Compton 238

379   Stimson briefs Truman June 6: Sherwin 210; H&A 360; McCullough 392

381   Franck report to Harrison: Price (1995); H&A 366–367; Jones 533

381   Harrison asks Scientific Panel for comments on Franck report: Stoff 148–149; H&A 367 gives this as June 16

383   Bard meets with Truman: H&A 370, 693n40; Smith 52–53

384   Szilard memorandum: Lanouette 260–274

386   Compton sends petition results to Nichols: Lanouette 274; H&A 399–400

386   "To me the obvious answer": Peierls 204–5

387   Twentieth Air Force under Twining: Norris 413

387   Hiroshima bombed May 7 and June 2, Nagasaki July 22 and Aug. 1: Weintraub 350, 397

387   Projector jammed: Norris 417; Weintraub 386, 396–398; Polmar 31

387   LeMay authorizes mission order August 5: Norris 417

387   Hiroshima population 340,000: USSBS 6

388   *Little Boy* preparation: Weintraub 414; Coster-Mullen (2010) 275

389   Farrell cables Groves re take-off time: Norris 417

389   Farrell cables Groves with change in bomb-arming plan; crew briefings begin: Weintraub 414–415; Norris 417

389   Weather planes depart 1:37 a.m.: Weintraub 415

389   Like the opening of a drugstore: Russ 61; Ramsey 15; Weintraub 415

389   *Enola Gay* take-off: Coster-Mullen (2010) 34; Russ 62

390   *Little Boy* arming procedure: Coster-Mullen (2010) 107–109, 159; M1109(1), 363

391   Robert Lewis' log: Weintraub 417

392   Lawrence Johnston: Coster-Mullen (2010) 37, 93–98; HHMW 171–172

393   Parsons' log: Ramsey 15; Coster-Mullen (2010) 107–108

393   "I had a feeling": Weintraub 418

393   "There'll be a short intermission": Coster-Mullen (2010) 36

394   Aioi bridge as aiming point: Rhodes 709

395   Tibbets executes escape maneuver: Coster-Mullen (2010) 38

395   "If I live a hundred years": Weintraub 424; Rhodes 711

395   Cloud over Hiroshima: Ramsey (1982)

395   Groves receives 6:45 p.m. call: Norris 418–419; Groves 320–321

395   "In order to ease the growing tension": Groves 321–322

395   Groves receives Parson's coded message 11:35 p.m.: Norris 419, Groves 322. Curiously, in Groves' autobiography, "test" is plural. In his report to Marshall, M1109(1) 339, the word is in the singular.

396   Tibbets awarded DSC: Norris 419

396   Farrell sends Groves lengthier cable: Groves 323

397   Emulsion gone from camera plane film: HHMW 394, 397

397   Reconnaissance had to wait: Ramsey 16

397   Text of Truman release: Ferrell 48–51

400   Groves-Oppenheimer conversation: http://blog.nuclearsecrecy.com/2012/04/04/weekly-document-the-hiroshima-phone-call-1945/ (July 27, 2018); http://www.dannen.com/decision/opp-tel.html (July 27, 2018)

401   Sam Cohen's description of Oppenheimer: Cohen (1983) 21–22

401   Alice Smith: Smith 77

402 McAllister Hull quote: Hull 73
402 Hans Bethe quote: Palevsky 70
402 Six million leaflets: Norris 419–420
403 Russia declares war on Japan: Weintraub 473–477
403 Second strike originally scheduled for August 20: Groves 341–342; Ramsey 16
404 JANCFU: Russ 65–68; Coster-Mullen (2010) 53
405 Field Order 17: Coster-Mullen (2010) 321
406 Nagasaki crew briefing begins 00:30: Coster-Mullen (2010) 69
406 Inoperative fuel pump: Coster-Mullen (2010) 70, 404–405; Sweeney 203–204
406 *Bockscar* take-off time from Ramsey 17; Coster-Mullen (2010) 71
407 *Big Stink* nowhere to be seen; Coster-Mullen (2010) 72
407 50-mile dog-leg sweeps: Coster-Mullen (2010) 73; Sweeney 212
407 Good weather reported at both targets: Coster-Mullen (2010) 73–74; Sweeney 211
407 Robert Serber ejected from plane: Serber & Crease 113–114
407 *Bockscar* departs Kokura 11:30 a.m.: Sweeney 214–215
408 Clouds between 6,000 and 8,000 feet: Sweeney 216
408 Radar-bombing order: Coster-Mullen (2010) 76; Sweeney 217
409 Five shock waves: Coster-Mullen (2010) 77–78; Groves 346
409 Spitzer transmits strike report: Sweeney 220
409 Leonard Cheshire observation of bomb cloud: Coster-Mullen (2010) 80
409 Mayday call: Coster-Mullen (2010) 80
409 "I was so mentally and physically exhausted": Coster-Mullen (2010) 81–82; Sweeney 226
409 7 or 35 gallons of fuel: Coster-Mullen (2010) 81
409 More time over enemy territory: Coster-Mullen (2010) 84
409 Reconnaissance of Nagasaki unavailable for a week: Ramsey (1982)
411 Emperor in favor of ending war: Weintraub 499–506
411 "any demand for modification": Weintraub 508
411 "the authority of the emperor": Weintraub 513–514
411 "next bomb of the implosion type": Norris 413, 416; https://nsarchive2.gwu.edu//NSAEBB/NSAEBB162/67.pdf
412 Henry Wallace diary quote: Blum 473–474
412 Tactical use of bombs: Settle 132–135; Bernstein (1991)
412 Allied reply picked up in Tokyo: Weintraub 547, 553
412 Imperial rescript drafted: Weintraub 581–583, 591
412 11:48 p.m. coded message: Weintraub 596

412 B-29 daylight raids: Weintraub 604
413 Hirohito statement broadcast at noon: Weintraub 618; McCullough 461
413 "A new and most cruel bomb": Weintraub 594
413 Manhattan Project Atomic Bomb Investigating Group: Jones 543
413 Japanese survey estimates 71,000 dead: Weintraub 431
416 Robert Serber measures shadow: Serber & Crease 129, 137–138
418 Groves prepares Smyth report for release: Groves 348–351
418 Stimson and Truman clear Smyth report: Groves 351; Jones 561
419 Groves' views on secrecy: Norris 437
420 Farm Hall: The Farm hall story is related in a number of sources: Irving 1–8; Bernstein & Cassidy (1995); Bernstein (1996); Cassidy (2017); Logan (1996); Walker (1995). von Laue's comment on the *Lesart* appears in Bernstein & Cassidy (1995).
425 "I have no qualms": Norris 426
425 Oppenheimer speech: Hawkins 293–294; Pais 48–58; Groves 355; HHMW 401–402.

Chapter 10—Epilogue

427 *Crossroads* test: Weisgall (1994)
428 Tsar Bomba: http://en.wikipedia.org/wiki/Tsar_Bomba (July 27, 2018); Garwin & Charpak 64–65
429 Non-US nuclear programs: Reed (2015a) and references therein; Albright & Stricker
434 Hirschfelder: In Badash et al. (1980) 67–88.

Glossary

25 Manhattan Engineer District code for uranium-235 from 92-U-235.

49 Manhattan Engineer District code for plutonium-239 from 94-Pu-239.

Activation energy Generic term for energy that must be supplied to cause a reaction to happen; see also *Fission barrier* and *Coulomb barrier*. In nuclear reactions, activation energies are usually expressed in millions of electron volts (MeV).

AEC Atomic Energy Commission (United States). Succeeded by the Nuclear Regulatory Commission (NRC).

Alpha (α) decay Natural radioactive decay mechanism characteristic of heavy elements such as radium and uranium in which a nucleus ejects an alpha-particle, which is a nucleus of helium-4.

Alsos Code-name of a World War II Allied intelligence-gathering unit deployed to assess Italian and especially German work in the scientific and technical areas, including nuclear physics.

Ångstrom Unit of length equal to one ten-billionth of a meter. Characteristic of the effective sizes of atoms.

Atomic number (Z) Number of protons in the nucleus of an atom. Identifies the chemical element to which the atom corresponds.

Atomic weight (A) The weight of an atom in atomic mass units. The symbol *A* is also used to designate the mass number or nucleon number, the total number of protons plus neutrons within a nucleus.

Barn (bn) Unit of reaction cross-section equal to one trillionth of one trillionth of a square centimeter.

Becquerel (Bq) A unit of rate of radioactive decay; 1 Becquerel = 1 decay per second. See also *Curie*.

© Springer Nature Switzerland AG 2020
B. C. Reed, *Manhattan Project*,
https://doi.org/10.1007/978-3-030-45734-1

Beta (β^-) decay Natural radioactive decay mechanism of neutron-rich nuclei wherein a neutron spontaneously transmutes into a proton plus an electron, ejecting the latter to the outside world. The electron is also known as a β^- particle. The resulting nucleus is one element heavier in the Periodic Table than the parent nucleus.

Binding energy A form of energy which is created from mass, and which can be transformed back into mass. In reactions where the mass of the output product(s) is less than that of the input reactant(s), binding energy is said to be liberated ($E = mc^2$), and the energy appears in the form of kinetic energy of the products. If the mass of the output products is greater than that of the input reactants, kinetic energy from the input reactants is transmuted into mass. See also *Mass defect.*

Black oxide Uranium oxide: U_3O_8.

Bockscar Name of the B-29 bomber which carried the Nagasaki *Fat Man* nuclear weapon.

Brown oxide Uranium oxide: UO_2

B-Pile First large-scale (250 megawatts) nuclear reactor constructed at the Hanford Engineer Works (*HEW*, Washington) for the purpose of breeding plutonium. B-pile began operation in late 1944, and was soon followed by the D and F piles at the same site.

Calutron A device based on a *Cyclotron* which is used for separating isotopes of different atomic weights by ionizing them and passing them through a strong magnetic field; a contraction of *Cal*ifornia *U*niversity cyclo*tron*. See also *Mass spectroscopy.*

CEW Clinton Engineer Works, Tennessee. Location of Manhattan Project uranium enrichment facilities.

CIW Carnegie Institution of Washington.

Combined Policy Committee (CPC) American-British-Canadian committee established in August, 1943, to coordinate nuclear research and to serve as the focal point for interchanging information.

Control rod Device made of a neutron-capturing material that is used in a nuclear reactor to control the reaction rate. Cadmium and boron are excellent neutron capturers.

Coulomb barrier Amount of kinetic energy that an "incoming" particle or nucleus which is approaching a "target" nucleus must possess in order to overcome the repulsive electrical force between protons within the two nuclei in order to collide and induce a nuclear reaction with the target nucleus. Typically measured in millions of electron volts (MeV).

CP-1 Critical (or Chicago) Pile number 1, the first nuclear reactor to achieve a self-sustaining nuclear chain reaction. This uncooled, graphite-moderated device operated for the first time on December 2, 1942, under the direction of Enrico Fermi.

Critical mass Minimum mass of a fissile material necessary to achieve a self-sustaining fission chain reaction, taking into account loss of neutrons through the surface of the material. If the material is not surrounded by a neutron-reflecting tamper, the term "bare critical mass" is used. For uranium-235 and plutonium-239, the bare critical masses are about 45 and 17 kg.

Cross-section A quantity which measures the probability that a given *nuclide* will undergo a particular type of reaction (fission, scattering, capture …) when struck by an incoming particle. Cross-sections are expressed as areas in *barns*, and depend on the type of particle being struck, the type of striking particle, and the energy of the striking particle.

Curie (Ci) A unit of rate of radioactive decay equal to 37 billion decays per second, approximately the alpha-decay rate of one gram of freshly-isolated radium-226. See also *Becquerel.*

Cyclotron A modified mass spectrometer (see *Mass spectroscopy*) used for accelerating electrically charged particles to very great energies using electric and magnetic fields.

Degussa German Gold and Silver Exchange Corporation.

Diffusion Generic term for the passage of particles through space. The speed of the particles depends on their mass and the temperature of the environment. In the Manhattan Project, uranium was enriched by both gaseous and thermal diffusion processes.

Dragon machine Colloquial name for a device developed at Los Alamos wherein a slug of uranium-235 would be dropped through a hole in a plate of uranium-235 to momentarily induce a fast-neutron fission chain reaction.

Enola Gay Name of the B-29 bomber which carried the Hiroshima *Little Boy* nuclear weapon.

Enrichment Generic term for any process which alters the abundance ratio of isotopes in a sample of some input feed material. Usually used in the sense of a process which increases the number of fissile uranium-235 nuclei in comparison to the number of non-fissile uranium-238 nuclei. In the Manhattan Project, both electromagnetic and diffusion enrichment techniques were employed.

eV Electron-volt. A unit of energy used in atomic and nuclear physics research. Chemical reactions typically involve energy exchanges of a few eV. See also MeV.

Fat Man Code name for the Nagasaki implosion-type plutonium bomb, which achieved an explosive *yield* of about 22 kt.

First criticality Moment in the detonation of a nuclear weapon when the core first achieves conditions necessary for a self-sustaining chain reaction. Compare *Second criticality.*

Fissile A fissile material is one whose nuclei will undergo fission when struck by bombarding neutrons of any energy. Uranium-235 and plutonium-239 are both fissile. Fissile is a subset of *Fissionable.* See also *Fission barrier.*

Fission Nuclear reaction wherein a nucleus splits into two roughly equal fragments, typically accompanied by a significant release of energy ($\sim$200 MeV). Fission may be induced by striking the nucleus with an outside particle (usually a neutron), but also happens spontaneously in some heavy elements. Compare *Fusion.*

Fission barrier Minimum amount of kinetic energy a bombarding neutron must possess in order to induce fission in a target nucleus, typically measured in millions of electron volts. For elements in the middle of the Periodic Table, the fission barrier can be as high as ~55 MeV, but for heavy nuclei such as those of uranium atoms is on the order of only 5–6 MeV. In these latter cases the barrier may be low enough to be exceeded by the *binding energy* liberated upon neutron capture, rendering a nuclide *fissile*.

Fissionable A fissionable material is one whose nuclei can be made to fission when struck by bombarding neutrons. In practice, the term is usually reserved for materials that fission only under bombardment by "fast" neutrons, typically of kinetic energy ~1 MeV or greater. Compare to *Fissile* above. Uranium-238 is fissionable, but not fissile.

Franck report Document prepared by University of Chicago scientists in June, 1945, addressing political and social problems associated with nuclear weapons. Now considered a founding document of the nuclear non-proliferation movement. See also *Jeffries report*.

Frisch-Peierls memorandum Memorandum prepared in early 1940 by Otto Frisch and Rudolf Peierls at Birmingham University, which alerted British government authorities to the possibility of fission bombs.

Fusion Nuclear reaction wherein two nuclei "fuse" to form a heavier nucleus, typically accompanied by an energy release of a few or few tens of MeV. Used in fusion weapons, which are known colloquially as "hydrogen bombs." Fusion reactions liberate less energy than fission reactions, but yield more energy per mass of reactant nuclei and often generate particles which can catalyze further fission and fusion reactions. Compare to *Fission*.

Half-life Characteristic time required for one-half of the nuclei of a naturally-decaying isotope to undergo a specific decay process. Half-lives vary from tiny fractions of a second to billions of years.

Heavy water A form of water in which the hydrogen atoms are replaced with deuterium, an isotopic form of hydrogen. Chemical symbol D_2O. D designates a deuterium, or "heavy hydrogen" nucleus, ^2_1H. Heavy water occurs naturally, and can be extracted from ordinary water. Heavy water is of interest in nuclear power and research as it makes an excellent neutron *moderator*.

Heereswaffenamt War Office (Germany).

HEW Hanford Engineer Works, Washington state. Location of Manhattan Project plutonium production facilities.

Hex Colloquial term for uranium hexafluoride, UF_6.

Hibakusha Japanese term for people who survived both the Hiroshima and Nagasaki bombings.

Implosion A chemical explosion which is directed "inwards." In the context of nuclear weapons, used to crush an initially sub-critical mass to critical density.

Initiator Device at the core of a nuclear weapon that releases neutrons to initiate the chain reaction. In the Manhattan Project, initiators were also known as Urchins.

Interim Committee Advisory group established by Secretary of War Henry Stimson in May, 1945, to advise on postwar atomic-energy planning.

Isotope See also *Nuclide*. Nucleus or atom of an element that has the number of protons characteristic of the element (*Atomic number*), and some specific number of neutrons. All nuclei of a given element have the same number of protons, but different isotopes of an element have different numbers of neutrons. Different isotopes of a given element consequently have different *Atomic weights*.

Ivy Mike First true American thermonuclear (fusion) weapon, detonated November 1952. Yield ~10.4 megatons.

Jeffries report A document prepared by University of Chicago scientists in late 1944 describing anticipated postwar research and industrial applications of nuclear energy. Also known as the "Prospectus on Nucleonics." See also *Franck report*.

Joe-1 Western term for the first test of a Soviet nuclear weapon, 1949.

Jumbo Name of a 200-ton steel vessel that was intended to be used to contain the first test explosion of a nuclear weapon. Jumbo was never used, and parts of it still remain at the *Trinity* site.

K-25 Code name for the gaseous diffusion plant at the Clinton Engineer Works (*CEW*).

Kiloton (kt) A unit of energy equal to that released by the explosion of 1000 metric tons of conventional explosive (1 metric ton = 1000 kg), commonly used to quantify the energy *yield* of nuclear weapons. Equal to nearly 1.2 million kilowatt-hours. World War II-era nuclear weapons had yields in the 10–20 kt range.

KWIP Kaiser-Wilhelm Institute for Physics (Germany)

Lewis Committee There were various Lewis Committees during the Manhattan Project, all involving MIT chemical engineer Warren Lewis. The most important ones reviewed the entire atomic-energy program at the time the CP-1 reactor went critical in late 1942, and the proposed research program at Los Alamos in March/April 1943.

Little Boy Code name for the Hiroshima gun-type uranium fission bomb, which achieved a *yield* of about 13 kt.

Mass defect Difference in mass between an "assembled" nucleus and the sum of the masses of the individual protons and neutrons that comprise it; usually expressed as an equivalent amount of energy. All stable nuclei have masses less than the sum of the masses of their constituent *nucleons*.

Mass spectroscopy An experimental technique for determining masses of atoms to high precision. Ionized atoms or molecules are directed into a region of space containing a magnetic field; the trajectories of the particles consequently depend on their mass. By noting where particles "land," masses can be accurately measured. See also *Cyclotron* and *Calutron*.

MAUD committee British government committee established in response to the *Frisch-Peierls memorandum* to investigate possible military uses of nuclear fission. In its July, 1941, report the committee analyzed the possibilities for fission bombs.

MED Manhattan Engineer District of the United States Army.

Megaton (Mt) A unit of energy equal to that released by the explosion of one million metric tons of conventional explosive, commonly used to quantify the energy release of extremely powerful nuclear weapons. Equal to nearly 1.2 billion kilowatt-hours.

Metallurgical Laboratory Code name for the atomic research laboratory at the University of Chicago, directed by Arthur Compton. This laboratory had particular responsibility for development of nuclear reactors and plutonium-separation chemistry.

MeV Mega electron-volt; one million electron-volts. Nuclear reactions typically involve energy exchanges of a few MeV. See also electron-volt (eV).

Military Policy Committee (MPC) Established in September, 1943, by Secretary of War Henry Stimson to advise on development and use of nuclear weapons. The MPC acted as a sort of Board of Directors of the Manhattan Project.

Moderator Material within a nuclear reactor which slows high-energy neutrons to "thermal" velocities in order to increase their chance of fissioning U-235 nuclei. Graphite and heavy water make excellent moderators. Ordinary water can also be used, but requires a reactor fueled with enriched uranium.

MW Megawatt (one million Watts). A unit of power for quantifying the rate of generation or consumption of energy.

NAS National Academy of Sciences (United States).

NDRC National Defense Research Committee. Established by President Roosevelt in June, 1940, to support and coordinate research conducted by civilian scientists which might have military applications. The Uranium Committee was absorbed into the NDRC when the latter was established, and the NDRC was absorbed into the *OSRD* in July, 1941.

Neutron Electrically neutral constituent particle of atomic nuclei. Given the number of protons in the nucleus (*Atomic number*), the number of neutrons in a nucleus dictates the *isotope* of the element involved. Neutrons can be thought of as a form of "nuclear glue" which holds nuclei together against repulsive electrostatic forces that protons exert on each other.

Neutron number (N) Number of neutrons within a nucleus. The number of neutrons N plus the number of protons Z (*Atomic number*) totals to the *Nucleon number* (also known as mass number) A. See also *Atomic weight*.

NBS National Bureau of Standards (United States).

NRL Naval Research Laboratory (United States).

Nucleon Collective term for neutrons and protons.

Nucleon number (A) Total number of protons plus neutrons within a nucleus, always an integer number. See *atomic number* and *neutron number*.

Nuclide Generic term for a nucleus of a given number of protons and neutrons. Notation: $^{A}_{Z}X$, where X is the symbol for the element involved, Z is the number of protons (*Atomic number*), and A is the total number of protons plus neutrons (*Atomic weight*; sometimes known as mass number or *nucleon number*). Essentially synonymous with *Isotope*, except that the latter term is usually employed in the context of referring to nuclides of a given element, which will all have the same Z value but different atomic weights.

Nucleus Positively-charged core of an atom, comprising protons and neutrons.

Operation Freshman Ill-fated commando raid staged in November, 1942. The target of the raid was a heavy-water plant in Vemork, Norway, which had been seized by Germany. Commandos were to be landed in gliders, but both crashed, killing over thirty men. See also *Operation Gunnerside.*

Operation Gunnerside Successful commando raid mounted against a heavy-water plant in Vemork, Norway, February 1943. See also *Operation Freshman.*

OSRD Office of Scientific Research and Development. Established by President Roosevelt in July, 1941, to coordinate research and development of devices and processes that might be of military value (e.g., radar, proximity fuses, fission weapons, synthetic rubber, fuels).

Operation Olympic Plan for proposed invasion of Japan, November, 1945.

Overpressure Condition of atmospheric pressure above "normal" atmospheric pressure, caused by the detonation of a nuclear weapon, usually measured in pounds per square inch (psi).

P-5 The "primary five" nuclear weapons states: United States, Russia, Britain, France, China.

Parity Oddness or evenness of the number of protons and neutrons in a nucleus. In non-proliferation parlance, the relative numbers of nuclear weapons held by various countries.

Pile Historic term for a nuclear reactor.

Planning Board The Manhattan Project involved two Planning Boards. The first was established in November, 1941, to develop recommendations concerning plans for production of fissile materials and contracts for engineering studies. The second, at Los Alamos, was organized to coordinate technical work at the laboratory.

Predetonation Detonation of a nuclear explosive before the bomb core is fully assembled, resulting in an explosive *yield* less than intended. May be caused by neutron-emitting impurities or spontaneous fissions.

Project Alberta Code name for Los Alamos program to prepare bombs for combat.

Proton Constituent positively-charged particle of atomic nuclei. The number of protons in a nucleus is equal to the *Atomic number* of the nucleus.

Queen Marys Colloquial name for plutonium-processing facilities at the Hanford Engineer Works (*HEW*). These 800-foot buildings rivaled the ocean liner Queen Mary in length (1020 feet).

Reaction channel One of a number of possible outcomes in a reaction involving two (or more) input particles.

Reflector See *tamper.*

Rem Unit of radiation exposure; "Radiation Equivalent in Man." Synonymous with *Roentgen.* For humans, an acute dose on the order of 500 rems will often result in death.

Reproduction factor Measure of the net number of neutrons generated per each consumed in a nuclear reactor, designated by the symbol k. If $k \geq 1$, a self-sustaining reaction is in progress.

Roentgen See *Rem.*

RRC Reich Research Council (Germany).

S-1 Committee; S-1 Section Name acquired by the *Uranium Committee* after it was absorbed into the Office of Scientific Research and Development (*OSRD*) when the latter was established in July, 1941.

S-1 Executive Committee Successor to the S-1 Committee established June, 1942, within the *OSRD* to coordinate research into various methods of fissile-material production. Chaired by James Conant, the other members were Lyman Briggs, Ernest Lawrence, Harold Urey, Arthur Compton, and Eger Murphree.

S-50 Code name for the thermal diffusion plant at the Clinton Engineer Works (*CEW*).

Scientific Panel A subcommittee of the *Interim Committee* (1945) established to provide advice on technical issues related to the use and future development of nuclear weapons. Members were Robert Oppenheimer, Arthur Compton, Enrico Fermi, and Ernest Lawrence.

Second criticality Moment in the course of the detonation of a fission weapon where the core has expanded to the point where conditions necessary for a self-sustaining chain reaction no longer hold. Compare *First criticality*.

Section S-1 See S-1 Committee.

SED Special Engineer Detachment; a group of U. S. Army World War II personnel with technical and scientific training.

Smyth Report Colloquial title of a report authored by Henry Smyth and issued by the United States government just after the bombings of Hiroshima and Nagasaki in August, 1945. This document was the first public description of the Manhattan Project; its full tile was "Atomic Energy for Military Purposes: The Official Report on the Development of the Atomic Bomb under the Auspices of the United States Government, 1940–1945."

SODC Standard Oil Development Company.

Tamper A heavy (usually metallic) structure that surrounds the core of a nuclear weapon, designed to reflect escaping neutrons back into the core and briefly retard expansion of the core while it explodes; sometimes known as a *reflector*. Both effects act to increase weapon efficiency.

Target Committee Group of military officers and scientists established April, 1945, to advise on targeting of nuclear weapons against Japanese cities.

Top Policy Group Committee of government, military, and scientific personnel established by President Roosevelt, October, 1941, to advise on policy considerations raised by nuclear issues.

Trinity First test of a nuclear weapon, July 16, 1945, in southern New Mexico. This implosion device achieved a *yield* of about 22 kt.

TVA Tennessee Valley Authority, an agency of the United States government.

Uranium Committee Formally, the Advisory Committee on Uranium, established October, 1939, to investigate possible military applications of nuclear fission. This was the first United States government group convened to consider the possibility of fission weapons and nuclear power. The Uranium Committee was absorbed into the *NDRC* in June, 1940, and became known as *Section S-1* of the Office of Scientific Research and Development (*OSRD*) when it was established in July, 1941.

USSBS United States Strategic Bombing Survey.

X-10 Code name for the graphite reactor at the Clinton Engineer Works (*CEW*).

Xenon poisoning Xenon is a product of nuclear fissions; as it accumulates within a reactor, it "poisons" the reaction due to its tendency to capture neutrons. If not for the short half-life involved (9 hours), the responsible isotope, Xe-135, would continue to accumulate until the reaction could not longer proceed.

Y-12 Code name for the electromagnetic separation complex at the Clinton Engineer Works (*CEW*).

Yield Energy released by a nuclear weapon, usually measured in *kilotons* (kt) or *megatons* (Mt).

Bibliography

Books, Journal Articles, and Reports

P.H. Abelson, Cleavage of the uranium nucleus. Phys. Rev. **55**, 418 (1939)

P.H. Abelson, R. Gunn, A.H. Van Keuren, Progress report on Liquid Thermal Diffusion Research. NRL report 0-1977, January 4, 1943.

P.H. Abelson, N. Rosen, J.I. Hoover, Liquid Thermal Diffusion. NRL report TID-5229, September 10, 1946. http://www.osti.gov/energycitations/product.biblio.jsp?osti_id=4311423

J. Abelson, P.H. Abelson, *Uncle Phil and the Atomic Bomb* (Roberts and Company, Greenwood Village, Colorado, 2008)

G. Acocella, F. Guerra, N. Robotti, Enrico Fermi's discovery of neutron-induced artificial radioactivity: the recovery of his first laboratory notebook. Phys. Perspect. **6**(1), 29–41 (2004)

J.-J. Ahern, We had the hose turned on us! Ross Gunn and The Naval research laboratory's early research into nuclear propulsion, 1939–1946. Int. J. Naval Hist. **2**(1) (2003). http://www.ijnhonline.org/wp-content/uploads/2012/01/article_ahern_pdf_apr03.pdf

J. Albright, M. Kunstel, *Bombshell: The Secret Story of America's Unknown Atomic Spy Conspiracy* (Times Books, New York, 1997)

D. Albright, A. Stricker, *Revisiting South Africa's Nuclear Weapons Program: Its History, Dismantlement, and Lessons for Today* (Institute for Science and International Security, Washington, 2016)

L.W. Alvarez, *Adventures of a Physicist* (Basic Books, New York, 1987)

© Springer Nature Switzerland AG 2020
B. C. Reed, *Manhattan Project*,
https://doi.org/10.1007/978-3-030-45734-1

E. Amaldi, O. D'Agostino, E. Fermi, B. Pontecorvo, F. Rasetti, E. Segrè, Artificial radioactivity produced by neutron bombardment–II. Proc. R. Soc. Lond. Ser. A **149**(868), 522–558 (1935)

H.L. Anderson, E.T. Booth, J.R. Dunning, E. Fermi, G.N. Glasoe, F. Slack, The fission of uranium. Phys. Rev. **55**, 511–512 (1939)

H.L. Anderson, E. Fermi, H.B. Hanstein, Production of neutrons in uranium bombarded by neutrons. Phys. Rev. **55**, 797–798 (1939)

H.L. Anderson, E. Fermi, L. Szilard, Neutron production and absorption in uranium. Phys. Rev. **56**, 284–286 (1939)

H.L. Anderson, The legacy of Fermi and Szilard. Bull. At. Sci. **XXX**(7), 56–62 (1974)

H.L. Anderson, Fermi, Szilard and Trinity. Bull. At. Sci. **30**(8), 40–47 (1974)

F. Aston, Neon. Nature **104**(2613), 334 (1919)

F. Aston, A positive ray spectrograph. Phil. Mag. Ser. 6, **38**(228), 707–714 (1919)

F. Aston, The constitution of atmospheric neon. Phil. Mag. Ser. 6, **39**(232), 449–455 (1920)

F. Aston, The mass spectra of chemical elements. Phil. Mag. Ser. 6, **39**(233), 611–625 (1920)

F. Aston, Constitution of thallium and uranium. Nature **128**(3234), 725 (1931)

D.F. Babcock, The discovery of xenon-135 as a reactor poison. Nucl. News **7**, 38–42 (1964)

L. Badash, Radioactivity before the Curies. Am. J. Phys. **33**(2), 128–135 (1965)

L. Badash, The discovery of thorium's radioactivity. J. Chem. Educ. **43**(4), 219–220 (1966)

L. Badash, The discovery of radioactivity. Phys. Today **49**(2), 21–26 (1996)

L. Badash, J.O. Hirschfelder, H.P. Broida (eds.), *Reminiscences of Los Alamos 1943–1945* (Reidel, Dordrecht, 1980)

K.T. Bainbridge, Trinity. Los Alamos report LA-6300-H. https://fas.org/sgp/othergov/doe/lanl/docs1/00317133.pdf

K.T. Bainbridge, Orchestrating the test, in *All In Our Time: The Reminiscences of Twelve Nuclear Pioneers*, ed. by J. Wilson (The Bulletin of the Atomic Scientists, Chicago, 1975)

S.G. Bankoff, A. Weinberg, Notes on Hanford reactor startup. Phys. Today **57**(4), 17–19 (2004)

G.E. Barasch, Light Flash Produced by an Atmospheric Nuclear Explosion. Los Alamos Scientific Laboratory report LASL-79-84 (1979). https://fas.org/sgp/othergov/doe/lanl/docs4/00362363.pdf

N. Bascomb, *The Winter Fortress: The Epic Mission to Sabotage Hitler's Atomic Bomb* (Houghton Mifflin Harcourt, Boston, 2016)

H. Becquerel, Sur les radiations émises par phosphorescents. Comptes Rendus **122**, 420–421 (1896)

H. Becquerel, Sur les radiationes invisibles émises par les corps phosphorescents. Comptes Rendus **122**, 501–503 (1896)

H. Becquerel, Contribution à l'étude du rayonnement du radium. Comptes Rendus **130**, 120–126 (1900)

H. Becquerel, Déviation du rayonnement du radium dans un champ électrique. Comptes Rendus **130**, 809–815 (1900)

B. Bederson, SEDs at Los Alamos: a personal memoir. Phys. Perspect. **3**(1), 52–75 (2001)

B. Bernstein, The uneasy alliance: Roosevelt, Churchill, and the atomic bomb, 1940–1945. West. Polit. Q. **29**(2), 202–230 (1976)

B. Bernstein, Eclipsed by Hiroshima and Nagasaki: early thinking about tactical nuclear weapons. Int. Secur. **15**(4), 149–173 (1991)

J. Bernstein, D.C. Cassidy, Bomb apologetics: farm hall, August 1945. Phys. Today **48**(8), 32–36 (1995)

J. Bernstein, *Hitler's Uranium Club: The Secret Recordings at Farm Hall* (American Institute of Physics, New York, 1996)

J. Bernstein, *Plutonium: A History of the World's Most Dangerous Element* (Joseph Henry Press, Washington, 2007)

J. Bernstein, A memorandum that changed the world. Am. J. Phys. **79**(5), 440–446 (2011)

J. Bernstein, *Nuclear Iran* (Harvard University Press, Cambridge, 2014)

H.A. Bethe, R.F. Christy, Memorandum on the Immediate After Effects of the Gadget. http://blog.nuclearsecrecy.com/wp-content/uploads/2012/10/1944-Bethe-and-Christy-Memorandum-on-the-Immediate-After-Effects-of-the-Gadget.pdf

H.A. Bethe, Theory of the Fireball. Los Alamos report LA-3064 (1964). http://www.fas.org/sgp/othergov/doe/lanl/docs1/00367118.pdf

H. Bethe, The German uranium project. Phys. Today **53**(7), 34–36 (2000)

K. Bird, M.J. Sherwin, *American Prometheus: The Triumph and Tragedy of J. Robert Oppenheimer* (Knopf, New York, 2005)

J.M. Blum (ed.), *The Price of Vision: The Diary of Henry A. Wallace, 1942–1946* (Houghton Mifflin, Boston, 1973)

N. Bohr, Neutron capture and nuclear constitution. Nature **137**(3461), 344–348 (1936)

N. Bohr, Disintegration of heavy nuclei. Nature **143**(3617), 330 (1939)

N. Bohr, Resonance in uranium and thorium disintegrations and the phenomenon of nuclear fission. Phys. Rev. **55**, 418–419 (1939)

N. Bohr, J.A. Wheeler, The mechanism of nuclear fission. Phys. Rev. **56**, 426–450 (1939)

N. Bohr, *Collected Works. Volume 9: Nuclear Physics (1929–1952)*, ed. by R. Peierls (North-Holland, Amsterdam, 1986)

H.C. Børreson, Flawed nuclear physics and atomic intelligence in the campaign to deny norwegian heavy-water to Germany, 1942–1944. Phys. Perspect. **14**(4), 471–497 (2012)

E. Booth, J. Dunning, W.L. Laurence, A.O. Nier, W. Zinn, *The Beginnings of the Nuclear Age* (Newcomen Society in North America, New York, 1969)

W. Bothe, H. Becker, Künstliche Erregung von Kern-γ-Strahlen. Zeitschrift fur Physik **66**(5/6), 289–306 (1930)

A. Bramley, A.K. Brewer, A thermal method for the separation of isotopes. J. Chem. Phys. **7**, 553–554 (1939)

A. Brown, *The Neutron and the Bomb: A Biography of Sir James Chadwick* (Oxford University Press, Oxford, 1997)

A.A. Broyles, Nuclear explosions. Am. J. Phys. **50**(7), 586–594 (1982)

R.H. Campbell, *The Silverplate Bombers* (McFarland & Co., Jefferson, North Carolina, 2005)

S. Cannon, The Hanford Site Historic District–Manhattan Project 1943–1946, Cold War Era 1947–1990. Pacific Northwest National Laboratory (2002). DOE/RL-97-1047. http://www.osti.gov/bridge/product.biblio.jsp?osti_id=807939

R.P. Carlisle, J.M. Zenzen, *Supplying the Nuclear Arsenal: American Production Reactors, 1942–1992* (Johns Hopkins University Press, Baltimore, 1996)

C. Carson, D.A. Hollinger (eds.), *Remembering Oppenheimer: Centennial Studies and Reflections* (University of California, Berkeley, 2005)

D.C. Cassidy, A historical perspective on *Copenhagen*. Phys. Today **53**(7), 28–32 (2000)

D.C. Cassidy, *Beyond Uncertainty: Heisenberg, Quantum Physics, and the Bomb* (Bellevue Literary Press, New York, 2009)

D.C. Cassidy, *Farm Hall and the German Atomic Project of World War II: A Dramatic History* (Springer, 2017)

D.C. Cassidy, *A Short History of Physics in the American Century* (Harvard University Press, Cambridge, 2011)

J. Chadwick, Possible existence of a neutron. Nature **129**(3252), 312 (1932)

J. Chadwick, The existence of a neutron. Proc. R. Soc. Lond. **A136**, 692–708 (1932)

M.B. Chambers, Technically sweet Los Alamos: the development of a federally sponsored scientific community. Ph.D. thesis, University of New Mexico (1974)

S. Chapman, On the law of distribution of molecular velocities, and on the theory of viscosity and thermal conduction, in a non-uniform simple monatomic gas. Philos. Trans. R. Soc. Lond. **A216**, 279–348 (1916)

S. Chapman, F.W. Dootson, A note on thermal diffusion. Lond. Edinb. Dublin Philos. Mag. J. Sci. **33**, 248–253 (1917)

R.W. Clark, *The Birth of the Bomb: The Untold Story of Britain's Part in the Weapon that Changed the World* (Phoenix House, London, 1961)

R.W. Clark, *Tizard* (M.I.T. Press, Cambridge, MA, 1965)

K. Clusius, G. Dickel, Neus Verfahren zur Gasentmischung und Isotopentrennung. Naturwissenschaften **26**, 546 (1938)

K. Clusius, G. Dickel, Zur trennung der chlorisotope. Naturwissenschaften **27**, 148–149 (1939)

K.P. Cohen, S.K. Runcorn, H.E. Suess, H.G. Thode, Harold Clayton Urey, 29 April 1893–5 January 1981. Biogr. Memo. Fellows R. Soc. **29**, 622–659 (1983)

S. Cohen, *The Truth About the Neutron Bomb: The Inventor of the Bomb Speaks Out* (William Morrow, New York, 1983)

A.L. Compere, W.L. Griffith, The U. S. Calutron Program for Uranium Enrichment: History, Technology, Operations, and Production. Oak Ridge National Laboratory report ORNL-5928 (1991)

A.H. Compton, *Atomic Quest* (Oxford University Press, New York, 1956)

K.T. Compton, If the atomic bomb had not been used. The Atlantic Monthly **178**(6), 54–56 (1946)

J. Conant, *109 East Palace: Robert Oppenheimer and the Secret City of Los Alamos* (Simon and Schuster, New York, 2005)

J. Conant, *Man of the Hour: James B. Conant, Warrior Scientist* (Simon and Schuster, New York, 2017)

J. Coster-Mullen, *Atom Bombs: The Top Secret Inside Story of Little Boy and Fat Man* (Coster-Mullen, Waukesha, WI, 2010)

E. Crawford, R.L. Sime, M. Walker, A nobel tale of postwar injustice. Phys. Today **50**(9), 26–32 (1997)

I. Curie, F. Joliot, Émission de protons de grande vitesse par les substances hydrogénnés sous l'influence des rayons γ très pénétrants. Comptes Rendus **194**, 273–275 (1932)

I. Curie, F. Joliot, Un nouveau type de radioactivite. Comptes Rendus **198**, 254–256 (1934)

I. Curie, P. Savitch, Sur les radioéléments formes dans l'uranium irradié per les neutrons. J. Phys. Radium **8**(10), 385–387 (1937)

I. Curie, P. Savitch, Sur les radioéléments formes dans l'uranium irradié per les neutrons. II. J. Phys. Radium **9**(9), 355–359 (1938)

P. Curie, S. Curie, Sur une substance nouvelle radio-active, contenue dans la pitchblende. Comptes Rendus **127**, 175–178 (1898). Marie Curie's initial appears here as "S," for "Sklodowski."

P. Curie, P. Curie, Mme., G. Bémont, Sur une nouvelle substance fortement radioactive, contenue dans la pitchblende. Comptes Rendus **127**, 215–1217 (1898)

G. Dannen, Szilard's inventions patently halted. Phys. Today **54**(3), 102–103 (2001)

A.J. Dempster, Isotopic constitution of uranium. Nature **136**(3431), 180 (1935)

Department of Energy Nevada Operations Office: United States Nuclear Tests July 1945 through September 1992. (Report DOE/NV-209 REV 16); http://nnss. gov/docs/docs_LibraryPublications/DOE_NV-209_Rev16.pdf

F. Dyson, *Disturbing the Universe* (Harper and Row, New York, 1979)

D. Enskog, Über eine Verallgemeinerung der zweiten Maxwellschen Theorie der Gase. Physikalische Zeitschrift **12**, 56–60 (1911); Bemerkungen zu einer Fundamentalgleichung in der kinetischen Gastheorie, ibid., 533–539

D.C. Fakley, The British Mission. Los Alamos Science **4**(7), 186–189 (1983)

G. Farmelo, *Churchill's Bomb: How the United States Overtook Britain in the First Nuclear Arms Race* (Basic Books, New York, 2013)

B.T. Feld, G.W. Szilard, K. Winsor, *The Collected Works of Leo Szilard. Volume I–Scientific Papers* (MIT Press, London, 1972)

E. Fermi, Radioactivity Induced by Neutron Bombardment. Nature **133**(3368), 757 (1934)

E. Fermi, Possible production of elements of atomic number higher than 92. Nature **133**(3372), 898–899 (1934)

E. Fermi, E. Amaldi, O. D'Agostino, F. Rasetti, E. Segrè, Artificial radioactivity produced by neutron bombardment. Proc. R. Soc. Lond. Ser. A **146**(857), 483–500 (1934)

E. Fermi, Elementary theory of the chain-reacting pile. Science **105**(2751), 27–32 (1947)

E. Fermi, *Nuclear Physics* (University of Chicago Press, Chicago, 1949)

E. Fermi, Experimental production of a divergent chain reaction. Am. J. Phys. **20**(9), 536–558 (1952)

E. Fermi, *Collected Papers, Vol. 2, United States, 1939–1945* (University of Chicago Press, Chicago, 1965)

R.H. Ferrell, *Harry S. Truman and the Bomb* (High Plains Publishing Co., Worland, WY, 1996)

N.I. Fetisov, Spectra of neutrons inelastically scattered on U-238. At. Energy **3**, 995–998 (1957)

F.L. Fey, Health Physics Survey of Trinity Site. Los Alamos report LA-3719 (June, 1967) https://fas.org/sgp/othergov/doe/lanl/lib-www/la-pubs/00314894.pdf

L. Fine, J.A. Remington, *United States Army in World War II: The Technical Services–The Corps of Engineers: Construction in the United States* (Center of Military History, United States Army, Washington, 1989)

N.S. Finney, How F. D. R. planned to use the A-Bomb. Look **14**(6), 23–27 (1950)

G.N. Flerov, K.A. Petrzhak, Spontaneous fission of uranium. Phys. Rev. **58**, 89 (1940)

S. Flügge, Kann der Energieinhalt der Atomkerne technisch nutzbar gemacht werden? Naturwissenschaften **27**, 402–410 (1939)

S. Flügge, G. von Droste, Energetische Betrachtungen zu der Entstehung von Barium bei der Neutronenbestrahlung Uran. Zeitschrift fur Physikalische Chemie **42B**(1), 274–280 (1939)

K.W. Ford, John Wheeler's work on particles, nuclei, and weapons. Phys. Today **62**(4), 29–33 (2009)

R.D. Fowler, R.W. Dodson, Intensely ionizing particles produced by neutron bombardment of uranium and thorium. Phys. Rev. **55**, 417–418 (1939)

C. Frank, *Operation Epsilon: The Farm Hall Transcripts* (University of California Press, 1993)

M. Frayn, *Copenhagen* (Anchor Books, New York, 1998)

H. Friedman, L.B. Lockhart, I.H. Blifford, Detecting the Soviet bomb: Joe-1 in a rain barrel. Phys. Today **49**(11), 38–41 (1996)

O.R. Frisch, Physical evidence for the division of heavy nuclei under neutron bombardment. Nature **143**(3616), 276 (1939)

O.R. Frisch, Radioactivity and subatomic phenomena. Annu. Rep. Prog. Chem. **37**, 7–22 (1940)

O.R. Frisch, The discovery of fission. Phys. Today **20**(11), 43–52 (1967)

O.R. Frisch, *What Little I Remember* (Cambridge University Press, Cambridge, 1979)

R.L. Garwin, G. Charpak, *Megawatts and Megatons: A Turning Point in the Nuclear Age?* (Knopf, New York, 2001)

H. Geiger, E. Marsden, On a diffuse reflection of the α-particles. Proc. R. Soc. **A82**(557), 495–500 (1909)

M.S. Gerber, A Brief History of the T Plant Facility at the Hanford Site. Westinghouse Hanford Company, Richland, WA (1994). Report WHC-MR-0452. http://pdw.hanford.gov/arpir/index.cfm/viewDoc?accession=0071652H

M.S. Gerber, The Plutonium Production Story at Hanford Site: Process and Facilities History. Westinghouse Hanford Company, Richland, WA (1996). Report WHC-MR-0521. http://www.osti.gov/bridge/product.biblio.jsp?osti_id=664389

D.M. Giangreco, *Hell to Pay: Operation DOWNFALL and the Invasion of Japan, 1945–1947* (Naval Institute Press, Washington, DC, 2017)

D.M. Giangreco, Casualty projections for the U. S. invasions of Japan, 1945–1946: planning and policy implications. J. Mil. Hist. **61**(3), 521–581 (1997)

S. Glasstone, P.J. Dolan, *The Effects of Nuclear Weapons* (United States Department of Defense and Energy Research and Development Agency, Washington, 1977)

S. Goldberg, Inventing a climate of opinion: Vannevar Bush and the decision to build the bomb. Isis **83**(3), 429–452 (1992)

P.J. Goodchild, *Robert Oppenheimer: Shatterer of Worlds* (British Broadcasting Corporation, London, 1980)

F.G. Gosling, *The Manhattan Project: Making the Atomic Bomb* (U. S. Department of Energy, Washington, 2010) http://energy.gov/management/downloads/gosling-manhattan-project-making-atomic-bomb

S. Goudsmit, *Alsos* (Schuman, New York, 1947)

M. Gowing, *Britain and Atomic Energy 1939–1945* (St. Martin's Press, London, 1964)

H.G. Graetzer, Discovery of nuclear fission. Am. J. Phys. **32**(1), 9–15 (1964)

G. Grasso, C. Oppici, M. Sumini, A nucleonics study of the 1945 haigerloch B-VIII nuclear reactor. Phys. Persp. **11**, 318–335 (2009)

G.K. Green, L.W. Alvarez, Heavily ionizing particles from uranium. Phys. Rev. **55**, 417 (1939)

S. Groueff, *Manhattan Project: The Untold Story of the Making of the Atomic Bomb* (Little, Brown and Company, New York, 1967). Reissued by Authors Guild Backprint.com (2000)

L.R. Groves, *Now It Can be Told: The Story of the Manhattan Project* (Da Capo Press, New York, 1983)

F. Guerra, M. Leone, N. Robotti, Enrico Fermi's discovery of neutron-induced artificial radioactivity: neutrons and neutron sources. Phys. Perspect. **8**(3), 255–281 (2006)

F. Guerra, N. Robotti, Enrico Fermi's discovery of neutron-induced artificial radioactivity: the influence of his theory of beta decay. Phys. Perspect. **11**(4), 379–404 (2009)

F. Guerra, M. Leone, N. Robotti, The discovery of artificial radioactivity. Phys. Perspect. **14**(1), 33–58 (2012)

O. Hahn, F. Strassmann, Über die Enstehung von Radiumisotopen aus Uran beim Bestrahlen mit schnellen und verlangsamten Neutronen. Naturwissenschaften **26**, 755–756 (1938)

O. Hahn, F. Strassmann, Concerning the existence of alkaline earth metals resulting from neutron irradiation of uranium. Naturwissenschaften **27**, 11–15 (1939) [Title translated]

P. Halpern, Washington: a DC circuit tour. Phys. Perspect. **12**(4), 443–466 (2010)

D. Harvey, History of the Hanford Site, 1943–1990. Pacific Northwest National Laboratory, http://ecology.pnnl.gov/library/History/Hanford-History-All.pdf

W.R. Hansen, J.C. Rodgers, Radiological Survey and Evaluation of the Fallout Area from the Trinity Test. Los Alamos report LA-10256-MS (June, 1985). https://fas.org/sgp/othergov/doe/lanl/lib-www/la-pubs/00318776.pdf

S.K. Hanson, A.D. Pollington, C.R. Waidmann, W.S. Kinman, A.M. Wende, J.L. Miller, J.A. Berger, W.J. Oldham, H.D. Selby, Measurements of extinct fission products in nuclear bomb debris: determination of the yield of the Trinity test 70 y later. Proc. Nat. Acad. Sci. **113**(29), 8104–8108 (2016)

D. Hawkins, *Manhattan District History. Project Y: The Los Alamos Project. Volume I: Inception until August 1945* (Los Alamos Scientific Laboratory, Los Alamos, NM, 1947). Los Alamos publication LAMS-2532. https://fas.org/sgp/othergov/doe/lanl/docs1/00103803.pdf

W. Heisenberg, Research in Germany on the technical application of atomic energy. Nature **160**(4059), 211–215 (1947)

W. Heisenberg, D.C. Cassidy, W. Sweet, A lecture on bomb physics: February 1942. Phys. Today **48**(8), 27–30 (1995)

G. Hermann, Five decades ago: from the "Transuranics" to nuclear fission. Angew. Chem. Int. Ed. Engl. **29**(5), 481–508 (1990)

J. Hersey, *Hiroshima* (Vintage, New York, 1989)

R.G. Hewlett, O.E. Anderson, Jr., *A History of the United States Atomic Energy Commission, Vol. 1: The New World, 1939/1946* (Pennsylvania State University Press, University Park, PA, 1962)

L. Hoddeson, P.W. Henriksen, R.A. Meade, C. Westfall, *Critical Assembly: A Technical History of Los Alamos during the Oppenheimer Years* (Cambridge University Press, Cambridge, 1993)

F. Houtermans, Zur Frage der Auslösung von Kern-Kettenreaktionen (1941). https://www.mpiwg-berlin.mpg.de/Preprints/P414.PDF

J. Hughes, 1932: the annus mirabilis of nuclear physics. Phys. World **13**, 43–50 (2000)

M. Hull, A. Bianco, *Rider of the Pale Horse: A Memoir of Los Alamos and Beyond* (University of New Mexico Press, Albuquerque, 2005)

J. Hunner, *Inventing Los Alamos: The Growth of an Atomic Community* (University of Oklahoma Press, Norman, OK, 2004)

D. Irving, *The Virus House: Germany's Atomic Research and Allied Counter-Measures* (Focal Point Publications, 2010)

W. Isaacson, *Einstein: His Life and Universe* (Simon and Schuster, New York, 2007)

F. Joliot, I. Curie, Artificial production of a new kind of radio-element. Nature **133**(3354), 201–202 (1934)

V.C. Jones, *United States Army in World War II: Special Studies - Manhattan: The Army and the Atomic Bomb* (Center of Military History, United States Army, Washington, 1985)

R.L. Kathren, J.B. Gough, G.T. Benefiel, *The Plutonium Story: The Journals of Professor Glenn T. Seaborg 1939–1946* (Battelle Press, Columbus, Ohio, 1994). An abbreviated version prepared by Seaborg is available at http://www.escholarship.org/uc/item/3hc273cb?display=all

P.C. Keith, The role of the process engineer in the atom bomb project. Chem. Eng. **53**, 112–122 (1964)

C.C. Kelly (ed.), *Remembering the Manhattan Project: Perspectives on the Making of the Atomic Bomb and its Legacy* (World Scientific, Hackensack, NJ, 2004)

C.C. Kelly (ed.), *The Manhattan Project: The Birth of the Atomic Bomb in the Words of Its Creators, Eyewitnesses, and Historians* (Black Dog & Leventhal Press, New York, 2007)

C.C. Kelly, R.S. Norris, *A Guide to the Manhattan Project in Manhattan* (Atomic Heritage Foundation, Washington, 2012)

J.W. Kennedy, G.T. Seaborg, E. Segrè, A.C. Wahl, Properties of 94(239). Phys. Rev. **70**(7–8), 555–556 (1946)

I.B. Khriplovich, The eventful life of Fritz Houtermans. Phys. Today **45**(7), 29–37 (1992)

D. Kiernan, *The Girls of Atomic City: The Untold Story of the Women Who Helped Win World War II* (Touchstone/Simon and Schuster, New York, 2013)

G.B. Kistiakowsky, Trinity–a reminiscence. Bull. At. Sci. **36**(6), 19–22 (1980)

T. Koeth, M.E. Hiebert, Tracking the journey of a uranium cube. Phys. Today **72**(5), 36–43 (2019)

H. Kragh, Rutherford, Radioactivity, and the Atomic Nucleus. http://arxiv.org/abs/1202.0954

F. Kuhn, Jr., *Chadwick Calls Neutron 'Difficult catch'; His Find Hailed as Aid in Study of Atom.* New York Times, February 29, 1932, p. 1 and 8

L. Lamont, *Day of Trinity* (Antheneum, New York, 1965)

W. Lanouette, B. Silard, *Genius in the Shadows: A Biography of Leo Szilard, the Man Behind the Bomb* (Skyhorse Publishing, New York, 2013)

W.L. Laurence, Vast Power Source in Atomic Energy Opened by Science. The New York Times, May 5, 1940, p. 1 and 51

W.L. Laurence, The atom gives up. Saturday Evening Post **213**(10), 12–13, 60–63 (1940)

W.L. Laurence, Drama of the Atomic Bomb Found Climax in July 16 Test. The New York Times, September 26, 1945, p. 1 and 16

W.L. Laurence, *Dawn over Zero: The Story of the Atomic Bomb* (Knopf, New York, 1946)

E.O. Lawrence, N.E. Edlefsen, On the production of high speed protons. Science **72**(1867), 376 (1930)

E.O. Lawrence, D.H. Sloan, The production of high speed mercury ions without the use of high voltages. Phys. Rev. **37**, 231 (1931)

E.O. Lawrence, M.S. Livingston, A method for producing high speed hydrogen ions without the use of high voltages. Phys. Rev. **37**, 1707 (1931)

S. Lee, 'In no sense vital and actually not even important'? Reality and perception of Britain's contribution to the development of nuclear weapons. Contemp. Br. Hist. **20**(2), 159–185 (2006)

L.M. Libby, *The Uranium People* (Crane Russak, New York, 1979)

R.J. Lifton, *Death in Life: Survivors of Hiroshima* (Vintage, New York, 1969)

S.L. Lippincott, A conversation with Robert F. Christy–Part I. Phys. in Perspect. **8**(3), 282–317 (2006); A conversation with Robert F. Christy–Part II. Phys. Perspect. **8**(4), 408–450 (2006)

J. Logan, The critical mass. Am. Sci. **84**(3), 263–277 (1996)

Los Alamos Historical Society, *Los Alamos: Beginning of an Era 1943–1945* (2002). http://www.atomicarchive.com/Docs/ManhattanProject/la_index.shtml

R.J. Maddox, *Hiroshima in History: The Myths of Revisionism* (University of Missouri Press, Columbia, MO, 2007)

R. E. Malenfant, Experiments with the Dragon Machine. Los Alamos publication LA-14241-H (2005); http://www.osti.gov/energycitations/purl.cover.jsp?purl=/876514-I1Txj9/

Manhattan Engineer District: The Atomic Bombings of Hiroshima and Nagasaki (1946). http://www.atomicarchive.com/Docs/MED/index.shtml

J. C. Marshall, Chronology of District "X", 17 June 1942–28 October 1942.

K. Mayer, M. Wallenius, K. Lützenkirchen, J. Horta, A. Nicholl, G. Rasmussen, P. van Belle, Z. Varga, R. Buda, N. Erdmann, J.-V. Kratz, N. Trautmann. L.K. Fifield, G. Tims, M.B. Frölich, P. Steier, Uranium from German nuclear power projects of the 1940s–A nuclear forensic investigation. Angew. Chem. Int. Ed. **54**(45), 13452–13456 (2015)

M. McBride, *55 Years That Changed History–A Manhattan Project Timeline, 1894 to 1949* (The Secret City Store, Oak Ridge, TN, 2010)

D. McCullough, *Truman* (Simon and Schuster, New York, 1992)

M. McDonald, Survivor of 2 Atomic Bombs Dies at 93. New York Times, January 6, 2010. https://www.nytimes.com/2010/01/07/world/asia/07yamaguchi.html

E. McMillan, Radioactive recoils from uranium activated by neutrons. Phys. Rev. **56**, 510 (1939)

E. McMillan, P.H. Abelson, Radioactive element 93. Phys. Rev. **57**, 1185–1186 (1940)

L. Meitner, O.R. Frisch, Disintegration of uranium by neutrons: a new type of nuclear reaction. Nature **143**(3615), 239–240 (1939)

T. Merlan, *Life at Trinity Base Camp* (Human Systems Research, Las Cruces, NM, 2001)

D.E. Neuenschwander, Jumbo: Silent Partner in the Trinity Test. Radiations, 12–14 (Fall 2004)

D.E. Neuenschwander, The discovery of the nucleus. Radiations **17**(1), 13–22 (2011)

D.E. Neuenschwander, The discovery of the nucleus, Part 2: rutherford scattering and its aftermath. SPS Observer **XLV**(1), 11–16 (2011)

New York Times, Atom Explosion Frees 200,000,000 Volts; New Physics Phenomenon Credited to Hahn. January 29, 1939, p. 2

K.D. Nichols, *The Road to Trinity* (Morrow, New York, 1987)

J.W. Nicholson, The spectrum of nebulium. Mon. Not. R. Astron. Soc. **72**, 49–64 (1911)

A.O. Nier, The isotopic constitution of uranium and the half-lives of the uranium isotopes. I. Phys. Rev. **55**, 150–153 (1939)

A.O. Nier, Some reminiscences of mass spectroscopy and the Manhattan Project. J. Chem. Educ. **66**(5), 385–388 (1989)

A.O. Nier, E.T. Booth, J.R. Dunning, A.V. Grosse, Nuclear fission of separated uranium isotopes. Phys. Rev. **57**, 546 (1940)

A.O. Nier, E.T. Booth, J.R. Dunning, A.V. Grosse, Further experiments on fission of separated uranium isotopes. Phys. Rev. **57**, 748 (1940)

I. Noddack, Über das Element 93. Zeitschrift fur Angewandte Chemie **47**(37), 653–655 (1934). An English translation prepared by H.G. Graetzer is available at http://www.chemteam.info/Chem-History/Noddack-1934.html

R.S. Norris, *Racing for the Bomb: General Leslie R. Groves, the Manhattan Project's Indispensable Man* (Steerforth Press, South Royalton, VT, 2002)

R.S. Norris, H.M. Kristensen, U. S. Nuclear warheads, 1945–2009. Bull. At. Sci. **65**(4), 72–81 (2009)

Oak Ridge Operations, U. S. Atomic Energy Commission, A City is Born: The History of Oak Ridge (1961). Reprinted by Oak Ridge Heritage and Preservation Association (2009)

B.J. O'Keefe, *Nuclear Hostages* (Houghton Mifflin, Boston, 1983)

M. Oliphant, The beginning: Chadwick and the neutron. Bull. At. Sci. **38**(10), 14–18 (1982)

J.J. O'Neill, Enter atomic power. Harper's Mag. **181**, 1–10 (1940)

J.R. Oppenheimer, Physics in the contemporary world. Bull. At. Sci. **4**(3), 65–68 and 85–86 (1947)

A. Pais, R.P. Crease, *J. Robert Oppenheimer: A Life* (Oxford University Press, Oxford, 2006)

M. Palevsky, *Atomic Fragments: A Daughter's Questions* (University of California Press, Berkeley, 2000)

W.E. Parkins, The uranium bomb, the calutron, and the space-charge problem. Phys. Today **58**(5), 45–51 (2005)

G. Parshall, Shock Wave. U. S. News World Rep. **119**(5), 44–59 (1995)

R. Peierls, Critical conditions in neutron multiplication. Proc. Camb. Philos. Soc. **35**, 610–615 (1939)

R. Peierls, *Bird of Passage: Recollections of a Physicist* (Princeton University Press, Princeton, 1985)

W.G. Penney, D.E.J. Samuels, G.C. Scorgie, The nuclear explosive yields at Hiroshima and Nagasaki. Phil. Trans. Roy. Soc. London **A266**(1177), 357–424 (1970)

F. Perrin, Calcul relative aux conditions éventuelles de transmutation en chaine de l'uranium. Comptes Rendus **208**, 1394–1396 (1939)

B. Pierard, Japanese Balloon Bombs and the Hanford Engineer Works. B-Reactor Museum Association newsletter, Spring 2014, 6–7. http://b-reactor.org/wp-content/uploads/2016/06/mod2014-sprg.pdf

N. Polmar, *The Enola Gay: The B-29 That dropped The Atomic Bomb on Hiroshima* (Brassey's Inc., Dulles, VA, 2004)

R.F. Potter, Preserving the Hanford B-Reactor: a monument to the dawn of the nuclear age. Phys. Soc. **39**(1), 16–19 (2010)

D. Preston, *Before the fallout: from Marie Curie to Hiroshima* (Berkley Books, New York, 2005)

M. Price, Roots of dissent: the Chicago met lab and the origins of the Franck report. Isis **86**(2), 222–244 (1995)

A.S. Quist, Unclassified Controlled Nuclear Information and Restricted Data Concerning U. S. Calutrons. Oak Ridge Classification Associates report ORCA-3 (1999) https://www.osti.gov/scitech/servlets/purl/1291336

I.I. Rabi, *Science: the Center of Culture* (World Publishing, New York, 1970)

N. Ramsey, August 1945: the B-29 flight logs. Bull. At. Sci. **38**(10), 33–35 (1982)

N. Ramsey, History of Project A: http://www.alternatewars.com/WW2/WW2_Documents/War_Department/MED/History_of_Project_A.htm

B.C. Reed, Resource Letter MP-1: the Manhattan Project and related nuclear research. Am. J. Phys. **73**(9), 805–811 (2005)

B.C. Reed, Seeing the light: visibility of the July 1945 *Trinity* atomic bomb test from the inner solar system. Phys. Teach. **44**(9), 604–606 (2006)

B.C. Reed, Chadwick and the discovery of the neutron. Radiations **13**(1), 12–16 (2007)

B.C. Reed, Arthur Compton's 1941 report on explosive fission of U-235: a look at the physics. Am. J. Phys. **75**(12), 1065–1072 (2007)

B.C. Reed, Rudolf Peierls' 1939 analysis of critical conditions in neutron multiplication. Phys. Soc. **37**(4), 10–11 (2008)

B.C. Reed, The Bohr-Wheeler spontaneous fission limit: an undergraduate-level derivation. Eur. J. Phys. **30**, 763–770 (2009)

B.C. Reed, Bullion to B-Fields: the silver program of the Manhattan Project. Mich. Acad. **39**(3), 205–212 (2009)

B.C. Reed, A brief primer on tamped fission-bomb cores. Am. J. Phys. **77**(8), 730–733 (2009)

B.C. Reed, Predetonation probability of a fission-bomb core. Am. J. Phys. **78**(8), 804–808 (2010)

B.C. Reed, A desktop-computer simulation for exploring the fission barrier. Nat. Sci. **3**(4), 323–327 (2011)

B.C. Reed, Resource letter MP-2: the Manhattan Project and related nuclear research. Am. J. Phys. **79**(2), 151–163 (2011)

B.C. Reed, Liquid thermal diffusion during the Manhattan Project. Phys. Perspect. **13**(2), 161–188 (2011)

B.C. Reed, From treasury vault to the Manhattan Project. Am. Sci. **99**(1), 40–47 (2011)

B.C. Reed, The Manhattan Project, Phys. Scr. **89**(10), 108003 (26pp) (2014a)

B.C. Reed, The feed materials program of the Manhattan Project: a foundational component of the nuclear weapons complex. Phys. Perspect. **16**(4), 461–479 (2014)

B.C. Reed, Nuclear weapons at 70: reflections on the context and legacy of the Manhattan Project. Phys. Scr. **90**, 088001 (20pp) (2015a)

B.C. Reed, *Physics of the Manhattan Project*, 3rd edn. (Springer, Berlin, 2015)

B.C. Reed, *Atomic Bomb: The Story of the Manhattan Project. How nuclear physics became a global geopolitical game-changer* (Morgan and Claypool Publishers/IOP Concise Physics, San Rafael, California, 2015c)

B.C. Reed, Resource letter MP-3: the Manhattan Project and related nuclear research. Am. J. Phys. **84**(10), 734–745 (2016)

B.C. Reed, Chernobyl and Trinity-counting the curies. Fed. Am. Sci. Public Interest Rep. **69**(2), 12–15 (2016). https://fas.org/wp-content/uploads/2016/12/PIR-2016_v3.1_small.pdf

B.C. Reed, Revisiting the *Los Alamos Primer*. Phys. Today **70**(9), 42–49 (2017)

B.C. Reed, *The History and Science of the Manhattan Project*, 2nd edn. (Springer, Heidelberg, 2019)

R. Rhodes, *The Making of the Atomic Bomb* (Simon and Schuster, New York, 1986)

R. Rhodes, *Dark Sun: The Making of the Hydrogen Bomb* (Simon and Schuster, New York, 1995)

R.B. Roberts, R.C. Meyer, L.R. Hafstad, Droplet formation of uranium and thorium nuclei. Phys. Rev. **55**, 416–417 (1939)

G.O. Robinson, *The Oak Ridge Story* (Southern Publishers, Kingsport, Tennessee, 1950)

K. Ruane, *Churchill and the Bomb in War and Cold War* (Bloomsbury Academic, London, 2016)

H. Russ, *Project Alberta: The Preparation of Atomic Bombs for use in World War II* (Exceptional Books, Los Alamos, 1984)

E. Rutherford, Uranium radiation and the electrical conduction produced by it. Phil. Mag. Ser. 5, **xlvii**, 109–163 (1899)

E. Rutherford, A radio-active substance emitted from thorium compounds. Phil. Mag. Ser. 5, **49**(296), 1–14 (1900)

E. Rutherford, The cause and nature of radioactivity–Part I. Phil. Mag. Ser. 6, **4**(21), 370–396 (1902)

E. Rutherford, H. Geiger, An electrical method of counting the number of α particles from radioactive substances. Proc. R. Soc. (Lond.) **81**(546), 141–161 (1908)

E. Rutherford, T. Royds, The nature of the α particle from radioactive substances. Phil. Mag. Ser. 6, **xvii**, 281–286 (1909)

E. Rutherford, The scattering of α and β particles by matter and the structure of the atom. Phil. Mag. Ser. 6, **xxi**, 669–688 (1911)

E. Rutherford, Collision of α particles with light atoms. IV. An anomalous effect in nitrogen. Phil. Mag. Ser. 6, **37**, 581–587 (1919)

A. Sachs, *Early History Atomic Project in Relation to President Roosevelt, 1939–1940.* Unpublished manuscript, August 8–9, 1945.

D.N. Schwartz, *The Last Man Who New Everything: The Life and Times of Enrico Fermi, Father of the Nuclear Age* (Basic Books, New York, 2017)

S.I. Schwartz (ed.), *Atomic Audit: The Costs and Consequences of U. S. Nuclear Weapons Since 1940* (Brookings Institution Press, Washington, DC, 1995)

G.T. Seaborg, E.M. McMillan, J.W. Kennedy, A.C. Wahl, Radioactive element 94 from deuterons on uranium. Phys. Rev. **69**(7–8), 366–367 (1946)

G.T. Seaborg, A.C. Wahl, J.W. Kennedy, Radioactive element 94 from deuterons on uranium. Phys. Rev. **69**(7–8), 367 (1946)

E. Segrè, An unsuccessful search for transuranic elements. Phys. Rev. **55**, 1104–1105 (1939)

E. Segrè, *Enrico Fermi, Physicist* (University of Chicago Press, Chicago, 1970)

R.W. Seidel, *Los Alamos and the Development of the Atomic Bomb* (Otowi Crossing Press, Los Alamos, 1995)

T.M. Semkow, P.P. Parekh, D.K. Haines, Modeling the effects of the trinity test, in *Applied Modeling and Computations in Nuclear Science.* American Chemical Society Symposium Series 945 (2006), pp. 142–159.

R. Serber, *The Los Alamos Primer: The First Lectures on How To Build An Atomic Bomb* (University of California Press, Berkeley, 1992)

R. Serber, R.P. Crease, *Peace and War: Reminiscences of a Life on the Frontiers of Science* (Columbia University Press, New York, 1998)

F.A. Settle, *General George C. Marshall and the Atomic Bomb* (Praeger, Santa Barbara, CA, 2016)

M.J. Sherwin, *A World Destroyed: Hiroshima and the Origins of the Arms Race* (Vintage, New York, 1987)

R.L. Sime, Lise Meitner and the discovery of fission. J. Chem. Educ. **66**(5), 373–376 (1989)

R.L. Sime, *Lise Meitner: A Life in Physics* (University of California Press, Berkeley, 1996)

R.L. Sime, The search for transuranium elements and the discovery of nuclear fission. Phys. Perspect. **2**(1), 48–62 (2000)

A.K. Smith, *A Peril and a Hope: The Scientists' Movement in America, 1945–47* (University of Chicago Press, Chicago, 1965)

C.S. Smith, Some recollections of metallurgy at Los Alamos, 1943–45. J. Nucl. Mater. **100**(1–3), 3–10 (1981)

H.D. Smyth, *Atomic Energy for Military Purposes: The Official Report on the Development of the Atomic Bomb under the Auspices of the United States Government, 1940–1945* (Princeton University Press, Princeton, 1945)

A.H. Snell, Graveyard shift, Hanford, 28 September 1944–Henry W. Newson. Am. J. Phys. **50**(4), 343–348 (1982)

F. Soddy, Intra-atomic charge. Nature **92**(2301), 399–400 (1913)

G. Squires, Francis Aston and the mass spectrograph. J. Chem. Soc. Dalton Trans. **23**, 3893–3899 (1998)

H. Stimson, The decision to use the atomic bomb. Harper's Mag. **194**(1161), 97–107 (1947)

M.B. Stoff, J.F. Fanton, R.H. Williams, R.H., *The Manhattan Project: A Documentary Introduction to the Atomic Age* (McGraw-Hill, New York, 1991)

R. Stuewer, Bringing the news of fission to America. Phys. Today **38**(10), 49–56 (1985)

R. Stuewer, The origin of the liquid-drop model and the interpretation of nuclear fission. Perspect. Sci. **2**, 76–129 (1994)

R. Stuewer, Gamow, alpha decay, and the liquid-drop model of the nucleus, in *Astronomical Society of the Pacific Conference,* Series 129 (1997), pp. 30–43

R. Stuewer, *The Age of Innocence: Nuclear Physics Between the First and Second World Wars* (Oxford, 2018)

C. Sublette, Nuclear Weapons Frequently Asked Questions: http://nuclearweaponarchive.org/Nwfaq/Nfaq8.html

C.W. Sweeney, J.A. Antonucci, M.K. Antonucci, *War's End: An Eyewitness Account of America's Last Atomic Mission* (Avon, New York, 1997)

F.M. Szasz, *The Day the Sun Rose Twice* (University of New Mexico Press, Albuquerque, 1984)

F.M. Szasz, *British Scientists and the Manhattan Project: The Los Alamos Years* (Palgrave McMillan, London, 1992)

L. Szilard, W.H. Zinn, Instantaneous emission of fast neutrons in the interaction of slow neutrons with uranium. Phys. Rev. **55**, 799–800 (1939)

V.L. Telegdi, Szilard as inventor: accelerators and more. Phys. Today **53**(10), 25–28 (2000)

E. Teller, J.L. Shoolery, *Memoirs: A Twentieth-Century Journey in Science and Politics* (Perseus, Cambridge, MA, 2001)

H. Thayer, *Management of the Hanford Engineer Works in World War II* (American Society of Civil Engineers, New York, 1996)

G. Thomas, M. Morgan-Witts, *Enola Gay–Mission to Hiroshima* (White Owl Press, Loughborough, UK, 1995)

L.C. Thomas, *Polonium in the Playhouse: the Manhattan Project's Secret Chemistry Work in Dayton, Ohio* (Trillium, Columbus, OH, 2017)

E. Toomey, *Images of America: The Manhattan Project at Hanford Site* (Arcadia Publishing, Charleston, SC, 2015)

E.C. Truslow, R.C. Smith, *Manhattan District History: Project Y–The Los Alamos Project. Vol. II: August 1945 Through December 1946.* Los Alamos report LAMS-2532 (Vol. II). https://fas.org/sgp/othergov/doe/lanl/docs1/00103804.pdf

E. C. Truslow, R. C. Smith, *Manhattan District History: Nonscientific Aspects of Los Alamos Project Y 1942 Through 1946.* Los Alamos report LA-5200. https://fas.org/sgp/othergov/doe/lanl/docs1/00321210.pdf

L.A. Turner, Nuclear fission. Rev. Mod. Phys. **12**(1), 1–29 (1940)

L.A. Turner, Atomic energy from U^{238}. Phys. Rev. **69**(7–8), 366 (1946)

J.J. Thomson, Cathode rays. Phil. Mag. Ser. 5, **44**(269), 293–316 (1897)

J.J. Thomson, On rays of positive electricity. Phil. Mag. Ser. 6, **13**(77), 561–575 (1907)

S.M. Ulam, *Adventures of a Mathematician* (Charles Scribner's Sons, New York, 1976)

United States Strategic Bombing Survey: The Effects of the Atomic Bombings of Hiroshima and Nagasaki. Washington (1946). http://www.trumanlibrary.org/whistlestop/study_collections/bomb/large/documents/pdfs/65.pdf#zoom=100

H. von Halban, F. Joliot, L. Kowarski, Number of neutrons liberated in the nuclear fission of uranium. Nature **143**(3625), 680 (1939)

J.S. Walker, *Nuclear Energy and the Legacy of Harry S. Truman* (Truman State University Press, Kirksville, Missouri, 2016)

M. Walker, *German National Socialism and the Quest for Nuclear Power 1939–1949* (Cambridge U. S., 1989)

M. Walker, *Nazi Science: Myth, Truth, and the German Atomic Bomb* (Plenum, New York, 1995)

W. Warriner, Journey to Destiny: Train to the Manhattan Project. Atomic Heritage Foundation, Washington, DC (2013). http://www.atomicheritage.org/sites/default/files/resources/Train%20handout%20Final%20for%20Web.pdf

A. Wattenberg, December 2, 1942: the event and the people. Bull. At. Sci. **38**(10), 22–32 (1982)

A. Wattenberg, The birth of the nuclear age. Phys. Today **46**(1), 44–51 (1993)

S.R. Weart, Scientists with a secret. Phys. Today **29**(2), 23–30 (1976)

A.M. Weinberg, Eugene Wigner, nuclear engineer. Phys. Today **55**(10), 42–46 (2002)

S. Weinberg, *The Discovery of Subatomic Particles* (Cambridge University Press, Cambridge, 2003)

S. Weintraub, *The Last Great Victory: The End of World War II July/August 1945* (Dutton, New York, 1995)

J. Weisgall, *Operation Crossroads: The Atomic Tests at Bikini Atoll* (Naval Institute Press, Annapolis, MD, 1994)

V.F. Weisskopf, The Los Alamos years. Phys. Today **20**(10), 39–42 (1967)

R. Wideröe, Uber ein neues Prinzip zur Herstellung hoher Spannungen. Arkiv fur Electrotechnik **21**(4), 387–406 (1928)

E.P. Wigner, *Symmetries and Reflections: Scientific Essays* (Ox Bow Press, Woodbridge, CT, 1979)

W.J. Wilcox, *The Role of Oak Ridge in the Manhattan Project* (Privately Published, 2002)

W.J. Wilcox, *An Overview of the History of Y-12, 1942–1945*, 2nd edn. (The Secret City Store, Oak Ridge, TN, 2009)

H. Williams, *Made in Hanford. The Bomb that Changed the World* (Washington State University Press, Pullman, WA, 2011)

M.M.R. Williams, The development of nuclear reactor theory in the Montreal laboratory of the National Research Council of Canada (Division of Atomic Energy) 1943–1946. Prog. Nucl. Energy **36**(3), 239–322 (2000)

J. Wilson, *All in Our Time: The Reminiscences of Twelve Nuclear Pioneers* (Bulletin of the Atomic Scientists, Chicago, 1975)

A.L. Yergey, A.K. Yergey, Preparative scale mass spectrometry: a brief history of the calutron. J. Am. Soc. Mass Spectrom. **8**(9), 943–953 (1997)

G.P. Zachary, *Endless Frontier: Vannevar Bush, Engineer of the American Century* (MIT Press, Cambridge, MA, 1999)

Websites and Web-Based Documents

Alexander Sachs documents concerning the early history of the bomb project can be found at the site of the FDR library, http://www.fdrlibrary.marist.edu/archives/collections/franklin/index.php?p=collections/findingaid&id=309. Specifically, Sachs' cover letter to FDR of October 11 can be found on page 31 of http://www.fdrlibrary.marist.edu/_resources/images/atomic/atomic_04.pdf. There were actually three letters from Einstein to Roosevelt between August, 1939 and April, 1940, plus another in March, 1945. Texts of all four letters can be found at http://hypertextbook.com/eworld/einstein.shtml

American Institute of Physics Array of Contemporary American Physicists website on Manhattan Project: https://history.aip.org/acap/institutions/manhattan.jsp

Anderson, Sir John: http://en.wikipedia.org/wiki/John_Anderson,_1st_Viscount_Waverley

Argonne National Laboratory sites on Fermi and CP-1 witnesses: http://www.ne.anl.gov/About/legacy/unisci.shtml

Atomic Bomb casualty Commission: http://en.wikipedia.org/wiki/Atomic_Bomb_Casualty_Commission; http://www.nasonline.org/about-nas/history/archives/collections/abcc-1945-1982.html

Atomic Heritage Foundation: http://www.atomicheritage.org/

Atomic Heritage Foundation page on Canadian contributions to the Manhattan Project: https://www.atomicheritage.org/location/canada

B-Reactor Museum Association: http://www.b-reactor.org

British Mission to Los Alamos: http://www.atomicarchive.com/History/british/index.shtml

Bohr letters re Bohr-Heisenberg Copenhagen meeting: http://www.nbarchive.dk/collections/bohr-heisenberg/documents/

Department of Energy: The First Reactor (1982): https://www.energy.gov/ne/downloads/first-reactor

Federation of American Scientists index to Los Alamos reports: http://www.fas.org/sgp/othergov/doe/lanl/index1.html

Fermi's description of Trinity test: http://www.lanl.gov/history/story.php?story_id=13; http://www.atomicarchive.com/Docs/Trinity/Fermi.shtml

Franck Report: http://www.atomicarchive.com/Docs/ManhattanProject/FranckReport.shtml

Groves memo to Stimson, April 23, 1945: http://www.gwu.edu/%7Ensarchiv/NSAEBB/NSAEBB162/3a.pdf

Groves memo to Marshall, July 30, 1945: http://www.dannen.com/decision/bomb-rate.html

Harry Daghlian and Louis Slotin: http://en.wikipedia.org/wiki/Harry_K._Daghlian,_Jr; http://en.wikipedia.org/wiki/Louis_Slotin

Hiroshima and Nagasaki mission operations orders: http://time.com/3980421/hiroshima-nagasaki-operations-orders/

Historic American Engineering Record: B Reactor (105-B) Building, HAER No. WA-164. DOE/RL-2001-16. (Richland, WA: United States Department of Energy, 2001). http://wcpeace.org/history/Hanford/HAER_WA-164_B-Reactor.pdf

Indianapolis wreckage found: New York Times, Aug. 20, 2017; https://nyti.ms/2vP87GF

Interim Committee meetings: http://www.nuclearfiles.org/menu/key-issues/nuclear-weapons/history/pre-cold-war/interim-committee/index.htm

James Tuck: http://bayesrules.net/JamesTuckVitaeAndBiography.pdf

Jeffries Report: http://www.marshallfoundation.org/library/wp-content/uploads/sites/16/2015/05/xerox1482-45_opt.pdf

Kenneth Nichols on Clinton site codes: http://research.archives.gov/description/281585

Lawrence Berkeley National Laboratory: http://www.lbl.gov/Science-Articles/Research-Review/Magazine/1981/81fchp1.html

Manhattan Project National Historical Park legislation: http://www.gpo.gov/fdsys/pkg/CPRT-113HPRT91496/pdf/CPRT-113HPRT91496.pdf. The Park provision appears on pages 1245–1257.

McDonald Ranch House: http://en.wikipedia.org/wiki/McDonald_Ranch_House

McGill University website on Rutherford: http://www.physics.mcgill.ca/museum/emanations.htm

Monsanto Corporation Dayton operations: http://moundmuseum.com/

Mount Holyoke College site on documents relating to Hiroshima: http://www.mtholyoke.edu/acad/intrel/hiroshim.htm

Newseum story on top 100 news stories of twentieth century: https://tomprof.stanford.edu/posting/115

Nuclear Weapon Archive: http://nuclearweaponarchive.org/

Oak Ridge National Laboratory report on X-10 graphite reactor: https://www.ornl.gov/content/graphite-reactor

Operation Crossroads: http://en.wikipedia.org/wiki/Operation_Crossroads

Oppenheimer quotes: http://en.wikiquote.org/wiki/Robert_Oppenheimer

OSRD: A copy of the Executive Order establishing the OSRD can be found at http://www.presidency.ucsb.edu/ws/index.php?pid=16137#axzz1QbKXQHjp

Queen Mary buildings: http://www.atomicarchive.com/History/mp/p4s24.shtml

Robert Lewis log: http://en.wikipedia.org/wiki/Robert_A._Lewis

Roosevelt to Oppenheimer, June 29, 1943: http://lcweb2.loc.gov/mss/mcc/083/0001.jpg and 0002.jpg

Tinian Island: http://www.globalsecurity.org/military/facility/tinian.htm

Trinity eyewitness accounts: http://www.dannen.com/decision/trin-eye.html

Truman library documents on decision to drop the bomb: http://www.trumanlibrary.org/whistlestop/study_collections/bomb/large/index.php

Twenty-First Bomber Command and bombing of Tokyo: http://en.wikipedia.org/wiki/XXI_Bomber_Command, http://en.wikipedia.org/wiki/Bombing_of_Tokyo

ZEEP reactor: https://en.wikipedia.org/wiki/ZEEP

Index

© Springer Nature Switzerland AG 2020
B. C. Reed, *Manhattan Project*,
https://doi.org/10.1007/978-3-030-45734-1

Printed in the United States
by Baker & Taylor Publisher Services